W0245764

Hausdorff Approximations

Mathematics and Its Applications (*East European Series*)

Volume 50

Hausdorff Approximations

by

Bl. Sendov

Institute for Mathematics,
Bulgarian Academy of Sciences, Bulgaria

edited by

Gerald Beer

KLUWER ACADEMIC PUBLISHERS

DORDRECHT / BOSTON / LONDON

Library of Congress Cataloging-in-Publication Data

Sendov, Blagovest.
[Khausdorfovye priblizhenii͡a. English]
Hausdorff approximations / by Bl. Sendov.
p. cm. -- (Mathematics and its applications. East European series ; v. 50)
Translation of: Khausdorfovye priblizhenii͡a.
Includes bibliographical references and indexes.

1. Topological spaces. 2. Approximation theory. I. Title.
II. Series: Mathematics and its applications (Kluwer Academic Publishers). East European series ; v. 50.
QA611.3.S4613 1990
514'.3--dc20 90-43682

ISBN-13: 978-94-010-6787-4 e-ISBN-13: 978-94-009-0673-0
DOI: 10.1007/978-94-009-0673-0

Published by Kluwer Academic Publishers,
P.O. Box 17, 3300 AA Dordrecht, The Netherlands.

Kluwer Academic Publishers incorporates
the publishing programmes of
D. Reidel, Martinus Nijhoff, Dr W. Junk and MTP Press.

Sold and distributed in the U.S.A. and Canada
by Kluwer Academic Publishers,
101 Philip Drive, Norwell, MA 02061, U.S.A.

In all other countries, sold and distributed
by Kluwer Academic Publishers Group,
P.O. Box 322, 3300 AH Dordrecht, The Netherlands.

Printed on acid-free paper

SERIES EDITOR'S PREFACE

'Et moi, ..., si j'avait su comment en revenir, je n'y serais point allé.'

Jules Verne

The series is divergent; therefore we may be able to do something with it.

O. Heaviside

One service mathematics has rendered the human race. It has put common sense back where it belongs, on the topmost shelf next to the dusty canister labelled 'discarded non-sense'.

Eric T. Bell

Mathematics is a tool for thought. A highly necessary tool in a world where both feedback and non-linearities abound. Similarly, all kinds of parts of mathematics serve as tools for other parts and for other sciences.

Applying a simple rewriting rule to the quote on the right above one finds such statements as: 'One service topology has rendered mathematical physics ...'; 'One service logic has rendered computer science ...'; 'One service category theory has rendered mathematics ...'. All arguably true. And all statements obtainable this way form part of the raison d'être of this series.

This series, *Mathematics and Its Applications*, started in 1977. Now that over one hundred volumes have appeared it seems opportune to reexamine its scope. At the time I wrote

> "Growing specialization and diversification have brought a host of monographs and textbooks on increasingly specialized topics. However, the 'tree' of knowledge of mathematics and related fields does not grow only by putting forth new branches. It also happens, quite often in fact, that branches which were thought to be completely disparate are suddenly seen to be related. Further, the kind and level of sophistication of mathematics applied in various sciences has changed drastically in recent years: measure theory is used (non-trivially) in regional and theoretical economics; algebraic geometry interacts with physics; the Minkowsky lemma, coding theory and the structure of water meet one another in packing and covering theory; quantum fields, crystal defects and mathematical programming profit from homotopy theory; Lie algebras are relevant to filtering; and prediction and electrical engineering can use Stein spaces. And in addition to this there are such new emerging subdisciplines as 'experimental mathematics', 'CFD', 'completely integrable systems', 'chaos, synergetics and large-scale order', which are almost impossible to fit into the existing classification schemes. They draw upon widely different sections of mathematics."

By and large, all this still applies today. It is still true that at first sight mathematics seems rather fragmented and that to find, see, and exploit the deeper underlying interrelations more effort is needed and so are books that can help mathematicians and scientists do so. Accordingly MIA will continue to try to make such books available.

If anything, the description I gave in 1977 is now an understatement. To the examples of interaction areas one should add string theory where Riemann surfaces, algebraic geometry, modular functions, knots, quantum field theory, Kac-Moody algebras, monstrous moonshine (and more) all come together. And to the examples of things which can be usefully applied let me add the topic 'finite geometry'; a combination of words which sounds like it might not even exist, let alone be applicable. And yet it is being applied: to statistics via designs, to radar/sonar detection arrays (via finite projective planes), and to bus connections of VLSI chips (via difference sets). There seems to be no part of (so-called pure) mathematics that is not in immediate danger of being applied. And, accordingly, the applied mathematician needs to be aware of much more. Besides analysis and numerics, the traditional workhorses, he may need all kinds of combinatorics, algebra, probability, and so on.

In addition, the applied scientist needs to cope increasingly with the nonlinear world and the extra mathematical sophistication that this requires. For that is where the rewards are. Linear models are honest and a bit sad and depressing: proportional efforts and results. It is in the non-linear world that infinitesimal inputs may result in macroscopic outputs (or vice versa). To appreciate what I am hinting at: if electronics were linear we would have no fun with transistors and computers; we would have no TV; in fact you would not be reading these lines.

There is also no safety in ignoring such outlandish things as nonstandard analysis, superspace and anticommuting integration, p-adic and ultrametric space. All three have applications in both electrical engineering and physics. Once, complex numbers were equally outlandish, but they frequently proved the shortest path between 'real' results. Similarly, the first two topics named have already provided a number of 'wormhole' paths. There is no telling where all this is leading - fortunately.

Thus the original scope of the series, which for various (sound) reasons now comprises five subseries: white (Japan), yellow (China), red (USSR), blue (Eastern Europe), and green (everything else), still applies. It has been enlarged a bit to include books treating of the tools from one subdiscipline which are used in others. Thus the series still aims at books dealing with:

- a central concept which plays an important role in several different mathematical and/or scientific specialization areas;
- new applications of the results and ideas from one area of scientific endeavour into another;
- influences which the results, problems and concepts of one field of enquiry have, and have had, on the development of another.

Approximation of functions (by entities which can be coded in a finite way) is a topic of great importance, both theoretically and computationally. For instance, uniform approximation by polynomials on intervals, or various approximation methods where the error is measured by some integral (L^2, L^p,...), all subjects with a considerable literature.

However, when the function to be approximated is discontinuous, and many natural phenomena do involve modelling by discontinuous functions, the approximation schemes have disadvantages, as noted by Kolmogorov who, in fact, formulated some natural desirable 'axioms' in that setting.

Hausdorff approximation, which is defined by means of the Hausdorff distance between (completed) graphs of the functions involved, satisfies these axioms (and more).

In the seventies, a great deal of work was done on Hausdorff approximation, predominantly by Bulgarian mathematicians including, in particular, the author of the present volume. The original Russian version of this unique book appeared in 1978. Surprisingly, it was never translated. Therefore, it is a pleasure to now welcome a translation of this important book, updated to include references to the published literature up to and including 1988.

The shortest path between two truths in the real domain passes through the complex domain.

J. Hadamard

La physique ne nous donne pas seulement l'occasion de résoudre des problèmes ... elle nous fait pressentir la solution.

H. Poincaré

Never lend books, for no one ever returns them; the only books I have in my library are books that other folk have lent me.

Anatole France

The function of an expert is not to be more right than other people, but to be wrong for more sophisticated reasons.

David Butler

Amsterdam, July 1990

Michiel Hazewinkel

Table of Contents

Preface

This book is a translation of the monograph published in Russian in Bulgaria by the Publishing House of the Bulgarian Academy of Sciences in 1979. In the process of its translation, some additions were made with reference to the new publications in the area of Hausdorff approximations closely connected with the contents of the book.

A number of typographic and factual errors occuring in the Russian edition have been corrected, owing mainly to the editor of the English version, Dr. G. Beer, who has spared no effort in smoothing up the initial rough English translation. I am deeply grateful to Dr. Beer for his efforts and would like to emphasize that without his persistence, this book would not have come out in its English version.

I hope that in the English translation, the book will reach many more readers. The importance of Hausdorff approximations is growing with new applications, such as to signal processing, to curve fitting in the plane, etc., and with the further development of numerical methods of approximation with reference to Hausdorff distance.

Sofia
July 12, 1990

Bl. Sendov

Preface to the Russian Edition

The aim of this book is to present the basic results in the approximation of functions and point sets in the plane with respect to Hausdorff distance. Quite recently, the approximation theory with respect to Hausdorff distance has been developed by predominantly Bulgarian mathematicians. The joint reseach of B. I. Penkov and Bl. Sendov on the computation of ε-entropy and ε-capacity of sets of continuous functions with respect to Hausdorff distance initiated these studies. Various works on Hausdorff distance were reported and discussed in the Seminar on Approximation Theory in the Center for mathematics and mechanics at the Bulgarian Academy of Sciences and Sofia University "Kl. Ohridski".

I wish to thank my colleagues V. A. Popov, V. M. Vesselinov, T. P. Boyanov, B. D. Boyanov, A. S. Andreev, S. M. Markov, P. P. Petrushev, V. H. Hristov, G. L. Iliev, Sp. Tashev, and other participants in our Seminar on Approximation Theory for discussing certain questions that are included in the present book, as well as for their remarks on the manuscript.

Finally, I would like to thank D. Vachov and V. Andreev for their valuable assistance in preparing the book for publication.

Sofia
December 3, 1978

Bl. Sendov

Introduction

Approximation theory is an important field of modern mathematics. The necessity of representing complicated mathematical objects by more simple ones arises in purely theoretical problems of mathematics, as well as in concrete methods of computation. Approximation theory is especially valuable not only as the base of numerical analysis, but also in mathematical programming, control theory, and in other fields of mathematics and its applications.

The theory of approximation of functions originates from Chebyshev's problem on uniform (Chebyshev) approximation of continuous functions by algebraic polynomials. The set C_Δ of functions continuous on a closed interval Δ is metrizable by the uniform Chebyshev distance $R(f,g) = \|f - g\|_\Delta = \max\{|f(x) - g(x)| : x \in \Delta\}$ and the subset H_n of C_Δ consisting of algebraic polynomials of degree no higher than n is used as approximation means. Then the proximity between the approximated function $f \in C_\Delta$ and its approximating polynomial $P \in H_n$ is given by the Chebyshev distance $R(f,P) = \|f - P\|_\Delta$.

In many cases this way of measuring the distance between the approximated and approximating functions is very suitable, because if $R(f,P) \leq \varepsilon$, then every value $f(x)$ of the function f at the point $x \in \Delta$ can be recovered from the value of the polynomial $P(x)$ with an error that does not exceed ε. On the other hand, according to the classic Weierstrass theorem, for every function $f \in C_\Delta$ and for each $\varepsilon > 0$, there exists a positive integer n and a polynomial $P \in H_n$ such that $R(f,P) \leq \varepsilon$. This guarantees the possibility of representing each continuous function by algebraic polynomials with a preassigned accuracy for all values of the variable. On the other hand, every polynomial is defined by its coefficients, which are finite in number. This fact has stimulated profound and fruitful investigations in the theory of Chebyshev approximations and its generalizations.

Many natural phenomena are subject to mathematical modelling by using discontinuous functions. But the approximate representations of such functions by algebraic polynomials with respect to the uniform distance is impossible with arbitrary accuracy. Usually in those cases, weaker distances, e.g., integral distances, in particular the mean square distance

$$L_2(f,g) = \left(\int_a^b (f(x) - g(x))^2 \, dx\right)^{1/2}$$

are used.

An arbitrary square integrable function can be approximated by an algebraic polynomial with respect to the mean square distance with a preassigned accuracy. But for certain purposes the integral metric turns out to be quite weak, because it does not guarantee the geometric proximity of the graph of the approximated function and that of the approximating one. A. N. Kolmogorov [2] emphasized the necessity of defining a certain distance admitting the approximation of discontinuous functions with an arbitrary accuracy, and at the same time, being stronger than the integral distances:

"Describing the temporal course of a real process by use of the time function $f(t)$, assuming values in the 'phase set' X chosen in an appropriate way, it seems natural and legal to make the proposition that the function f possesses only discontinuities of the first kind (jumps). For the detailed research of such processes the introduction of a corresponding topology is very useful in the set D (of functions with discontinuities of the first kind).

The topology of uniform convergence, natural when studying continuous processes, turns out to be too strong when studying processes with discontinuities of the first kind. For example, it is desirable for the function sequence

$$f_n(t) = \begin{cases} x_1 & \text{if } t < t_n \\ x_2 & \text{if } t > t_n \end{cases},$$

where $t_n \to t_0$ when $n \to \infty$, to be convergent to the function

$$f(t) = \begin{cases} x_1 & \text{if } t < t_0 \\ x_2 & \text{if } t > t_0 \end{cases},$$

since the function f_n for large n deviates from the function f simply by a small shift at the moment of the jump from the state x_1 to the state x_2. As is well-known, uniform convergence fails when $x_1 \neq x_2$.

On the other hand, the topology of D should not be too weak for it is desirable that the most essential properties of the the function $f \in D$ be preserved at the boundary transition. For example, it is necessary that the convergence properties $f_n \to f$, $t_n \to t$, and $f_n(t_n + 0) - f_n(t_n - 0) \to C \neq 0$ jointly force $f(t + 0) - f(t - 0) = C$."

A distance, generating a topology with the properties suggested by A. N. Kolmogorov, was introduced by Yu. V. Prohorov [1]. When defining this distance between two functions $f, g \in D$, the Hausdorff distance between two closed point sets $\pi(f)$ and $\pi(g)$ in the plane is used, where $\pi(f)$ is the smallest closed point set containing the graph of f except the isolated points of the graph of f. The Hausdorff distance $r(F,G)$ between two bounded and closed point sets in the plane is defined by

$$r(F,G) = \inf\{\varepsilon : F \subset U(\varepsilon,G) \text{ and } G \subset U(\varepsilon,F)\}$$

where $U(\varepsilon,A)$ denotes the ε-neighborhood, with respect to the distance between points in the plane, of the set A.

With respect to certain applications, it seems natural to add to A. N. Kolmogorov's requirements the following one: The topology should be such that the sequence of continuous functions

$$g_n(x) = \begin{cases} -1 & \text{if } x \le -1/n \\ nx & \text{if } -1/n \le x \le 1/n \\ 1 & \text{if } x \ge 1/n \end{cases}$$

should converge to the function

$$g(x) = \begin{cases} -1 & \text{if } x < 0 \\ \lambda & \text{if } x = 0 \\ 1 & \text{if } x > 0 \end{cases},$$

where $\lambda \in [-1,1]$.

A distance, generating a topology with the above mentioned properties, can be defined as follows. With every bounded function f, defined on a closed interval Δ, is associated a minimal closed point set $F(f)$ convex with respect to the y-axis which contains the graph of f. The point set $F(f)$ is called the completed graph of the function f. Then the Hausdorff distance $r(F(f),F(g))$ between the point sets $F(f)$ and $F(g)$ in the plane will be used as a distance $r(f,g)$ between the functions f and g. The distance $r(f,g)$ is called the Hausdorff distance between functions.

The aim of the present book is to give an account of the main results in the theory of approximation of functions and point sets with respect to the Hausdorff distance in the plane. The Hausdorff distance has numerous and diverse applications when measuring the deviation of two sets. But it should be remembered that the Hausdorff distance between functions is defined in terms of the completed graphs of these functions. Our method of obtaining the completed graphs is not at all the only one possible.

The theory of approximation of functions has been well-developed for metric function spaces that are Banach spaces. The linearity and the existence of a norm, generating a distance, play an essential role in the techniques of approximation theory problems. It should be emphasized that the function space, metrized by Hausdorff distance, is not a Banach space. Therefore, the theory of function spaces, topologized by a Hausdorff distance, has some peculiarities and in many problems completely new concepts and approaches should be taken into account. For instance, the unit sphere in C_Δ is totally bounded with respect to Hausdorff distance, so that the calculation of the ε-entropy of the unit sphere or the calculation of the widths of all continuous functions, are permissible for that space.

The fact that function spaces, topologized by Hausdorff distance, are not Banach spaces, is often accepted as an offense against good manners in society, where everything takes place in normed linear spaces. But as E. Dolzenko and E. Sevastianov [2] noted, "the notion of proximity of two functions in the sense of the Hausdorff metric is a more general natural *visual* notion than the notion of proximity in the sense of any other metric, and in particular, it is more natural than their proximity in the sense of uniform deviation $\|f_1 - f_2\|_\Delta$. For instance, the proximity of two signals on the screen of an electronic oscilloscope is estimated visually as proximity in Hausdorff metric. Another example: in transforming by $y = f(x)$ the real signal x into a signal y, we usually know these signals with a precision ε, depending on the available measurement devices, so that the transformations f_1 and f_2 that differ by less than ε in Hausdorff metric, are indistinguishable."

There exists a series of engineering problems, e.g., the design of antennas and the design of electron filters, where the problem of approximation with respect to Hausdorff distance arises in a natural way.

In a way, the Hausdorff distance treats the coordinate axes equivalently. While a neighborhood of a function, with respect to the uniform distance, is obtained varying the points of its graph only in the direction of the ordinate axis, a neighborhood of a function with respect to Hausdorff distance is obtained by varying points of its graph in all directions in the plane. Since in real computations we not only find the function value with error, but the value of the argument for which the function value is calculated is given with certain error, too. Thus, it seems more natural to use in this situation the Hausdorff distance.

Since the Hausdorff distance between functions is defined with respect to their completed graphs, it is natural to consider multivalued functions whose values are segments, i.e., segment-valued functions. Thus, the problems of Hausdorff approximation come close to interval analysis (see Moore [1]). But we need a more specialized version of the interval analysis, which we shall call segment analysis, in order to distinguish it from the already formed interval analysis.

In the first chapter of this book, we present the necessary background information on segment analysis. The definition and the properties of the Hausdorff distance and its relation to uniform distance are established in the second chapter.

The presentation of the theory of Hausdorff approximations begins the third chapter. It is necessary to emphsize that it is assumed that the reader is familiar with the main results of the classical theory of uniform approximation. The contents are constructed by analogy with this classical theory, since it is possible to consider the uniform distance as a limiting case of Hausdorff distance (see §2.4). The third chapter is principally focused on the approximation of functions by means of the values of positive linear operators. First, we present the natural generalization of P. P. Korovkin's classical theory of convergence of sequences of linear operators with respect to Hausdorff distance. Then, we consider in detail the orders of approximation that a number of classical operators realize with respect to Hausdorff distance, for example, the operators of Fejer, Jackson, Vallée-Pousin, Landau, and Bernstein polynomials. A sufficiently general theorem is proved on the convergence of sequences of arbitrary linear operators. Finally, we look at the problem of convergence of partial sums of Fourier series with respect to Hausdorff distance. To this end, another way of obtaining the completed graph of a function, connected with the Gibbs' effect, is introduced.

In the fourth chapter we consider best approximation by means of algebraic polynomials, trigonometric polynomials, rational functions, spline functions, and piecewise monotone functions with respect to the Hausdorff distance. A characteristic peculiarity of the best approximation with respect to Hausdorff distance is the existense of universal estimates for the order of approximation. We consider particular classes of functions for which the order of approximation is better than the universal estimate. In many cases well-known estimates for the best approximation can be obtained as a consequence of the estimates for the best Hausdorff approximations.

The fifth chapter is devoted to converse theorems for the best Hausdorff approximation. In this direction, we observe new phenomena that are characteristic of the Hausdorff approximations. For example, Berstein's theorem for the existence of a function with preassigned best approximation does not hold here, in contrast to uniform and other classical distances. Another interesting phenomenon is that the characterization of classes of functions by their best Hausdorff approximations is connected not only with the order of this approximation, but also with the constant in front of the order. So to say, here we have not only separating orders, but also separating constants. It turns out that in some cases it is possible to compute the values of these separating constants.

The sixth chapter considers the problem of computing the ε-entropy and ε-capacity of the set of all bounded functions and the set of all continuous curves that are contained in a given rectangle with respect to Hausdorff distance. In connection with these problems, interesting combinatorial problems arise. Also, we compute the widths of the set of all bounded functions with respect to Hausdorff distance, and in this case, they coincide with the widths of the set of all continuous functions.

The seventh chapter deals with the best approximation of point sets in the plane by polynomial and other curves. Exact estimates are found with respect to order for the best approximation of a rectifiable curve by polynomial curves and also by piecewise monotone curves. Still, the question of finding the exact constants remains open. It turns out that the order of best approximation of a bounded, closed, connected set in the plane by polynomial curves is determined by the metric dimension of the set.

In the last chapter, chapter eight, we present a numerical method for finding the polynomial of best Hausdorff approximation.

approximation by means of partial sums of Fourier series with respect to Hausdorff distance. [illegible] the Fourier [illegible] of a function, [illegible] with [illegible].

In the fourth chapter we consider best approximation by means of algebraic polynomials, trigonometric polynomials, rational functions, spline functions, and [illegible] with respect to the Hausdorff distance. A [illegible] of best approximation with respect to Hausdorff distance [illegible] of [illegible]. We consider particular classes of [illegible] which the order of [illegible] the [illegible] estimate. In many cases [illegible] that the best approximation [illegible] of the [illegible] and the [illegible] normed spaces.

The fifth chapter is devoted to converse theorems for the best Hausdorff approximation. In this direction we observe how [illegible] of the Hausdorff approximations. For example, [illegible] for the existence of a [illegible] both [illegible] and [illegible] Hausdorff distance. [illegible] approximation [illegible] is connected not only with the order of [illegible] but also [illegible] in [illegible]. [illegible] we have not [illegible] converse [illegible] functions.

The sixth chapter considers the [illegible] of some classes of [illegible] and [illegible] capacity [illegible] the [illegible] of all continuous curves that are contained in a given rectangle with respect to Hausdorff distance. In connection with these problems, [illegible] with respect to Hausdorff distance, and in this case, [illegible] with the [illegible] of all [illegible] functions.

The seventh chapter deals with the [illegible] approximation [illegible] and other curves. [illegible] with respect to [illegible] approximation of [illegible] curves [illegible] Hausdorff distance, [illegible] the plane by [illegible].

In the last chapter, chapter eight, we [illegible] method for finding the polynomials of [illegible] Hausdorff approximation.

Chapter 1

Elements of Segment Analysis

We shall consider multivalued functions with special attention paid to appropriate notation and operations. Actually, we shall need only a very special type of real multivalued functions with closed and bounded convex images, i.e., each image is a closed interval (segment) in the extended real line. Functions that take segment values will be called **segment functions.**

At the end of the 1950s the so called "interval analysis" was introduced in connection with the development of numerical methods for computers (T. Sunaga [1], R. E. Moore and C. T. Yang [1], and R. E. Moore [1]). We shall use the interval arithmetic actually in the aspect that it was developed in the interval analysis, but some new definitions of "limit" and "derivative" will be introduced. In the interval analysis, the derivative is introduced by using the general theory of multivalued functions (H. Ratschek and G. Schröder [1]). We shall need another type of derivative that is naturally connected with the Hausdorff metric. This derivative is based on a certain extension of the notion of limit so that every sequence has a limit. In particular, the derivative we introduce is close to the sub-differential widely used in convex analysis (Bl.Sendov [25,26], R. Rockafellar [1]). The definition of segment function in our presentation differs fundamentally from the traditionally accepted one in the interval analysis for interval function.

In connection with the mentioned distinctions between the already established notions in interval analysis and our presentation, we shall call our treatment **segment analysis.** The elements of this analysis, which will be used in the consideration of different approximation problems, will be presented in what follows.

§ 1.1 Segment Arithmetic

Let R be the set of real numbers, and $\bar{R}$ the set of extended reals obtained by adjoining ∞ and $-\infty$ to R. A **segment** $[a,b]$ with $a, b \in \bar{R}$ is the following set of real numbers:

$$[a,b] = \{ x \in \bar{R} : a \leq x \leq b \}.$$

The set of all such segments will be denoted by $S(\bar{R})$ and the set of all **finite segments** with real endpoints will be denoted by $S(R)$. Obviously $\bar{R} \subset S(\bar{R})$ and $R \subset S(R)$, since there are admissible segments with coinciding endpoints, that is, point segments. If $a \in R$, then the segment $[a,a]$ will be denoted by a. The empty segment, denoted by $\emptyset$, will be considered as an element of $S(\bar{R})$ as well as of $S(R)$ (these may be obtained from taking $a > b$ in the definition of $[a,b]$).

1.1.1. Partial Orderings

Now we shall introduce order relations in the set $S(\bar{R})$ and hence in $S(R)$. Let $a, b \in S(\bar{R})$ where $a = [a_1, a_2]$ and $b = [b_1, b_2]$ and denote by $*$ one of the inequality symbols $<, \leq, <, \geq$. We shall write $a * b$ provided for each $x \in a$ and each $y \in b$ the inequality $x * y$ holds. Obviously, for example, $a \leq b$ if and only if $a_2 \leq b_1$. If $a, b \in S(\bar{R})$ and the inequalities $a \leq b$ and $b \leq a$ hold simultaneously, then it follows not only that $a = b$ but also that a and b are point segments, that is, $a, b \in \bar{R}$. The relation $\leq$ introduces a partial ordering of $S(\bar{R})$. The ordering is not linear; for example, if $a = [0, 2]$ and $b = [1, 3]$, then neither $a \leq b$ nor $b \leq a$ holds.

Inclusion is another method of partial ordering $S(\bar{R})$ that we will use. We write as usual $a \subset b$ and say a is **contained in** or **included in** b provided each element of a is an element of b. If both $a \subset b$ and $b \subset a$, then $a = b$ (this follows directly from the definition of inclusion).

1.1.2. Lattice Operations

It is convenient to use set operations when dealing with segments, since they are sets of real numbers. The partial order $\leq$ on segments is a lattice order. The meet of two segments a and b is the minimal segment c contained in both a and b. Clearly, the meet operation is simply set intersection. Notice that if $\emptyset$ is the empty segment, then $a \cap \emptyset = \emptyset$ for all segments a. If A denotes a set of real numbers and if to every $\alpha \in A$ there corresponds a segment a_α then the maximal segment c contained in all the segments a_α, $\alpha \in A$, is again their intersection in the set theoretic sense. This segment will be denoted by

$$c = \bigcap \{a_\alpha : \alpha \in A\} = \bigcap_{\alpha \in A} a_\alpha$$

The join of two segments a and b is the minimal segment c that contains both a and b. We call this segment c the **union** of a and b. Consistent the standard lattice notation, we write

$$c = a \vee b = b \vee a$$

If $\emptyset$ is the empty segment, then $a \vee \emptyset = a$ for each segment a. Generally, the union of two segments differs from the union of the points of these segments in a set-theoretic sense. For example, if $a = [0, 1]$ and $b = [3, 4]$, then $c = a \vee b = [0, 4]$. The operation of union of two segments was considered by T. Sunaga [1], but it was hardly used in the further development of interval analysis.

If A denotes a set of real numbers and if to every $\alpha \in A$ there corresponds a segment a_α then we call the minimal segment c containing all the segments a_α, $\alpha \in A$, their **union**. This segment will be denoted by

$$c = \vee \{a_\alpha : \alpha \in A\} = \bigvee_{\alpha \in A} a_\alpha$$

According to the above definition, c must also be a segment. For example, if $A = \{2, 3, 4, \ldots\}$ and $a_\alpha = [1/\alpha, 1 - 1/\alpha]$, then $c = \vee \{a_\alpha : \alpha = 2, 3, 4, \ldots\} = [0,1]$, in spite of the fact that all the segments a_α belong to the open interval (0,1). That is another distinction between our definition for union of segments and the set-theoretic notion of union of sets.

We have these relations for the operations of union and intersection:

(1.1) $$(a \vee b) \vee c = a \vee (b \vee c) = a \vee b \vee c$$

(1.2) $$(a \cap b) \cap c = a \cap (b \cap c) = a \cap b \cap c$$

(1.3) $$(a \cap c) \vee (b \cap c) \subset (a \vee b) \cap c$$

(1.4) $$(a \vee c) \cap (b \vee c) \supset (a \cap b) \vee c$$

where $a, b, c \in S(\overline{R})$. The equalities (1.1) and (1.2) are obvious. The inclusion (1.3) can also be proved easily. If $x \in a \cap c$, then $x \in (a \vee b) \cap c$. Similarly, if $x \in b \cap c$, then $x \in (a \vee b) \cap c$, hence $(a \cap c) \vee (b \cap c) \subset (a \vee b) \cap c$ holds. The inclusion (1.4) is proved similarly.

1.1.3. ARITHMETIC OPERATIONS

Let $*$ denote one of the arithmetic operations $+, -, \cdot, /$ (addition, subtraction, multiplication, division), and let $a, b \in S(\bar{R})$. By $a * b$ we mean the segment $c \in S(\bar{R})$ defined by

$$c = a * b = \{x : x = \xi * \eta \text{ where } \xi \in a \text{ and } \eta \in b\}. \tag{1.5}$$

This kind of definition for arithmetic operations is standard in interval analysis (R. Moore [1]). We shall introduce only the following convention: whenever the arithmetic operation $\xi * \eta$ is undefined in $\bar{R}$, we set by definition $c = a * b = [-\infty, \infty]$. For example, if the segment b contains zero, then $a/b = [-\infty, \infty]$ and if a and b both contain ∞, then $a - b = [-\infty, \infty]$. In particular, $\infty - \infty = [-\infty, \infty]$. Similarly, $0 \cdot \infty = [-\infty, \infty]$, $\infty/\infty = [-\infty, \infty]$, and so on.

In the interval analysis only finite intervals are usually used, and the arithmetic operations are not defined for all pairs of intervals. Thus, for example, the division by an interval containing zero is excluded. Infinite intervals have been used in the interval arithmetic by S. Laveuve [1] in a different way.

With the above convention, we obtain a closed segment arithmetic. All arithmetic operations are defined for all pairs of segments in $S(\bar{R})$ and the result of an operation again belongs to $S(\bar{R})$. Let us consider each arithmetic operation in some detail.

1.1.3.1 *Addition and Subtraction* . According to (1.5) if $a = [a_1, a_2]$ and $b = [b_1, b_2]$, then

$$c = a + b = [a_1 + b_1, a_2 + b_2]$$

so that if $c = [c_1, c_2]$ we have $c_1 = a_1 + b_1$ and $c_2 = a_2 + b_2$. Similarly we have

$$d = a - b = [a_1 - b_2, a_2 - b_1]$$

i.e., $d_1 = a_1 - b_2$ and $d_2 = a_2 - b_1$ where $d = [d_1, d_2]$. For example,

$$[-2, 4] + [3, 5] = [1, 9] \qquad [-1, 3] - [1, 2] = [-3, 2]$$

Obviously, the associative and commutative laws hold for addition.

An important peculiarity of the segment arithmetic is that subtraction is not the inverse operation of addition. From $a = b + c$, it does not follow at all that $b = a - c$. This means that we may not transfer a term from one side of the equality to the other with opposite sign. S. Markov [1] has developed an interval arithmetic, where subtraction is the inverse operation of addition.

Let us pay attention to the fact that if $a = c$ and $b = d$, where $a, b, c, d \in S(\bar{R})$, then $a + b = c + d$, i.e., equal segments can be added to and subtracted from both sides of the segment equality. Actually, the cancellation law for addition holds, provided the fixed summand is a finite segment. Adding the two segment equalites $a + b = c$ and $-b = -b$, we obtain

(1.6) $$a + b - b = c - b,$$

but $b - b$ is not equal to zero in general, for if $b = [b_1, b_2]$ then $b - b = [b_1 - b_2, b_2 - b_1]$, a symmetric segment with respect to the origin. The equality $b - b = 0$ holds if and only if $b_1 = b_2$, that is, when b is a point segment. It follows from (1.6) that

(1.7) $$a + b = c \text{ implies } a \subset c - b$$

From (1.7), we obtain a more general statement:

(1.8) $$a + b \subset c \text{ implies } a \subset c - b$$

1.1.3.2. *Multiplication and Division* . In view of (1.5), if $a = [a_1, a_2]$ and $b = [b_1, b_2]$, then

$$c = ab = [\min\{a_1b_1, a_1b_2, a_2b_1, a_2b_2\}, \max\{a_1b_1, a_1b_2, a_2b_1, a_2b_2\}]$$

and

$$d = a/b = \begin{cases} [a_1, a_2][1/b_1, 1/b_2] & \text{if } 0 \notin [b_1, b_2] \\ [-\infty, \infty] & \text{if } 0 \in [b_1, b_2] \end{cases}.$$

Let us consider the case when one of the factors is a point-segment. Let $a \in S(\bar{R})$ and $\alpha \in \bar{R}$; then the form of the product depends on the sign of α:

$$\text{if } \alpha \geq 0, \text{ then } \alpha a = [\alpha, \alpha][a_1, a_2] = [\alpha a_1, \alpha a_2];$$

if $\alpha < 0$, then $\alpha a = [\alpha, \alpha][a_1, a_2] = [\alpha a_2, \alpha a_1]$.

For $a, b \in S(R)$ and $\alpha, \beta \in R$ with $\alpha\beta \geq 0$, the following equalites hold:

(1.9) $$a + b = b + a;$$

(1.10) $$\alpha(a + b) = \alpha a + \alpha b;$$

(1.11) $$(\alpha + \beta)a = \alpha a + \beta a.$$

The equalities (1.9) - (1.11) show that $S(R)$ is a quasi-linear space over the set of real numbers, since these equalities hold only if $\alpha\beta \geq 0$ (G. Schröder [1]).

We note this important fact: the distributive law does not hold for multiplication and addition of segments. Instead, we have the so-called **subdistributive law** (H. Ratschek [1]):

(1.12) $$(a + b)c \subset ac + bc \quad \text{for all } a, b, c \in S(\bar{R}).$$

In (1.12) we have equality instead of inclusion if $c \in R$ or if $a, b \in R$ with $ab \geq 0$. The proof of (1.12) is very simple. If $x \in a$, $y \in b$, and $z \in c$, then $xz \in ac$, $yz \in bc$ and $x + y \in a + b$, from which (1.12) is immediate. The following example shows that in general we cannot replace the inclusion by equality in (1.12):

$$([-1, 1] + [0, 1])[-1,2] = [-1,2][-1, 2] = [-2, 4] \subset [-3, 4]$$

$$= [-2, 2] + [-1, 2] = [-1, 1][-1, 2] + [0, 1][-1, 2].$$

Let us establish one more relation.

Lemma 1.1. If $a, b, c, d \in S(R)$ and $*$ is an arbitrary arithmetic operation, then $(a * c) \vee (b * d) \subset (a \vee b) * (c \vee d)$.

Proof. Let $\xi \in (a * c)$; then $\xi = x * y$ with $x \in a$ and $y \in c$. Moreover, $x \in a \vee b$ and $y \in c \vee d$. Hence $\xi = x * y \in (a \vee b) * (c \vee d)$. Similarly, for every $\eta \in b * d$ we have $\eta \in (a \vee b) * (c \vee d)$. The lemma is proved.

Corollary 1.1. If $<a_n>_1^\infty$ and $<b_n>_1^\infty$ are two sequences and $*$ is an arithmetic operation, then

$$\bigvee_{n=1}^{\infty} (a_n * b_n) \subset \left(\bigvee_{n=1}^{\infty} a_n\right) * \left(\bigvee_{n=1}^{\infty} b_n\right)$$

1.1.4. DISTANCE AND NORM.

In S(R) we may define a distance $\rho(a,b)$ between segments a and b in the following way. If $a = [a_1, a_2]$ and $b = [b_1, b_2]$, then

$$\rho(a,b) = \max \{|a_1 - b_1|, |a_2 - b_2|\} \tag{1.13}$$

It is easy to check that

$$\rho(a,b) = \max \left[\max_{x \in a} \min_{y \in b} |x - y| , \max_{x \in b} \min_{y \in a} |x - y|\right]$$

i.e., $\rho(a,b)$ is the **Hausdorff distance** between the closed subsets a and b of the metric space of real numbers R. As a special case of E. Michael's theorem [1] for the space of closed subsets of a complete metric space, we obtain

Lemma 1.2. The set S(R) metrized by the distance ρ is a complete metric space.

Through the distance ρ one can define the **norm** $\|a\|$ of every segment in S(R), namely

$$\|a\| = \rho(a,0) = \max \{|a_1|, |a_2|\} \qquad \text{where } a = [a_1, a_2]. \tag{1.14}$$

An important remark is in order. If we use the norm (1.14) to define a distance $\delta(a,b)$ between two elements a and b of S(R) as it is done in normed spaces, i.e.,

$$\delta(a,b) = \|a - b\| = \max \{|a_1 - b_2|, |a_2 - b_1|\}$$

where $a = [a_1, a_2]$ and $a = [b_1, b_2]$ then the obtained distance differs from the initial distance ρ. Moreover, δ does not even satisfy the axioms for a distance, for $\delta(a,a) > 0$ for all segments a of positive length. This fact is not surprising, since S(R) is not a linear space, but only a quasi-linear one.

As distinct from the norm of a segment which is a number, we define the **absolute value** of a segment to be a segment, namely $|a| = [0, \|a\|]$.

§ 1.2. Segment sequences.

The theory of sequences of segments is the natural generalization of the analogous theory for sequences of real numbers, defining in a suitable way convergence and limit. One natural approach is to consider the segment sequence as a special case of a sequence of subsets, where the subsets belong to a given metric space. In this case, the theory of segment sequences is well-known.

Let $<a_n>_1^\infty$ with $a_n \in S(R)$ for $n = 1, 2, \ldots$ be a given segment sequence. We say $<a_n>_1^\infty$ **converges** to a segment $a \in S(R)$ provided $\lim_{n\to\infty} \rho(a_n,a) = 0$. The segment a is called the **limit** of the sequence $<a_n>_1^\infty$, and we write $\lim_{n\to\infty} a_n = a$. Obviously the segment sequence $<a_n>_1^\infty$ converges if and only if the number sequences $<a_n'>_1^\infty$ and $<a_n''>_1^\infty$ converge, where $a_n = [a_n'',a_n']$; moreover, $\lim_{n\to\infty} a_n = [\lim_{n\to\infty} a_n', \lim_{n\to\infty} a_n'']$.

Lemma 1.3. Let $<a_n>_1^\infty$ be a sequence in $S(R)$ such that for each n, $a_{n+1} \subset a_n$. Then $<a_n>_1^\infty$ is convergent and

$$\lim_{n\to\infty} a_n = \bigcap_{n=1}^{\infty} a_n$$

Proof . Write $a_n = [a_n',a_n'']$; for every positive integer n, the inequalities $a_{n+1}' \geq a_n'$ and $a_{n+1}'' \leq a_n''$ both hold. But then the number sequences $<a_n'>_1^\infty$ and $<a_n''>_1^\infty$ converge. Let $a' = \lim_{n\to\infty} a_n'$ and let $a'' = \lim_{n\to\infty} a_n''$. It is easy to check that

$$\lim_{n\to\infty} a_n = [a', a''] = \bigcap_{n=1}^{\infty} a_n .$$

The lemma is proved.

1.2.1 SEGMENT LIMITS

Definition 1.1. The segment a is called the **segment limit** (S-limit) of the segment sequence $<a_n>_1^\infty$ if a is the intersection of all segments that contain all a_n from a certain n on. We shall denote the segment limit by $a = \mathrm{Slim}_{n\to\infty} a_n$.

Formally, the S-limit a of a sequence $\langle a_n \rangle_1^\infty$ is

$$(1.15) \qquad a = \lim_{n \to \infty} b_n = \bigcap_{n=1}^{\infty} \bigvee_{i=n}^{\infty} a_i$$

Obviously, every segment sequence $\langle a_n \rangle_1^\infty$ with $a_n \in S(\bar{R})$ has an S-limit that also lies in $S(\bar{R})$.

The notion of S-limit is a generalization of the notion of Hausdorff metric limit for a finite segment (and hence for a number) sequence, as evidenced by the following statement.

Theorem 1.1. If the finite segment sequence $\langle a_n \rangle_1^\infty$ converges and has a limit a, then $\operatorname{Slim}_{n \to \infty} a_n = \lim_{n \to \infty} a_n = a$.

Proof. In view of the definition of S-limit it follows that

$$(1.16) \qquad a = \lim_{n \to \infty} a_n \subset \operatorname{Slim}_{n \to \infty} a_n ,$$

since every segment that contains all terms of $\langle a_n \rangle_1^\infty$ from a certain n on will contain the segment $a = \lim_{n \to \infty} a_n$ too. On the other hand, every open interval that contains the segment a will also contain all the terms of the sequence $\langle a_n \rangle_1^\infty$ from a certain n on, hence

$$(1.17) \qquad \operatorname{Slim}_{n \to \infty} a_n = \lim_{n \to \infty} a_n \subset a.$$

The conclusion now follows from (1.16) and (1.17).

Actually, as long as $\operatorname{Slim}_{n \to \infty} a_n$ is a finite interval, it must coincide with its Hausdorff metric limit. Write $b_n = \bigvee_{i=n}^{\infty} a_i$. According to the definition of the union of segments, b_n is also a segment and $b_{n+1} \subset b_n$ for $n = 1, 2, \dots$. If $a = \bigcap_{n=1}^{\infty} b_n$ is a finite segment, then according to Lemma 1.3 the sequence $\langle b_n \rangle_1^\infty$ converges and its (Hausdorff metric) limit is

$$a = \lim_{n \to \infty} b_n = \bigcap_{n=1}^{\infty} \bigvee_{i=n}^{\infty} a_i = \operatorname{Slim}_{n \to \infty} a_n .$$

Let us consider some examples. The point-interval sequence $a_n = (-1)^n$ $n = 1, 2, \ldots$ diverges with respect to Hausdorff distance, but it has an S-limit, namely $[-1, 1]$. The sequence $b_n = [n, n+1]$, $n = 1, 2, \ldots$, also diverges, since it is not bounded, but $\text{Slim}_{n\to\infty} b_n = \infty$. For the sequence $c_n = 2^n - (-2)^n$, $n = 1, 2, \ldots$, one has $\text{Slim}_{n\to\infty} c_n = [0,\infty]$.

1.2.2. THEOREMS ON SEGMENT LIMITS

The term-by-term addition, subtraction, multiplication and division of two segment sequences results in a segment sequence, too. The following statement gives the connection between S-limits of these sequences.

Theorem 1.2. Let $\langle a_n\rangle_1^\infty$ and $\langle b_n\rangle_1^\infty$ be two segment sequences. If $*$ denotes an arithmetic operation, then

$$\text{Slim}_{n\to\infty} (a_n * b_n) \subset (\text{Slim}_{n\to\infty} a_n) * (\text{Slim}_{n\to\infty} b_n).$$

Proof . Let c_p be a segment that contains all terms of $\langle a_n\rangle_1^\infty$ for $n \geq p$ and d_q be a segment that contains all terms of $\langle b_n\rangle_1^\infty$ for $n \geq q$. Then from the definition of the arithmetic operations with segments, we have

$$a_n * b_n = \{x : x = \xi * \eta \text{ where } \xi \in a_n \text{ and } \eta \in b_n\},$$

and hence all the terms of the sequence $\langle a_n * b_n\rangle_1^\infty$ for $n \geq \max\{p,q\}$ are contained in the segment $c_p * d_q$, i..e.,

$$a_n * b_n \subset c_p * d_q \qquad (n \geq \max\{p,q\}). \tag{1.18}$$

Since c_p and d_q were chosen arbitrarily, with the only condition imposed on them being that they contain all the terms of the corresponding sequences from a certain index on, the result follows.

In the special case of convergent number sequences, from Theorem 1.2 follows all the known theorems for the limit of a sum, difference, product, or quotient of two convergent number sequences. It is necessary only to note that if $a, b, c \in \mathbb{R}$ and $c \subset a * b$, then $c = a * b$ since $a * b$ is a number. If $*$ denotes division, the number b must of course be different from zero.

The inclusion in Theorem 1.2 cannot be replaced by equality in the general case. For instance, if $a_n = (-1)^n$ and $b_n = (-1)^n$, and $c_n = a_n b_n = 1$, we have

$$\text{Slim}_{n\to\infty} a_n = \text{Slim}_{n\to\infty} b_n = \text{Slim}_{n\to\infty} (-1)^n = [-1,1]$$

$$\text{Slim}_{n\to\infty} c_n = 1$$

whereas $[-1, 1][-1, 1] = [-1, 1] \neq 1$. We could supply similar examples for all arithmetic operations. However, the inclusion in Theorem 1.2 can be replaced by equality if one of the the sequences converges.

Theorem 1.3. Let $\langle a_n\rangle_1^\infty$ and $\langle b_n\rangle_1^\infty$ be two sequences such that one of them converges, say $\langle a_n\rangle_1^\infty$, converges: $\lim_{n\to\infty} a_n = \text{Slim}_{n\to\infty} a_n = a$. Suppose $a * b \neq [-\infty, \infty]$ where $*$ is an arithmetic operation, and $b = \text{Slim}_{n\to\infty} b_n$. If $c_n = a_n * b_n$ and $\text{Slim}_{n\to\infty} c_n = c$, then $c = a * b$.

Proof. According to Theorem 1.2 we have the inclusion $c \subset a * b$. It remains to prove

(1.19) $\qquad a * b \subset c.$

Let d be an arbitrary open interval that contains the segment c. According to the definition of S-limit, d contains all the terms of $\langle c_n\rangle_1^\infty$ from a certain n on, i.e., for $n \geq n(d)$ we have $a_n * b_n \subset d$. Since $a * b \neq [-\infty, \infty]$ and $\langle a_n\rangle_1^\infty$ converges, then from a certain n on, all a_n are finite, and if $*$ denotes division, then from a certain n on, all b_n do not contain zero. If $\xi \in a_n$ and $\eta \in b_n$, then the operation $\xi * \eta$ is defined and $\xi * \eta$ is a continuous function of ξ and η. Since the sequence $\langle a_n\rangle_1^\infty$ converges to the segment a, for sufficiently large n, the inclusion

(1.20) $\qquad a * b_n \subset d$

holds. But if (1.20) holds for all sufficiently large n then $a * b \subset d$ holds, in view of the above remarks. Hence, every interval containing the segment c contains $a * b$ as well, i.e., (1.19) is fulfilled and the theorem is proved.

Corollary 1.2. If $c \in S(\bar{R})$ and $\langle a_n\rangle_1^\infty$ is a segment sequence, then $\text{Slim}_{n\to\infty} (c * a_n) = c * \text{Slim}_{n\to\infty} a_n$.

§1.3 Segment functions

Let $\Omega \subset \bar{R}$. The set of all segment functions defined on Ω will be denote by A_Ω, i.e., every $f \in A_\Omega$ is a certain mapping of Ω into $S(\bar{R})$. By A_Ω will be deonoted the set of all extended real (single valued) functions defined on Ω, i.e., every $f \in A_\Omega$ is a certain mapping of Ω onto $\bar{R}$. Since $\bar{R} \subset S(\bar{R})$, we have $A_\Omega \subset A_\Omega$.

To every $\delta > 0$ there correspond two operators defined on A_Ω with values in $A_{\bar{\Omega}}$ (where $\bar{\Omega}$ is the closure of the set Ω), such that

$$I(\delta, f; x) = I(\Omega, \delta, f; x) = \inf \{ y : y \in f(t),\ t \in [x - \delta, x + \delta] \cap \Omega \},$$

$$S(\delta, f; x) = S(\Omega, \delta, f; x) = \sup \{ y : y \in f(t),\ t \in [x - \delta, x + \delta] \cap \Omega \}.$$

It is directly seen that for the **modulus of continuity**

(1.21) $$\omega(f; \delta) = \sup \{ |f(x') - f(x'')| : |x' - x''| \le \delta,\ x', x'' \in \Omega \}$$

of every function $f \in A_\Omega$ the following equality holds:

(1.22) $$\omega(f; \delta) = \sup_{x \in \Omega} (S(\delta/2, f; x) - I(\delta/2, f; x))$$

The **lower and upper Baire functions** for $f \in A_\Omega$ are

$$I(f; x) = I(\Omega, f; x) = \lim_{\delta \to +0} I(\delta, f; x),$$

$$S(f; x) = S(\Omega, f; x) = \lim_{\delta \to +0} S(\delta, f; x).$$

According to the definition of I and S, they are operators defined on A_Ω with values in $A_{\bar{\Omega}}$.

Definition 1.2. The **completed graph** of a function $f \in A_\Omega$ is the segment function $F(f) \in A_{\bar{\Omega}}$ defined by

$$F(f; x) = F(\Omega, f; x) = [I(f; x), S(f; x)].$$

According to the definition F is an operator defined on A_Ω with values in $A_{\bar{\Omega}}$. Let us denote by $\mathbf{F}_\Omega$ the set of those functions f in A_Ω for which $F(f; x) = f(x)$ for all $x \in \Omega$. Clearly, $\mathbf{F}_\Omega$ consists of all the fixed points of the operator F in A_Ω. As usual, by C_Ω we shall denote all the continuous single valued functions defined on Ω. Obviously, $C_\Omega \subset A_\Omega$ and for every $f \in C_\Omega$ the equality $F(f) = f$ holds, i.e., $C_\Omega \subset \mathbf{F}_\Omega$. It is not difficult to see that

$$C_\Omega = A_\Omega \cap \mathbf{F}_\Omega, \tag{1.23}$$

that is, the single valued function f is continuous on Ω if and only if $F(f) \in A_\Omega$.

If $f \in A_\Omega$, then $I(f)$ (resp. $S(f)$) is lower (resp. upper) semicontinuous. Let us recall that a function $f \in A_\Omega$ is lower (resp. upper) semicontinuous if and only if there exists a sequence $\langle f_n \rangle_1^\infty$ in C_Ω such that for each n $f_n(x) \le f(x)$ (resp. $f_n(x) \ge f(x)$) and $\lim_{n\to\infty} f_n(x) = f(x)$ for all x in Ω.

According to the definition of the set $\mathbf{F}_\Omega$, for every $f \in \mathbf{F}_\Omega$ there exist two functions $\varphi(x) = I(f; x)$ and $\psi(x) = S(f; x)$ where φ is lower semicontinuous and ψ is upper semicontinuous, such that $f(x) = [\varphi(x), \psi(x)]$. The converse statement also holds, as we now show.

Lemma 1.4. If $\varphi, \psi \in A_\Omega$, where φ is lower semicontinuous and ψ is upper semicontinuous, and the inequality $\varphi(x) \le \psi(x)$ holds for every $x \in \Omega$, then the function $f(x) = [\varphi(x), \psi(x)]$ belongs to $\mathbf{F}_\Omega$.

Proof . It is easy to note that the lower semicontinuos functions are the fixed points of the operator I in A_Ω and the upper semicontinuous functions are the fixed points of the operator S in A_Ω. According to the hypotheses of the lemma, we obtain

$$F(f; x) = [I(f; x), S(f; x)] = [I(\varphi; x), S(\psi; x)] = [\varphi(x), \psi(x)] = f(x).$$

This completes the proof.

1.3.1. THE SEGMENT LIMIT OF A SEGMENT FUNCTION

Definition 1.3. Let $f \in A_\Omega$ and $x_0 \in \bar{\Omega}$. The segment a is called the **segment limit (S-limit) of the function f at the point x_0** if a is the minimal segment that contains all the segments of the form $\mathrm{Slim}_{n\to\infty} f(x_n)$ where $x_n \in \Omega, x_n \neq x_0$, and $\lim_{n\to\infty} x_n = x_0$. The segment a will be denoted by

$$a = \underset{x\to x_0}{\mathrm{Slim}}\ f(x).$$

From the defintion of the set F_Ω directly follows

Lemma 1.5. The relation

$$\underset{x\to x_0}{\mathrm{Slim}}\ f(x) \subset f(x_0) \qquad (x_0 \in \Omega) \tag{1.24}$$

is a necessary and sufficient condition for $f \in A_\Omega$ to belong to F_Ω.

Proof. If for a certain x_0 in Ω the inclusion (1.24) fails, then the equality

$$f(x_0) = [I(f; x_0), S(f; x_0)]$$

also does not hold at this point, so that $f \notin F_\Omega$. Thus, necessity is established. Conversely, suppose (1.24) holds for every $x_0 \in \Omega$. Then the function $\varphi(x) = \min\{y : y \in f(x)\} \in A_\Omega$ is lower semicontinuous and the function $\psi(x) = \max\{y : y \in f(x)\} \in A_\Omega$ is upper semicontinuous. Since $f(x) = [\varphi(x), \psi(x)]$, by Lemma 1.4 we have $f \in F_\Omega$. The lemma is proved.

Let us show that set F_Ω is closed with respect to all arithmetic operations.

Theorem 1.4. If $f, g \in F_\Omega$ and $*$ is one of the arithmetic operations, then

$$\underset{x\to x_0}{\mathrm{Slim}}\ (f(x) * g(x)) \subset f(x_0) * g(x_0) \tag{1.25}$$

and hence $f * g \in F_\Omega$.

Proof. Let $\langle x_n \rangle_1^\infty$ be an arbitrary sequence such that $x_n \in \Omega$, $x_n \neq x_0$, and $\lim_{n\to\infty} x_n = x_0$. By Theorem 1.2,

(1.26) $$\underset{x \to x_0}{\mathrm{Slim}} \, (f(x_n) * g(x_n)) = \underset{x \to x_0}{\mathrm{Slim}} \, f(x_n) * \underset{x \to x_0}{\mathrm{Slim}} \, g(x_n) \subset f(x_0) * g(x_0).$$

The last inclusion follows from Lemma 1.5, since $f, g \in F_\Omega$. But since $\langle x_n \rangle_1^\infty$ was chosen arbitrarily, we see that from (1.26) follows (1.25).

1.3.2. SEGMENT DERIVATIVES

Using the notion of S-limit, we shall define the S-derivative for every function from A_Ω and this derivative is a generalization of the usual derivative.

Definition 1.4. The segment function

(1.27) $$D(f; x) = \underset{h \to 0}{\mathrm{Slim}} \frac{f(x + h) - f(x)}{h}$$

is called the **segment derivative** (S-derivative) of the function $f \in A_\Omega$.

Obviously the operation of segment differentiation is well-defined for every function in A_Ω and its values also belong to A_Ω. On the other hand, according to (1.27), if $f \in A_\Omega$ has a derivative $f'(x_0)$ at the point $x_0 \in \Omega$, then

(1.28) $$D(f; x_0) = f'(x_0).$$

Bl. Sendov [26] established a series of properties of S-derivative. Here we shall give only those that will be necessary later on.

Theorem 1.5. If $f, g \in A_\Omega$ then for every $x \in \Omega$,

(1.29) $$D(f + g; x) \subset D(f; x) + D(g; x).$$

and if the function f has a derivative at the point x_0 in the usual sense, then

(1.30) $$D(f + g; x_0) = D(f; x_0) + D(g; x_0) = f'(x_0) + D(g; x_0).$$

Proof. By Theorem 1.2 we have

$$D(f + g; x) = \underset{h \to 0}{\mathrm{Slim}} \frac{f(x + h) + g(x + h) - f(x) - g(x)}{h}$$

$$\subset \underset{h\to 0}{\mathrm{Slim}} \frac{f(x+h) - f(x)}{h} + \underset{h\to 0}{\mathrm{Slim}} \frac{g(x + h) - g(x)}{h}$$

$$= D(f; x) + D(g; x).$$

Thus, (1.29) is proved.

Let us assume that the function f has a derivative at the point x_0, so that $D(f; x_0) = f'(x_0) \in R$. This means that f is continuous in a neighborhood $\Delta \subset \Omega$ of the point x_0. Then from $h(x) = f(x) + g(x)$, it follows that $g(x) = h(x) - f(x)$ for every $x \in \Delta$. Hence, according to (1.29), we have

$$(1.31) \qquad D(g; x_0) \subset D(h; x_0) + D(-f; x_0) = D(h; x_0) - f'(x_0) = D(h; x_0) - D(f; x_0).$$

But since $D(f; x_0) = f'(x_0) \in R$, it now follows from (1.31) that

$$D(f; x_0) + D(g; x_0) \subset D(h; x_0) = D(f + g; x_0).$$

This last inclusion and (1.29) give (1.30), completing the proof.

Let us note that in general it is impossible to replace (1.29) by (1.30). We shall consider as an example the function $f(x) = |x|$. From the definition of S-derivative, it follows that

$$D(|t|; x) = \begin{cases} -1 & \text{if } x < 0 \\ [-1, 1] & \text{if } x = 0 \\ 1 & \text{if } x > 0 \end{cases} .$$

Then,

$$0 = D(0; x) = D(|t| - |t|; x) \subset D(|t|; x) - D(|t|; x)$$

$$= \begin{cases} 0 & \text{if } x \neq 0 \\ [-2, 2] & \text{if } x = 0 \end{cases},$$

and it is not possible to replace the inclusion by equality everywhere.

We can also prove laws for the S-derivative of a product, quotient, and a composition; they are similar to the corresponding laws for the usual derivative, where equality is replaced by inclusion.

The notion of convexity of k-th order for segment functions is introduced in the expected way. The k-th **divided difference** $[x_0, x_1, x_2, \ldots, x_k; f]$ for $f \in A_\Omega$ at the points $x_0, x_1, x_2, \ldots, x_k \in \Omega$ with $x_0 < x_1 < x_2 < \cdots < x_k$ is defined recusively as follows:

$$[x_i; f] = f(x_i)$$

$$[x_i, x_{i+1}; f] = ([x_{i+1}; f] - [x_i; f])/(x_{i+1} - x_i)$$

$$[x_i, x_{i+1}, \ldots, x_{i+l}; f] = \frac{[x_{i+1}, x_{i+2}, \ldots, x_{i+l}; f] - [x_i, x_{i+1}, \ldots, x_{i+l-1}; f]}{x_{i+l} - x_i}$$

Definition 1.5. The segment function $f \in A_\Omega$ is called **convex of k-th order** on Ω if for every $k + 1$ distinct points $x_0, x_1, x_2, \ldots, x_k \in \Omega$, we have the inequality $[x_0, x_1, x_2, \ldots, x_k; f] \geq 0$.

According to this definition, for single valued functions, convexity of zero order means non-negativity, convexity of first order means monotonicity, convexity of second order means usual convexity, etc.

Let us denote by D^k the k-fold application of the operator of segment differentiation. The following statement is valid.

Theorem 1.6. A necessary and sufficient condition for a function $f \in A_\Omega$ to be convex of k-th order on Ω is the inequalty $D^k(f; x) \geq 0$ for each x in Ω.

The proof follows directly from the definition of convexity of order k and the use of S-limit when points tend to a single point.

1.3.3. Segment Continuity

In view of Lemma 1.5, it is natural to name the functions in F_Ω **segment continuous** (S-continuous). Let us consider certain properties of S-continuous functions, when Ω is a segment Δ. We first have an intermediate value theorem.

Lemma 1.6. If $f \in F_\Delta$ where $\Delta = [a, b]$ and for $x_1, x_2 \in \Delta$ and $y_1 \in f(x_1)$ and $y_2 \in f(x_2)$, the inequalities $x_1 \leq x_2$ and $y_1 < y_2$ hold, then for each $y_0 \in (y_1, y_2)$ there exists $x_0 \in [x_1, x_2]$ such that $y_0 \in f(x_0)$.

Proof. If $S(f; x_1) \geq y_0$, then we can set x_0 equal to x_1. Otherwise, $S(f; x_1) < y_0$ and we set $x_0 = \sup\{x : x \in [x_1, x_2]$ and $S(f; x) \leq y_0\}$. Then by the lower semicontinuity of $I(f)$ we get $S(f; x_0) \geq y_0 \geq I(f; x_0)$, and the lemma is proved.

The proof of the following lemma is similar to the proof of the previous one.

Lemma 1.7. If $f, g \in \mathbf{F}_\Delta$, and $x_1, x_2 \in \Delta$ are such that $I(f; x_1) > S(g; x_1)$ and $I(g; x_2) > S(f; x_2)$, then there exists $x_0 \in [x_1, x_2]$ for which the set $f(x_0) \cap g(x_0) \neq \emptyset$.

If $\Delta = [a, b]$ is a finite closed interval, then the graph of the function $f \in \mathbf{F}_\Delta$ is a bounded and closed point set in the plane R_2 and it may have positive area. We have the following result of Bl. Sendov and V. Popov [1].

Theorem 1.7. Let f be a real valued bounded function with domain Δ where $\Delta = [a, b] \in S(R)$. A necessary and sufficient condition for the function f to be integrable in the Riemann sense is that the completed graph $F(f)$ of this function has planar measure zero .

Proof . The completed graph $F(f)$ of the function f is a bounded and closed set and hence this set is measurable. Its measure is

$$\mu(F(f)) = \iint_{F(f)} dx\, dy$$

By the Fubini Theorem,

$$\iint_{F(f)} dx\, dy = \int_a^b (S(f; x) - I(f; x))\, dx = \int_a^b S(f; x)\, dx - \int_a^b I(f; x)\, dx,$$

where $S(f)$ and $I(f)$ are the upper and lower Baire functions for f. They are semicontinuous and are therefore integrable in the Lebesgue sense. Hence, it follows that if $\mu(F(f)) = 0$, then

$$\int_a^b S(f; x)\, dx = \int_a^b I(f; x)\, dx \tag{1.32}$$

and vice versa. But it follows from (1.32) that f is integrable in the Riemann sense. The converse is also true, as Riemann integrability of f means that the points of discontinuity of f have measure zero and $S(f; x) = I(f; x)$ almost everywhere, so that (1.32) holds. The theorem is proved.

1.3.4. H-CONTINUITY

Definition 1.6. Let us denote by $\mathbf{H}_\Omega$ the set of those functions $f \in \mathbf{F}_\Omega$ for which the conditions $g \in \mathbf{F}_\Omega$ and $g(x) \subset f(x)$ for all x jointly imply $f = g$. A function $f \in \mathbf{H}_\Omega$ is called **Hausdorff continuous** (H-continuous).

The name "Hausdorff continuous" will be justified later on when we supply a characterization of H-continuity in terms of Hausdorff distance.

Theorem 1.8. The equalities

$$S(I(f)) = S(f), \qquad I(S(f)) = I(f) \tag{1.33}$$

are necessary and sufficient conditions for the function $f \in \mathbf{F}_\Omega$ to be H-continuous.

Proof. For necessity, if (1.33) fails then $g(x) = [I(S(f; x)), S(I(f; x))]$ would be properly contained in $f(x) = [I(f; x), S(f; x)]$ for some x. Thus, for this g, we would have $g \in \mathbf{F}_\Omega$, $g(x) \subset f(x)$ for all x, but $f \neq g$.

For sufficiency, suppose (1.33) holds and $F(g) \subset F(f)$, i.e.,

$$I(f; x) \leq I(g; x) \leq S(g; x) \leq S(f; x) \tag{1.34}$$

for all $x \in \Omega$. According to the monotonicity of the operators S and I, we obtain from (1.33) and (1.34) that $S(I(g)) = S(f)$ and $I(S(g)) = I(f)$. On the other hand, from the definition of the Baire functions, it follows that

$$S(I(g); x) \leq S(g; x), \qquad I(S(g); x) \geq I(g; x),$$

and therefore $S(g) = S(f)$ and $I(g) = I(f)$, i.e., $F(g) = F(f)$.

Corollary 1.3 (V. Veselinov [12]). The equality $F(S(f)) = F(I(f))$ is a necessary and sufficient condition for $f \in \mathbf{H}_\Omega$.

Corollary 1.4. To every function $f \in \mathbf{H}_\Omega$ there corresponds $g \in H_\Omega$ such that $g \subset f$ and such that $F(g) = F(f) = f$.

Proof. For every $x \in \Omega$ we have $f(x) = [I(f; x), S(f; x)]$. Let $g(x) = S(f; x)$. By the upper semicontinuity of $S(f)$ we have $S(g; x) = S(f; x)$. By Theorem 1.8, we have $I(g; x) = I(f; x)$. This says that the completed graph of g is f.

Let us note that H-continuity is not preserved by shrinking Ω. Precisely, if $\Delta \subset \Omega$ and $f \in H_\Omega$, then we cannot conclude that $f \in H_\Delta$ if we consider the restriction of f to Δ. For example, let

$$\sigma(x) = \begin{cases} 0 & \text{if } -1 \leq x < 0 \\ [0, 1] & \text{if } x = 0 \\ 1 & \text{if } 0 < x \leq 1 \end{cases}$$

Obviously, $\sigma \in H_{[-1, 1]}$ but $\sigma \notin H_{[0, 1]}$ although $[0, 1] \subset [-1, 1]$.

Definition 1.7. We shall denote by H_Ω the set of those single valued functions $f \in A_\Omega$ for which $F(f) \in H_\Omega$.

The set H_Ω is sufficiently extensive. It is obvious that $C_\Omega \subset H_\Omega$, and it is not difficult to show that all the monotone functions of A_Δ with $\Delta = [a, b]$ belong to H_Δ if they are continuous at the endpoints of the segment, to the right at a and to the left at b, respectively. P. Korovkin [2] has shown that H_Δ contains functions that are not Lebesgue measurable. We find it worthwhile to demonstrate the existence of such functions here.

Let P be a perfect set of positive measure without interior points in the closed interval [0, 1] containing 0 and 1. We define a set K_P of functions whose domain is the closed interval [0, 1] in the following way. It is well known that $[0, 1] \setminus P$ is the union of a disjoint denumerable set of open intervals $\Delta_i = (a_i, b_i)$. On each subinterval $[a_i, b_i]$, all the functions $f \in K_P$ are defined in the same way:

$$f(a_i) = f(b_i) = 1, \qquad f((a_i + b_i)/2) = 0,$$

and f is linear on each of the closed subintervals $[a_i, (a_i + b_i)/2]$, $[(a_i + b_i)/2, b_i]$.

On the set P every function $f \in K_P$ is defined in arbitrary way with values in [0,1]. Since the measure of P differs from zero, then among the functions that belong to K_P there are nonmeasurable ones. Let us prove that $K_P \subset H_{[0, 1]}$.

Lemma 1.8. If $f \in K_P$ then for each $x \in [0, 1] \setminus P$,

$$f(x) = S(f; x) = I(f; x) \tag{1.35}$$

and for $x \in P$

(1.36) $I(f; x) = 0, \quad S(f; x) = 1$

Proof. The equalities (1.35) follow directly from the condition that for $x \in [0, 1]\backslash P$ then x is an interior point of some (a_i, b_i), and on such intervals f is continuous and (1.35) holds. The equalities (1.36) follow from the condition that P has no interior points and that on every subinterval (a_i, b_i), the range of the function f is $[0, 1]$. The lemma is proved.

From the last lemma, it follows that if $f, g \in K_P$ then $F(f) = F(g)$ and therefore from $g \subset F(f)$ it follows that $F(g) = F(f)$. Thus it is proved that all functions in K_P belong to H_Ω.

Let us prove one necessary condition for H-continuity. For $f \in F_\Omega$ and x_0 an interior point of Ω we introduce the following notation:

$$S(f; x_0 - 0) = \limsup_{x \to x_0 - 0} S(f; x), \qquad S(f; x_0 + 0) = \limsup_{x \to x_0 + 0} S(f; x)$$

and similarly,

$$I(f; x_0 - 0) = \limsup_{x \to x_0 - 0} I(f; x), \qquad I(f; x_0 + 0) = \limsup_{x \to x_0 + 0} I(f; x)$$

The point $x_0 \in \Omega$ will be called a **left boundary point** of Ω if there exists $\delta > 0$ such that the open interval $(x_0 - \delta, x_0)$ does not contain points of Ω. The notion of **right boundary point** is defined similarly.

Theorem 1.9. Necessary conditions for $f \in F_\Omega$ to be H-continuous on Ω, that is $f \in H_\Omega$, are the following:

(a) If x_0 is an interior point of Ω, then for each $y \in f(x_0)$ we have

$$\min\{I(f; x_0 - 0), I(f; x_0 + 0)\} \le y \le \max\{S(f; x_0 - 0), S(f; x_0 + 0)\};$$

(b) If x_0 is a left boundary point of Ω, then for each $y \in f(x_0)$ we have

$$I(f; x_0 + 0) \le y \le S(f; x_0 + 0);$$

(c) If x_0 is a right boundary point of Ω, then for each $y \in f(x_0)$ we have

$$I(f; x_0 - 0) \le y \le S(f; x_0 - 0).$$

Proof. Let $f \in F_\Omega$ and let the hypotheses of the theorem be violated at a certain $x_0 \in \Omega$. We define a function $g \in F_\Omega$ as follows. For $x \neq x_0$, set $g(x) = f(x)$. If x_0 is an interior point of Ω

$$g(x_0) = \frac{1}{2}(\min \{I(f; x_0 - 0), I(f; x_0 + 0)\} + \max \{S(f; x_0 - 0), S(f; x_0 + 0)\}).$$

If x_0 is a left boundary point of Ω, then

$$g(x_0) = \frac{1}{2}(I(f; x_0 + 0) + S(f; x_0 + 0)),$$

and if x_0 is a right boundary point of Ω, then

$$g(x_0) = \frac{1}{2}(I(f; x_0 - 0) + S(f; x_0 - 0)$$

Then $F(g) \subset F(f)$ and $F(g) \neq F(f)$. Hence, according to the definition of H-continuity, it follows that $f \notin H_\Omega$.

Remarks . For another way to define the derivative of a segment function using a nonstandard operation of subtraction, see S. Markov [2]. A definition of derivative with respect to Hausdorff distance was given by Bl. Sendov, B. Penkov, V. Popov and S. Markov [1]. A detailed characterization of the segment derivative of a Lipschitz function is given in Bl. Sendov, S. Tashev, and P. Petrushev [1].

A sufficient condtion for the existence of a solution of a differential equation, that involves a segment derivative is considered in S. Tashev [2]. Segment derivation of a function is used in S. Tashev [4].

The notion of the complete graph of a function is used in many situations, even when Hausdorff distance is not considered. As an example, see P. Petrushev [6]. A. P. Petukhov introduced the canonical graph of a function for Hausdorff approximation of measurable functions.

General limits of sequences of sets in a metric space without convexity are considered in K. Kuratowski [1]. In the theory of convex multivalued functions (see e.g., R. Phelps [1]), segment continuity appears in the guise of upper semicontinuity, whereas H-continuity is closely related to the notion of convex usco mapping. Continuity properties for multivalued mappings in a general setting are discussed in C. Berge [1], K. Kuratowski [1], R. Smithson [1], and in E. Klein and A. Thompson [1].

Chapter 2

Hausdorff Distance

§ 2.1. Hausdorff distance between subsets of a metric space

Let (M,ρ) be a metric space, i.e., to every pair $a, b \in M$ there corresponds a nonnegative number $\rho(a, b)$ such that for all $a, b, c \in M$

$$\rho(a, b) = \rho(b, a) \geq 0, \tag{2.1}$$

$$\rho(a, b) = 0 \Leftrightarrow a = b, \tag{2.2}$$

$$\rho(a, b) \leq \rho(a, c) + \rho(c, b). \tag{2.3}$$

Usually the distance between the point $b \in M$ and the set $A \subset M$ is defined as

$$\rho(b, A) = \inf_{a \in A} \rho(a, b),$$

and the "distance" between two sets $A, B \subset M$ is defined as

$$\rho(A, B) = \inf_{a \in A} \rho(a, B) = \inf_{b \in B} \rho(b, A). \tag{2.4}$$

The "distance" between two subsets of a given metric space as defined by (2.4) is not a true distance, since it does not satisfy the condition (2.2). It is not difficult to see that $\rho(A, B) = 0$ when A and B have a common point or a common limit point.

In his famous book, F. Hausdorff [1] defined the distance between the subsets of a given metric space in another way. Let $A \subset M$. We shall denote by $U(\alpha, A)$ where $\alpha \geq 0$ the set of all points $x \in M$ such that $\rho(x; A) \leq \alpha$, i.e.,

$$U(\alpha, A) = \{x : x \in M \text{ and } \rho(x; A) \leq \alpha\}.$$

The infimimum of those α for which both $U(\alpha, A) \supset B$ and $U(\alpha, B) \supset A$ is called the **Hausdorff distance** or **distance in the sense of deviation of sets** between the sets A and B induced by the distance ρ. We denote this distance by $r(\rho; A, B)$. The distance $r(\rho; A, B)$ does not satisfy condition (2.2) in the totality of nonempty subsets of the space M. For instance, if a is an arbitrary point of the line R with the usual metric, and we have $A = \{x : \rho(x, a) < 1\}$ and $B = \{x : \rho(x, a) \leq 1\}$, then $A \neq B$, but $r(\rho; A, B) = 0$.

It is easy to see that if F(M) is the set of all closed subsets of the metric space M, then $r(\rho; A, B)$ is an infinite valued metric on F(M), which satisfies the conditions (2.1) - (2.3). Clearly, $r(\rho; A, B)$ can become ∞, for example, when A is bounded and B is unbounded. We shall admit as a distance also the symbol ∞. Of course, we can eliminate it by considering only bounded sets, but this is inappropriate, since later on we shall need to deal with unbounded sets. There are many cases when A and B are unbounded, yet the number $r(\rho; A, B)$ is finite, e. g., this happens when the sets are parallel lines in the plane.

In the sequel, the following equivalent form of Hausdorff distance will be often more convenient:

$$r(\rho; A, B) = \max \{ \sup_{a\in A} \inf_{b\in B} \rho(a, b), \sup_{b\in B} \inf_{a\in A} \rho(a, b)\}. \tag{2.5}$$

The following two simple lemmas are useful for the estimation of Hausdorff distance.

Lemma 2.1. Let $A, B \subset M$. If for a certain point $a \in A$ we have $U(\alpha, a) \cap B = \emptyset$, then $r(\rho; A, B) \geq \alpha$.

Proof. From the definition of $U(\alpha, a)$ and the hypotheses of the lemma, it follows that for every point $b \in B$ we have $\rho(a, b) \geq \alpha$. As a result, $\inf_{b\in B} \rho(a, b) \geq 0$ so that

$$\sup_{a\in A} \inf_{b\in B} \rho(a, b) \geq \alpha.$$

But from the last inequality, according to (2.5), the assertion of the lemma follows.

Lemma 2.2. Let $A, B \subset M$, satisfying the following conditions:

(a) for every point $a \in A$ there exist points $a' \in M$ and $b \in B$ such that $a, b \in U(\alpha, a')$;

(b) for every point $b \in B$ there exist points $b' \in M$ and $a \in A$ such that $a, b \in U(\alpha, b')$.

Then $r(\rho; A, B) \leq 2\alpha$.

Proof. According to the hypotheses of the lemma, for every point $a \in A$ there exist $a' \in M$ and $b \in B$ such that both $\rho(a, a') \leq \alpha$ and $\rho(a', b) \leq \alpha$. By (2.3)

$$\rho(a, b) \leq \rho(a, a') + \rho(a', b) \leq 2\alpha.$$

Thus for every $a \in A$ the inequality $\inf_{b \in B} \rho(a, b) \leq 2\alpha$ holds, and consequently

$$\sup_{a \in A} \inf_{b \in B} \rho(a, b) \leq 2\alpha. \tag{2.6}$$

In a similar way we obtain

$$\sup_{b \in B} \inf_{a \in A} \rho(a, b) \leq 2\alpha. \tag{2.7}$$

Combining (2.5), (2.6), and (2.7), the result follows.

§ 2.2. The metric space F_Ω

We shall now define a distance in the set of functions F_Ω, which will be called Hausdorff distance or H-distance.

Having in mind the definition of Hausdorff distance between two subsets of a metric space, let us consider the plane R_2 equipped with the usual metric as a metric space (R_2, ρ) and the graphs of two functions $f, g \in F_\Omega$ as subsets of R_2. Then identifying segment functions with their graphs we have

$$r(\rho; f, g) = \max \{ \sup_{A \in f} \inf_{B \in g} \rho(A, B),\ \sup_{A \in g} \inf_{B \in f} \rho(A, B) \}. \tag{2.8}$$

Theorem 2.1. The distance (2.8) in F_Ω meets the axioms for a metric:

(2.9) $r(\rho; f, g) = r(\rho; g, f)$,

(2.10) $r(\rho; f, g) = 0 \Leftrightarrow f = g$,

(2.11) $r(\rho; f, g) \leq r(\rho; f, h) + r(\rho; h, g)$.

Proof. Hausdorff distance satisfies the metric axioms on the closed subsets of the plane, and its restriction to any family of closed subsets will also be a metric. Thus, it suffices to show that the graph of each function $f \in F_\Omega$ is a closed subset of the plane, because $F(f) = f$.

To this end, let $\langle x_n \rangle_1^\infty$ be a sequence in Ω and for each n let $y_n \in f(x_n)$. Suppose $\langle A_n(x_n, y_n) \rangle_1^\infty$ is convergent to $A(x_0, y_0)$. We must show $y_0 \in f(x_0)$. Since $f(x) = F(f; x) = [I(f; x), S(f; x)]$, for each n we have

$$I(f; x_n) \leq y_n \leq S(f; x_n).$$

From the upper (resp. lower) semicontinuity of $S(f; x)$ (resp. $I(f; x)$) at x_0, we get $I(f; x_0) \leq y_0 \leq S(f; x_0)$, so that $y_0 \in F(f; x_0) = f(x_0)$. This completes the proof.

For our aims it is conveneint to introduce a parametrized family of distances as follows: for each $\alpha > 0$,

$$\rho_\alpha(A(x,y), B(\xi, \eta)) = \max \{ \alpha^{-1}|x - \xi|, |y - \eta| \}$$

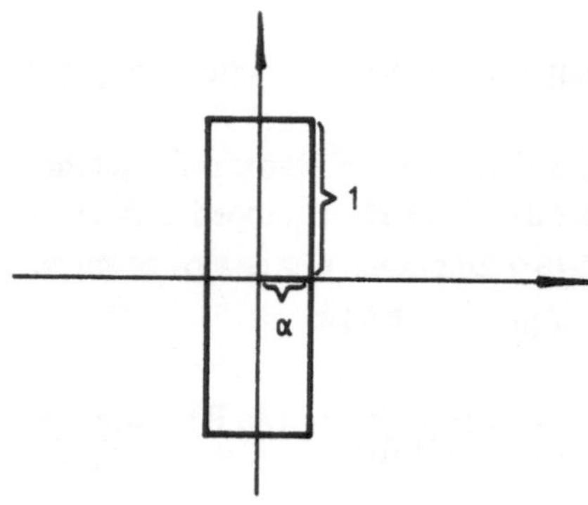

FIGURE 2.1

With respect to each metric, the unit sphere is a rectangle with sides that are parallel to the coordinate axes and are in the ratio $1:\alpha$ (see Figure 2.1).

The Hausdorff distance $r(\rho_\alpha; f, g)$ between two functions $f, g \in \mathbf{F}_\Omega$ will be denoted by $r(\Omega, \alpha; f, g)$. If Ω is understood, we shall abbreviate this by $r(\alpha; f, g)$. The distance $r(\alpha; f, g)$ is called **Hausdorff distance** (H-distance) **with parameter** α.

For $f, g \in \mathbf{F}_\Omega$, we can write

$$(2.12) \qquad r(\alpha; f, g) = \max \{ \sup_{x\in\Omega} \sup_{y\in f(x)} \inf_{\xi\in\Omega} \inf_{\eta\in g(\xi)} \max \{ \alpha^{-1}|x-\xi|, |y-\eta|\}, \sup_{x\in\Omega} \sup_{y\in g(x)} \inf_{\xi\in\Omega} \inf_{\eta\in f(\xi)} \max \{ \alpha^{-1}|x-\xi|, |y-\eta|\}\}.$$

Now let us present variants of Lemmas 2.1 and 2.2 for the Hausdorff distance $r(\alpha; f, g)$ with a parameter $\alpha > 0$.

Lemma 2.3. Let $f, g \in \mathbf{F}_\Omega$. If for a certain $x_0 \in \Omega$ and for all $x \in [x_0 - \alpha\delta, x_0 + \alpha\delta] \cap \Omega$ we have $|y - \eta| > \delta$ for every $y \in f(x)$ and every $\eta \in g(x_0)$, then $r(\alpha; f, g) \geq \delta$.

Lemma 2.4. Let $f, g \in \mathbf{F}_\Omega$, and suppose the following conditions hold:

(1) for every $x \in \Omega$ and for every $y \in f(x)$ there exists $\xi \in [x - \alpha\delta, x + \alpha\delta] \cap \Omega$ and $\eta \in g(\xi)$ such that $|y - \eta| \leq \delta$;

(2) for every $x \in \Omega$ and for every $y \in g(x)$ there exists $\xi \in [x - \alpha\delta, x + \alpha\delta] \cap \Omega$ and $\eta \in f(\xi)$ such that $|y - \eta| \leq \delta$.

Then $r(\alpha; f, g) \leq \delta$.

We now focus our attention on the subset of $\mathbf{F}_\Omega$ consisting of those segment functions with bounded range. In order to prove completeness of this class with respect to H-distance when the domain is compact (Bl. Sendov and B. Penkov [2]), we need some auxilliary statements that will be proved in advance.

Lemma 2.5. Let $f \in A_\Omega$ have bounded range. Let $M = \sup \{|y| : y \in f(x), x \in \Omega\} < \infty$. Then for every $\delta > 0$ there exists a continuous function $\psi \in C_\Omega$ such that

$$(2.13) \qquad S(\delta, f; x) \leq \psi(x) \leq S(4\delta, f; x)$$

for all $x \in \Omega$ and such that the modulus of continuity of ψ satisfies the inequality

(2.14) $$\omega(\psi; t) \le 2\delta^{-1}Mt\,.$$

Proof. Let $\delta > 0$, $x_i = 4i\delta$ and $\Delta_i = [x_{i-1}, x_i] \cap \Omega$. Let us denote by $\Omega(\delta)$ the set of all integers i such that Δ_i is not empty. For $i \in \Omega(\delta)$ we define

$$M_i = \sup\{y : y \in f(t),\ t \in \Delta_i\},$$

$$M_i' = \begin{cases} \max\{M_{i-1}, M_i\} & \text{if } i-1 \in \Omega(\delta) \\ M_i & \text{if } i-1 \notin \Omega(\delta) \end{cases}$$

$$M_i'' = \begin{cases} \max\{M_i, M_{i+1}\} & \text{if } i+1 \in \Omega(\delta) \\ M_i & \text{if } i+1 \notin \Omega(\delta) \end{cases}.$$

Let ψ be the following piecewise linear function defined on Δ_i (and consequently on Ω, since $\cup\{\Delta_i : i \in \Omega(\delta)\} = \Omega$) by:

(2.15) $$\psi(x) = \begin{cases} M_i' & \text{if } x_{i-1} \le x \le x_{i-1} + \delta \\ \delta^{-1}(M_i - M_i')(x - x_{i-1} - \delta) + M_i' & \text{if } x_{i-1} + \delta \le x \le x_{i-1} + 2\delta \\ \delta^{-1}(M_i - M_i'')(x_i - \delta - x) + M_i'' & \text{if } x_i - 2\delta \le x \le x_i - \delta \\ M_i'' & \text{if } x_i - \delta \le x \le x_i \end{cases}$$

It is easy to verify that ψ is continuous and has a modulus of continuity $\omega(\psi; t)$ that satisfies (2.14), since the inequality $|M_i| \le M$ holds for all $i \in \Omega(\delta)$. On the other hand, according to the definition of $S(\delta, f)$, from (2.15) follows (2.13).

The proof of the next lemma is the same.

Lemma 2.6. Let $f \in A_\Omega$ have bounded range. Let $M = \sup\{|y| : y \in f(x), x \in \Omega\} < \infty$. Then for every $\delta > 0$ there exists a continuous function $\varphi \in C_\Omega$ such that

(2.16) $$I(4\delta, f; x) \le \varphi(x) \le I(\delta, f; x)$$

for all $x \in \Omega$ and such that the modulus of continuity of φ satisfies the inequality $\omega(\varphi; t) \le 2\delta^{-1}Mt\,.$

Theorem 2.2. Let Ω be compact. Then the subset of F_Ω consisting of the functions with finite segment values equipped with H-distance is a complete metric space.

Proof . Let $\langle f_n \rangle_1^\infty$ be a Cauchy sequence of functions in F_Ω consisting of the functions whose values are finite segments. Then for every $\varepsilon > 0$ there exists $n(\varepsilon)$ such that for $p, q > n(\varepsilon)$ the inequality

$$(2.17) \qquad r(\alpha; f_p, f_q) < \varepsilon$$

holds. It is necessary to prove that there exists f_0 in F_Ω with finite segment values such that

$$(2.18) \qquad \lim_{n \to \infty} r(\alpha; f_n, f_0) = 0.$$

From the compactness of Ω it is easy to check that each function in the sequence actually has bounded range, for an upper (resp. lower) semicontinuous real function with compact domain has a largest (resp. smallest) value. Also, it follows from the definition of Hausdorff distance and the fact the functions have bounded range that $\sup\{|y| : y \in f_n(x),\ x \in \Omega,\ n = 1, 2, 3, \ldots\} = M < \infty$.

For every f_n and $\delta > 0$ we choose a function $\psi_{n,\delta}$ as in Lemma 2.5. All the functions of the sequence $\langle \psi_{n,\delta} \rangle_1^\infty$ have modulus of continuity $\omega(\psi_{n,\delta}; t) \le 2\delta^{-1}Mt$. Consequently the function ψ_δ defined by

$$\psi_\delta(x) = \liminf_{n \to \infty} \psi_{n,\delta}(x)$$

has the same modulus of continuity. The same can be said for

$$\varphi_\delta(x) = \limsup_{n \to \infty} \varphi_{n,\delta}(x)$$

where the function $\varphi_{n,\delta}$ is chosen for f_n as in Lemma 2.6. From the way we have defined the functions $\psi_{n,\delta}$, by the equality (2.15) it follows that

$$\psi_{n,2\delta}(x) \ge \psi_{n,\delta}(x) \quad \text{and} \quad \varphi_{n,2\delta}(x) \le \varphi_{n,\delta}(x) ,$$

and hence we have

(2.19) $\varphi_{2\delta}(x) \le \varphi_\delta(x) \le \psi_\delta(x) \le \varphi_{2\delta}(x).$

Then for $\delta = 1/2, 1/4, 1/8, \ldots$ we obtain two sequences of functions $<\varphi_n>_1^\infty$ and $<\psi_n>_1^\infty$ such that $\varphi_m(x) \le \varphi_{m+1}(x) \le \psi_{m+1}(x) \le \varphi_m(x)$ and hence, if we write

(2.20) $\varphi(x) = \lim_{m \to \infty} \varphi_m(x), \qquad \psi(x) = \lim_{m \to \infty} \psi_m(x),$

then φ is lower semicontinuous and ψ is upper semicontinuous. Thus, by Lemma 1.4, the function

(2.21) $f_0(x) = [\varphi(x), \psi(x)]$

belongs to F_Ω. Clearly, the values of the segment function f_0 are finite segments.

Let us now show that (2.18) holds for the function f_0. This will complete the proof of the theorem. To this end, it is sufficient to establish for every $\varepsilon > 0$ the existence of $n_0 = n_0(\varepsilon)$ such that for $n > n_0$, we have

(2.22) $r(\alpha; f_n, f_0) < \varepsilon.$

Since Ω is compact and the functions $\psi_{n,\delta}$ are uniformly continuous (because $\omega(\psi_{n,\delta}; t) \le 2\delta^{-1}Mt$), for each $\varepsilon > 0$ and $x \in \Omega$ there exists $n_0 = n_0(\varepsilon)$ such that for all x in Ω and $n > n_0$ we have

(2.23) $|\psi_{n,\delta}(x) - \psi_\delta(x)| < \varepsilon.$

According to Lemma 2.5,

(2.24) $S(\delta, f_n; x) \le \psi_{n,\delta}(x) \le S(4\delta, f_n; x).$

We obtain from (2.23) and (2.24) that

(2.25) $S(\delta, f_n; x) - \varepsilon \le \psi_\delta(x) \le S(4\delta, f_n; x) + \varepsilon.$

The left-side inequality of (2.25) gives $S(f_n; x) - \varepsilon \le \psi_\delta(x)$ and according to (2.20) and (2.21) we have

(2.26) $S(f_0; x) \ge S(f_n; x) - \varepsilon\,.$

The right-side inequality of (2.25) according to (2.19) gives

(2.27) $S(f_0; x) \le S(4\delta, f_n; x) + \varepsilon.$

We can prove in similar way that

(2.28) $I(4\delta, f_n; x) - \varepsilon \le I(f_0; x) \le I(f_n; x) + \varepsilon.$

From (2.26) - (2.28) and $\delta = \alpha\varepsilon/4$ it follows that

(2.29) $I(\alpha\varepsilon, f_n; x) - \varepsilon \le I(f_0; x) \le S(f_0; x) \le S(\alpha\varepsilon, f_n; x) + \varepsilon,$

(2.30) $I(f_0; x) - \varepsilon \le I(f_n; x) \le S(f_n; x) \le S(f_0; x) + \varepsilon.$

The inequality (2.29) shows that for every $x \in \Omega$ and for every $y \in f_0(x)$ there exists $\xi \in [x - \alpha\varepsilon, x + \alpha\varepsilon] \cap \Omega$ and $\eta \in f_n(\xi)$ such that $|y - \eta| < \varepsilon$. The inequality (2.30) ensures that for every $x \in \Omega$ and for every $y \in f_n(x)$ there exists $\eta \in f_0(x)$ such that $|y - \eta| < \varepsilon$. Applying Lemma 1.7, we obtain (2.22). The theorem is proved.

§ 2.3. H-distance in A_Ω and its properties

We defined the H-distance between functions that belong to F_Ω. If $f, g \in A_\Omega$ then we define the H-distance between them to be the H-distance betwen their completed graphs. If $f, g \in A_\Omega$ then by definition

$$r(\Omega, \alpha; f, g) = r(\alpha; f, g) = r(\alpha; F(f), F(g)).$$

Of course, the H-distance in A_Ω does not meet all the axioms of the metric. The condition $r(\alpha; f, g) = 0 \Rightarrow f = g$ is violated, since two different functions may have the same completed graphs. We shall consider some properties of H-distance in A_Ω and in F_Ω.

First, we observe that from the inclusion $\Delta_1 \subset \Delta_2$ for intervals, it does not follow that $r(\Delta_1, \alpha; f, g) \le r(\Delta_2, \alpha; f, g)$. For example, if φ, ψ are these real functions

$$\varphi(x) = \begin{cases} |x| & \text{if} \quad |x| \le 1 \\ 1 & \text{if} \quad 1 \le |x| \le 2 \end{cases}$$

$$\psi(x) = \begin{cases} 0 & \text{if} \quad |x| \le 1 \\ |x| - 1 & \text{if} \quad 1 \le |x| \le 2 \end{cases},$$

then it is directly verified (see Figure 2.2) that $r([-1, 1], \alpha; \varphi, \psi) = 1$ whereas $r([-2, 2], \alpha; \varphi, \psi) = 1/(1 + \alpha) < 1$.

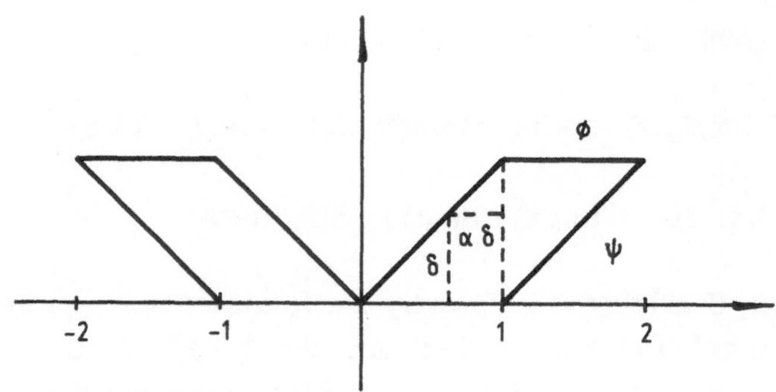

FIGURE 2.2

This example demonstrates one of the essential distinctions between the Hausdorff distance and the classical distances of analysis. It might happen that the sequence $\langle f_n \rangle_1^\infty$ converges with regard to Hausdorff distance on Δ_2, but it diverges on $\Delta_1 \subset \Delta_2$. This property is illustrated in the next example. Let

$$f_n(x) = \max\{0, 1 - |(-1)^n - nx|\}, \quad x \in [-1, 1], \; n = 1, 2, 3, \ldots$$

It is directly calculated that $r([-1, 1], \alpha; f_n, f_0) = 2/(1 + \alpha n)$ where

$$f_0(x) = \begin{cases} 0 & \text{if} \quad 0 < |x| \le 1 \\ [0, 1] & \text{if} \quad x = 0 \end{cases},$$

and

$$\lim_{n \to \infty} r([-1, 1], \alpha; f_n, f_0) = 0,$$

i.e., the sequence $\langle f_n \rangle_1^\infty$ converges to f_0 with respect to Hausdorff distance on the closed interval $[-1, 1]$. On the other hand, for every n we have $r([0, 1], \alpha; f_n, f_{n+1}) = 1$

and, consequently, the sequence $<f_n>_1^\infty$ diverges with respect to Hausdorff distance on the closed interval [0, 1].

In connection with the above examples we shall prove the following statement.

Lemma 2.7. If f and g are continuous and 2π-periodic functions, then $r((-\infty,\infty), \alpha; f, g) \leq r([-\pi,\pi], \alpha; f, g)$, and moreover, if f and g are even functions, then $r((-\infty,\infty), \alpha; f, g) = r([-\pi,\pi], \alpha; f, g)$.

Proof. We shall introduce some simplifying notation:

$$r((-\infty,\infty), \alpha; f, g) = p, \qquad r([-\pi,\pi], \alpha; f, g) = q \tag{2.31}$$

By virtue of the first equality in (2.31) there exists a point $X_0(x_0,y_0)$ of the graph of one of the two functions such that the open rectangle

$$D_0 = \{(x,y) : |x - x_0| < \alpha p, \ |y - y_0| < p\}$$

does not contain points of the graph of the other function. Since the functions are 2π-periodic, then we may assume that $x_0 \in [-\pi, \pi]$. But then the equality

$$p \leq q \tag{2.32}$$

holds, and thus the first part of the lemma is proved.

By virtue of the second equality (2.31) there exists a point $X_1(x_1,y_1)$ with $x_1 \in [-\pi, \pi]$ which, without loss of generality, we can take from the graph of the function f, that is, $y_1 = f(x_1)$, and this point is such that the rectangle

$$D_1^* = \{(x,y) : x \in (x_1 - \alpha q, x_1 + \alpha q) \cap [-\pi, \pi], \ |y - y_1| < q\} \tag{2.33}$$

does not contain points of the graph of the function g. We claim that the open rectangle

$$D_1 = \{(x,y) : \ |x - x_1| < \alpha q, \ |y - y_1| < q\} \tag{2.34}$$

also does not contain points of the graph of g. Hence, it will follow that $q \leq p$, which together with (2.32) gives $p = q$.

Let us assume to the contrary that the rectangle D_1 defined by (2.34) contains a point $(x_2, g(x_2))$ of the graph of the function g. Then

$$x_2 \in (x_1 - \alpha q, x_1 + \alpha q), \qquad x_2 \notin [-\pi, \pi]. \tag{2.35}$$

Without loss of generality, we may assume $x_2 > \pi$ so that there exists a positive integer s such that $x_2 - 2s\pi \in [-\pi, \pi]$. But then $-x_2 + 2s\pi \in [-\pi, \pi]$. From the periodicity and the evenness of g, it follows that $g(-x_2 + 2s\pi) = g(x_2)$. On the other hand it is not dificult to verify that $-x_2 + 2s\pi \in (x_1 - \alpha q, x_1 + \alpha q)$, so that the point $(-x_2 + 2s\pi, g(-x_2 + 2s\pi))$ belongs to the rectangle D_1^* defined by (2.33). This contradicts our assumption. The lemma is proved.

The next statement follows directly from the definition of Hausdorff distance.

Lemma 2.8. If f and g are arbitrary bounded real functions, defined on the closed interval Δ, and c is an arbitrary constant, then

(2.36) $$r(\Delta, \alpha; cf, cg) = |c|r(\Delta, |c|\alpha; f, g),$$

(2.37) $$r(\Delta, \alpha; f, g) \le \max\{1, \alpha_1/\alpha\}r(\Delta, \alpha_1; f, g).$$

Let us prove the following statement involving change of parameter.

Lemma 2.9. Let $\varphi(t)$ be a monotone function on the closed interval $\Delta_1 = [a_1, b_1]$ which satisfies a Lipschitz condition with a constant L, i.e., $|\varphi(t') - \varphi(t'')| \le L|t' - t''|$, and such that $\varphi(a_1) = a$, $\varphi(b_1) = b$. If we write $f_1(t) = f(\varphi(t))$ and $g_1(t) = g(\varphi(t))$, where $f, g \in A_\Delta$ with $\Delta = [a, b]$, then

$$r(\Delta, L; f, g) \le r(\Delta_1, 1; f_1, g_1).$$

Proof. This follows directly from the inequalities

$$\max\{L^{-1}|x - \xi|, |f(x) - g(\xi)|\} = \max\{L^{-1}|\varphi(t) - \varphi(\zeta)|, |f(\varphi(t)) - g(\varphi(\zeta))|\}$$

$$\le \max\{|t - \eta|, |f_1(t) - g_1(\zeta)|.$$

From (2.37) and the above lemma, we obtain

(2.38) $$r(\Delta, L; f, g) \le \max(1,\alpha)r(\Delta_1, \alpha; f_1, g_1).$$

Lemma 2.19. Let $f, g \in A_\Omega$. If for all $x \in \Omega$, the inequalities

(2.39) $\varphi(x) \leq f(x) \leq \psi(x)$

(2.40) $\varphi(x) - c \leq g(x) \leq \psi(x) + c$

hold, where c is a constant, then

(2.41) $r(\Omega, \alpha; f, g) \leq r(\Omega, \alpha; \varphi, \psi) + c.$

Proof. It is sufficient to note that for arbitrary $x, \xi \in \Omega$ it follows from (2.39) and (2.40) that

$$-(\psi(\xi) - \varphi(x) + c) \leq f(x) - g(\xi) \leq \psi(x) - \varphi(\xi) + c.$$

§ 2.4. Relationships between uniform distance and the Hausdorff distance

The uniform distance in the set C_Ω of all continuous functions on Ω is defined as follows:

$$R(f, g) = \|f - g\|_\Omega = \sup_{x \in \Omega} |f(x) - g(x)|$$

It is possible to extend this definition to all the functions in A_Ω, where the expression $\sup_{x \in \Omega} |f(x) - g(x)|$ should be understood as

(2.42) $\sup_{x \in \Omega} |f(x) - g(x)| = \sup \{|y - \eta| : x \in \Omega, y \in f(x), \eta \in g(x)\}$

In the spirit of (2.42), we are able to recast the modulus of continuity of a function f in A_Ω defined by (1.2.1) by

(2.43) $\omega(f; \delta) = \sup \{|y - \eta| : y \in f(x_1), \eta \in f(x_2), x_1, x_2 \in \Omega, |x_1 - x_2| \leq \delta\}.$

For single valued functions, (2.43) of course agrees with the usual definition.

Theorem 2.3. If $f, g \in A_\Omega$ and $r(\Omega, \alpha; f, g) = r(\alpha; f, g)$ is the Hausdorff distance between f and g, then

(2.44) $r(\alpha; f, g) \leq \|f - g\|_\Omega \leq r(\alpha; f, g) + \omega(\alpha r(\alpha; f, g),$

where $\omega(\delta) = \min\ \{\omega(f;\delta), \omega(g;\delta)\}$.

Proof. The inequality on the left side of (2.44) follows from the defintion of H-distance in (2.12), since $r(\alpha; f, g)$ is at most the Hausdorff distance between the graphs of f and g, and for $y \in f(x)$ and $\eta \in g(x)$, we have

$$\inf_{\xi\in\Omega}\ \inf_{\eta'\in g(\xi)}\ \max\ \{\alpha^{-1}|x-\xi|, |y-\eta'|\} \le |y-\eta| \le R(f,g).$$

To establish the inequality on the right hand side of (2.44), we may assume that f and g are in $\mathbf{F}_\Omega$, as Hausdorff distance between segment functions is defined in terms of completed graphs, and uniform distance is not decreased by replacing segment functions by their completed graphs. It is directly seen that for every $x \in \Omega$ and $y \in f(x)$ there exists $\xi_x \in \Omega$ and $\eta' \in g(\xi_x)$ such that

$$\inf_{\xi\in\Omega}\ \inf_{\eta\in g(\xi)}\ \max\ \{\alpha^{-1}|x-\xi|, |y-\eta|\} = \max\ \{\alpha^{-1}|x-\xi_x|, |y-\eta'|\}.$$

On the other hand, according to (2.12), for every $x \in \Omega$ and $y \in f(x)$, the inequalities

$$\alpha^{-1}|x-\xi_x| \le r(\alpha; f, g), \qquad |y-\eta'| \le r(\alpha; f, g)$$

hold. But then, in view of

$$|y-\eta| \le |y-\eta'| + |\eta'-\eta|,$$

the inequality

$$(2.45)\qquad |y-\eta| \le r(\alpha; f, g) + \omega(g; |x-\xi_x|)$$

$$\le r(\alpha; f, g) + \omega(g; \alpha r(\alpha; f, g))$$

holds for every $x \in \Omega$. Similarly, we obtain

$$(2.46)\qquad |y-\eta| \le r(\alpha; f, g) + \omega(f; \alpha r(\alpha; f, g)).$$

We get the right hand side of (2.44) from (2.45) and (2.46), completing the proof.

The following corollaries are obtained from Theorem 2.3.

Corollary 2.1. Let $<f_n>_1^\infty$ be a sequence in A_Ω with Ω compact, and let $f \in C_\Omega$. Then a necessary and sufficient condition for the uniform convergence of the sequence $<f_n>_1^\infty$ to f is its Hausdorff convergence.

Proof. This follows directly from (2.44), which can be written as

$$r(\alpha; f, f_n) \le \|f - f_n\|_\Omega \le r(\alpha; f, f_n) + \omega(\alpha r(\alpha; f, f_n)).$$

Since $f \in C_\Omega$, from $\lim_{n \to \infty} r(\alpha; f, f_n) = 0$, it follows

$$\lim_{n \to \infty} \omega(f; \alpha r(\alpha; f, f_n)) = 0.$$

Corollary 2.2. The uniform distance and the Hausdorff distance are topologically equivalent on every set of equicontinuous functions.

Since $\omega(g; \delta) = 0$ for each δ when g is a constant function, we also have

Corollary 2.3. If $g(x) = c$ where c is a constant then $r(\alpha; f, g) = R(f, g)$.

Let us note that the ratio of the uniform distance to the Hausdorff distance for two given functions may be arbitrarily large. For instance, if f and g are defined on $\Delta = [0, 1]$ in the following manner

$$f(x) = \begin{cases} x/\lambda & \text{if } 0 \le x \le \lambda \\ 1 & \text{if } 1 < \lambda \le 1 \end{cases},$$

$$g(x) = \begin{cases} 0 & \text{if } 0 \le x \le \lambda \\ (x - \lambda)/\lambda & \text{if } \lambda \le x \le 2\lambda \\ 1 & \text{if } 2\lambda \le x \le 1 \end{cases},$$

then by Lemma 2.4, we obtain that $r(\alpha; f, g) \le \lambda/(\alpha + \lambda)$. On the other hand $\|f - g\|_\Omega = R(f, g) \ge |f(\lambda) - g(\lambda)| = 1$, and hence

$$R(f, g)/r(\alpha; f, g) \ge 1/\alpha\lambda$$

where λ can be chosen arbitrarily small.

Corollary 2.4. If $f, g \in C_\Omega$ with Ω compact, then

$$\lim_{\alpha \to +0} r(\alpha; f, g) = R(f, g).$$

Proof. The proof follows directly from (2.44), letting $\alpha \to +0$.

The last corollary says that the uniform distance may be considered as a limiting case of Hausdorff distance with parameter $\alpha = 0$, that is $r(0; f, g) = R(f, g)$.

The consideration of Hausdorff distance as a generalization of the uniform distance enables us to obtain a series of classical theorems on uniform approximation of continuous functions as corollaries of the corresponding theorems for Hausdorff approximations of wider classes of functions. A series of such results will be considered later on.

In order to clarify the relationship of Hausdorff distance to the uniform distance, we make some further remarks. It is well-known that the space C_Ω of functions continuous on the compact set Ω, metrized by uniform distance, is a complete metric space. The set of continuous functions defined on Ω, is also metrizable by Hausdorff distance with parameter $\alpha > 0$. But the obtained metric space will not be complete. Perhaps the simplest example is the following Cauchy sequence in $C_{[0, 1]}$:

$$f_n(x) = \begin{cases} \sin(1/x) & \text{if } x \geq 1/n\pi \\ 0 & \text{if } x \leq 1/n\pi \end{cases}.$$

According to Theorem 2.2, the set of functions in $\mathbf{F}_\Omega$ whose values are finite segments is a complete metric space that clearly contains the closure of the set of continuous functions on the compact set Ω, metrized by Hausdorff distance. If Ω is a closed interval Δ, this subspace of $\mathbf{F}_\Delta$ can be shown to be the completion of C_Δ , that is, the continuous functions are dense in this subspace. The key condition is the intermediate value property expressed in Lemma 1.7.

Suppose $f \in \mathbf{F}_\Delta$ has finite segments as values. Fix $\varepsilon > 0$, and choose a regular partition $\{x_0, x_1, \ldots, x_n\}$ of Δ whose mesh is less than $\varepsilon\alpha$. Let $y_i \in f(x_i)$ be arbitrary for $0 \leq i \leq n$. By Lemma 1.7 and the fact that $\Delta_i = [x_{i-1}, x_i]$ is compact, $\Delta_i^* = \bigcup \{f(x) : x \in [x_{i-1}, x_i]\}$ is a finite segment. Let $\varphi \in C_\Delta$ be a function whose restriction to Δ_i has range Δ_i^* and that satisfies $\varphi(x_i) = y_i$. Evidently, we have $r(\Delta, \alpha; \varphi, f) \leq \varepsilon$.

We note that the continuous functions are not dense in this subspace of $\mathbf{F}_\Omega$ whenever Ω has an isolated point, for a necessary condition for $f \in \mathbf{F}_\Omega$ to admit Hausdorff approximations by continuous functions is that f map isolated points of Ω to point segments.

It is well-known that the uniform distance generates a norm on C_Ω. If $f \in C_\Omega$, then

$$\|f\|_\Omega = R(f, 0) = \max_{x \in \Omega} |f(x)|.$$

The uniform distance can be recovered from the uniform norm according to

$$R(f, g) = \|f - g\|_\Omega .$$

As for the Hausdorff distance, according to Corollary 1.8,

$$r(\Omega, \alpha; f, 0) = R(f, 0) = \|f\|_\Omega ,$$

so that the Hausdorff distance induces the same norm as the uniform distance. But this norm does not generate the initial Hausdorff distance. It is very important to remember that F_Ω, metrized by the Hausdorff distance, is not a normed space. This fact explains a sequence of new phenomena that occur in the study of function spaces metrized by Hausdorff distance.

Let us prove another statement that relates the convergence with respect to Hausdorff distance to pointwise convergence.

Theorem 2.4. Let $\langle f_n \rangle_1^\infty$ be a sequence of functions in A_Ω with $\lim_{n \to \infty} r(\Omega, \alpha; f_n, f_0) = 0$ where $f_0 \in A_\Omega$. Then if the function f_0 is single valued and continuous at the point $x_0 \in \Omega$, we have

(2.47) $$\lim_{n \to \infty} f_n(x_0) = f_0(x_0).$$

Proof. If the functions f_n are multivalued, then the equality (2.47) may be understood in the following sense: if for every n we choose an arbitrary $y_n \in f_n(x_0)$ then $\lim_{n \to \infty} y_n = f(x_0)$. By continuity of f at the point x_0 then for every $x \in [x_0 - 2\delta, x_0 + 2\delta] \cap \Omega$ and $y \in f(x)$ we have $|y - f(x_0)| < \varepsilon$. Then for a sufficiently large n the inequalities

$$f(x_0) - 2\varepsilon \le I(\delta, f_n ; x_0) \le S(\delta, f_n ; x_0) \le f(x_0) + 2\varepsilon$$

hold, so that for each $y_n \in f_n(x_0)$ we have $|y_n - f(x_0)| < 2\varepsilon$. Since $\varepsilon > 0$ is arbitrary, the equality (2.47) is fulfilled and the theorem is proved.

§ 2.5. The modulus of H-continuity

Let $f \in A_\Omega$. We define the **modulus of H-continuity** (with parameter α) of f by the formula

$$\tau(\Omega, \alpha, f; \delta) = \tau(\alpha, f; \delta) = r(\alpha; S(\delta/2, f), I(\delta/2, f)).$$

Directly from the definition of H-continuity one obtains these facts:

(a) The modulus of H-continuity does not exceed the modulus of continuity

$$\tau(\alpha, f; \delta) \leq \omega(f; \delta), \tag{2.48}$$

where $\omega(f; \delta)$ is the modulus of continuity of f. Indeed, by definition,

$$\tau(\alpha, f; \delta) = r(\Omega, \alpha; S(\delta/2, f), I(\delta/2, f))$$

$$\leq \sup_{x \in \Omega} |S(\delta/2, f; x) - I(\delta/2, f; x)|$$

$$= \sup \{|y - \eta| : y \in f(x_1),\ \eta \in f(x_2),\ x_1, x_2 \in \Omega,\ |x_1 - x_2| \leq \delta\}$$

$$= \omega(f; \delta);$$

(b) The modulus of H-continuity tends to the modulus of continuity as the parameter tends to zero:

$$\lim_{\alpha \to +0} \tau(\alpha, f; \delta) = \omega(f; \delta). \tag{2.49}$$

Indeed, $\displaystyle\lim_{\alpha \to +0} r(\Omega, \alpha; f, g) = \sup_{x \in \Omega} |f(x) - g(x)|.$

(c) The modulus of H-continuity is a monotone nondecreasing function of δ, i.e, if $\delta_1 < \delta_2$, then $\tau(\alpha, f; \delta_1) \leq \tau(\alpha, f; \delta_2)$.

The modulus of H-continuity does not possess certain properties that are standard for the modulus of continuity. It is well-known that for every positive integer k the inequality $\omega(f; k\delta) \le k\omega(f; \delta)$ holds. This fails in general for the modulus of H-continuity. For example, if we take the function

$$f(x) = \begin{cases} 0 & \text{if } |x| \le \varepsilon \\ 1 & \text{if } \varepsilon < |x| \le 1 \end{cases},$$

then it is directly seen that

$$\tau(\alpha, f; \delta) = \begin{cases} \alpha^{-1}\delta & \text{if } 0 \le \delta < 2\varepsilon \\ 1 & \text{if } \delta \ge 2\varepsilon \end{cases}$$

Obviously, the inequality $\tau(\alpha, f; 2\delta) \le 2\tau(\alpha, f; \delta)$ will be violated for $\delta \in [\varepsilon/2, \varepsilon]$ provided $2\varepsilon/\alpha < 1$.

In §1.3.4 we gave a definition of H-continuous function. We shall show now that it is possible to define this notion through the modulus of H-continuity.

Theorem 2.5. The function $f \in A_\Omega$ is H-continuous if and only if

$$\lim_{\delta\to+0} \tau(\alpha, f; \delta) = \tau(\alpha, f; 0) = 0.$$

Proof. According to the defintion of the modulus of H-continuity, it follows that

$$\lim_{\delta\to+0} \tau(\alpha, f; \delta) = \lim_{\delta\to+0} r(\alpha; S(\delta/2, f), I(\delta/2, f))$$

$$= r(\alpha; S(f), I(f)),$$

or,

$$\lim_{\delta\to+0} \tau(\alpha, f; \delta) = r(\alpha; F(S(f)), F(I(f))).$$

But since both $F(S(f))$ and $F(I(f))$ are in F_Ω we see that $\lim_{\delta\to+0} \tau(\alpha, f; \delta) = 0$ holds if and only if $F(S(f)) = F(I(f))$. According to Corollary 1.3, this is a necessary and sufficient condition for the H-continuity of the function f. The theorem is proved.

Theorem 2.6. (P. P. Korovkin [3]). Let f have bounded range. A necessary and sufficient condition for f to be H-continuous is the following: for every $\varepsilon > 0$ there exists two functions φ and ψ, uniformly continuous on Ω, such that $\varphi(x) \le f(x) \le \psi(x)$ for all $x \in \Omega$ and $r(\alpha; \varphi, \psi) < \varepsilon$.

Proof. Let f be H-continuous. According to Lemma 2.5 and Lemma 2.6, for every $\delta > 0$ we can find two uniformly continuous φ and ψ such that

$$I(\delta/2, f; x) \le \varphi(x) \le f(x) \le \psi(x) \le S(\delta/2, f; x),$$

and hence

$$\tau(\alpha, f; \delta) = r(\alpha; S(\delta/2, f), I(\delta/2, f)) \ge r(\alpha; \varphi, \psi).$$

But since $\lim_{\delta\to+0} \tau(\alpha, f; \delta) = 0$, then $r(\alpha; \varphi, \psi)$ can be made arbitrarily small. This proves necessity.

For sufficiency, choose for each $\varepsilon > 0$ uniformly continuous functions φ_ε and ψ_ε such that for each x, $\varphi_\varepsilon(x) \le f(x) \le \psi_\varepsilon(x)$, and $r(\alpha; \varphi_\varepsilon, \psi_\varepsilon) < \varepsilon$. Clearly,

$$\varphi_\varepsilon(x) \le I(f; x) \le S(f; x) \le \psi_\varepsilon(x)$$

and $r(\alpha; S(f), I(f)) < \varepsilon$. Since ε was arbitrary, we get $r(\alpha; S(f), I(f)) = 0$. Thus $F(I(f)) = F(S(f))$ which, according to Corollary 1.3, completes the proof.

§ 2.6. The order of the modulus of H-continuity

Let us consider the various classes of functions that are obtained for a given order of convergence to zero of the modulus of H-continuity.

Theorem 2.7. (V. Veselinov [11]). If $f \in \mathbf{F}_\Omega$ where Δ is a closed interval and

$$\tau(\alpha, f; \delta) = o(\delta), \tag{2.50}$$

then f is a constant.

Proof. We shall prove that if the condition (2.50) holds, then for each x in Δ,

$$\lim_{h\to 0} \frac{|f(x+h) - f(x)|}{|h|} = 0, \tag{2.51}$$

so that $f'(x) = 0$ at each x, i.e., f is constant.

Let us assume that (2.51) is not true. Then there exists $x_0 \in \Delta$, $c_0 > 0$ and a sequence $\langle h_k \rangle_1^\infty$ with $\lim_{k\to\infty} h_k = 0$ such that $|f(x_0 + h_k) - f(x_0)| \geq c_0|h_k|$ for each k. Without loss of generality, we can assume that all h_k are positive, and for all k the inequality

$$f(x_0 + h_k) - f(x_0) \geq c_0 h_k$$

holds. Then $S(2h_k, f; t) - f(x_0) \geq c_0 h_k$, or $S(2h_k, f; t) \geq I(2h_k, f; t) + c_0$ for all $t \in [x_0 - h_k, x_0 + h_k] \cap \Delta$. Hence, we have

$$\tau(\alpha, f; 4h_k) = r(\alpha; S(2h_k, f), I(2h_k, f)) \geq c_0 h_k/2\alpha,$$

which contradicts (2.50). This completes the proof.

Theorem 2.7 shows that the modulus of continuity, as well as the modulus of H-continuity (for every value of the parameter α), tends to zero not faster than $O(\delta)$ for functions differing from a constant.

It is natural to study those functions $f \in \mathbf{F}_\Delta$ for which $\tau(\alpha, f; \delta) = O(\delta)$.

Theorem 2.8. If $f \in \mathbf{F}_\Omega$ and

$$\limsup_{\delta\to+0} \tau(\alpha, f; \delta)/\delta < 1/\alpha, \tag{2.52}$$

then f is continuous, $f \in \mathbf{C}_\Omega$. The inequality (2.52) is sharp.

Proof. Let us assume that f is discontinuous at some $x_0 \in \Omega$. Then

$$S(f; x_0) - I(f; x_0) = d > 0. \tag{2.53}$$

According to Theorem 1.9 either

$$\limsup_{x\to x_0-0} f(x) = S(f; x_0) \tag{2.54}$$

or

$$\limsup_{x\to x_0+0} f(x) = S(f; x_0). \tag{2.55}$$

We consider only the case (2.54) (the case (2.55) is analagous). There exists a monotone decreasing sequence $<h_n>_1^\infty$ of positive numbers convergent to zero for which $f(x_0 - 2h_n) \geq S(f; x_0) - d/2$, and hence

$$(2.56) \qquad S(h_n, f; t) \geq I(h_n, f; x_0 - h_n) + d/2$$

for all $t \in [x_0 - 3h_n, x_0 + h_n] \cap \Omega$. But from (2.56), if follows from Lemma 2.10 that

$$(2.57) \qquad \tau(\alpha, f; 2h_n) = r(\alpha; S(h_n, f), I(h_n, f)) \geq \min\{2h_n/\alpha, d/2\}.$$

For sufficiently large n, from (2.57) we obtain $\tau(\alpha, f; 2h_n)/2h_n \geq 1/\alpha$, and hence, $\limsup_{\delta \to 0} \tau(\alpha, f; \delta)/2\delta \geq 1/\alpha$, which contradicts (2.52).

To conclude the proof, it is sufficient to give an example of a discontinuous function σ such that $\tau(\alpha, \sigma; \delta) = \delta/\alpha$. For instance, such a function is $\sigma(x) = \operatorname{sgn} x$ for $-1 \leq x \leq 1$.

It should be noted that (2.52) is a sufficient condition for continuity, but not a necessary one. It is possible to construct a continuous function for which the H-modulus of continuity tends arbitrarily slowly to zero together with δ.

§ 2.7. H-continuity on a subset

The notion of H-continuity is strictly connected with the set on which the function is defined. As noted earlier, if $f \in \mathbf{H}_\Omega$ and Δ is a subset of Ω, then in general f does not belong to $\mathbf{H}_\Delta$. Special conditions for boundary points of Δ are noted in Theorem 1.9. For instance, let us consider the function $\sigma(x) = \operatorname{sgn} x$ for $-1 \leq x \leq 1$. Obiously, $\sigma \in \mathbf{H}_{[-1, 1]}$ since $\tau([-1, 1], \alpha, \sigma; \delta) = \delta/\alpha$, but $\sigma \notin \mathbf{H}_{[0, 1]}$, as $\tau([0, 1], \alpha, \sigma; \delta) = 2$ for all $\delta > 0$.

Defintion 2.1. Let Ω be an arbitrary set of real numbers and let $\Delta \subset \Omega$. The set of functions defined on Ω and H-continuous on Δ will be denoted by $\mathbf{H}_\Delta^\Omega$.

Let $f \in \mathbf{H}_\Delta^\Omega$. To determine the modulus of H-continuity of f, we take $S(\delta, f; x)$ and $I(\delta, f; x)$ for $x \in \Delta$, but

$$S(\Omega, \delta, f; x) = S(\delta, f; x) = \sup\{f(t) : t \in [x - \delta, x + \delta] \cap \Omega\}$$

$$I(\Omega, \delta, f; x) = I(\delta, f; x) = \sup\{f(t) : t \in [x - \delta, x + \delta] \cap \Omega\},$$

and hence

$$\tau(\alpha, f; \delta) = \tau(\Omega, \Delta, \alpha, f; \delta) = r(\Delta, \alpha; S(\Omega, \delta/2, f), I(\Omega, \delta/2, f)).$$

It follows from the definition of H_Δ^Ω that $H_\Delta^\Delta = H_\Delta$.

The proof of the next result is similar to that of Theorem 1.9 and is left to the reader.

Theorem 2.9. Let $f \in H_\Delta^\Omega$. Then the following conditions hold:

(a) If x_0 is a left boundary point of Δ and is an interior point of Ω, then

$$\liminf_{x\to x_0+0} I(f; x) \le \liminf_{x\to x_0-0} I(f; x) \le \limsup_{x\to x_0-0} S(f; x) \le \limsup_{x\to x_0+0} S(f; x),$$

and if $\liminf_{x\to x_0+0} f(x) = f(x_0+0)$ exists, then f is continuous at x_0;

(b) If x_0 is a right boundary point of Δ and is an interior point of Ω then

$$\liminf_{x\to x_0-0} I(f; x) \le \liminf_{x\to x_0+0} I(f; x) \le \limsup_{x\to x_0+0} S(f; x) \le \limsup_{x\to x_0-0} S(f; x),$$

and if $\liminf_{x\to x_0-0} f(x) = f(x_0-0)$ exists, then f is continuous at x_0.

Definition 2.2. The function f defined on the set Ω will be called H-continuous at the point $x_0 \in \Omega$ if $f \in H_{x_0}^\Omega$.

Lemma 2.10. If the function f is H-continuous at the point x_0, then f is single valued (and continuous) at this point.

Proof. According to the definitions, we have

$$\tau(\Omega, x_0, \alpha, f; \delta) = S(\Omega, \delta/2, f; x_0) - I(\Omega, \delta/2, f; x_0)$$

and if $\lim_{\delta\to 0} \tau(\Omega, x_0, \alpha, f; \delta) = 0$, then $S(f; x_0) = I(f; x_0)$, i e., $F(f)$ is single-valued and hence f is continuous at x_0.

§ 2.8. H-distance with weight

Let θ be a positive continuous function, defined on Ω, with $\theta(x) \geq 1$ for all $x \in \Omega$, and such that there exists $x_0 \in \Omega$ with $\theta(x_0) = 1$. Denote by $A_{\Omega,\theta}$ the set of all segment functions defined on Ω for which

$$\sup_{x\in\Omega} \frac{|f(x)|}{\theta(x)} = M_f^\theta < \infty.$$

The **modulus of continuity with weight** θ of the function $f \in A_{\Omega,\theta}$ is given by

$$\omega(\theta, f; \delta) = \sup_{x\in\Omega} \{\frac{|f(\xi) - f(t)|}{\theta(x)} : |\xi - x| \leq \delta/2,\ |t - x| \leq \delta/2,\ \xi, t \in \Omega\} \tag{2.58}$$

It should be pointed out that for $\Omega = (-\infty, \infty)$, the modulus of continuity with weight can be defined in another way and it is called the **generalized modulus of continuity** (N. Ahieser [1]):

$$\omega_\theta(f; \delta) = \sup \{(f(x + h) - f(x))/\overline{\theta}(\delta)\theta(x) : |h| < \delta,\ x \in \Omega\},$$

where $\overline{\theta}(\delta) = \sup \{\theta(x + \xi)/\theta(x) : |\xi| < \delta,\ x \in \Omega\}$.

As in §1.3 we define

$$S(\theta, \delta, f; x) = \sup \{\frac{f(t)}{\theta(t)} : t \in [x - \delta,\ x + \delta] \cap \Omega\}, \tag{2.59}$$

$$I(\theta, \delta, f; x) = \inf \{\frac{f(t)}{\theta(t)} : t \in [x - \delta,\ x + \delta] \cap \Omega\}, \tag{2.60}$$

and also

$$S(\theta, f; x) = \lim_{\delta\to+0} S(\theta, \delta, f; x), \quad I(\theta, f; x) = \lim_{\delta\to+0} I(\theta, \delta, f; x).$$

Then the completed graph $F(\theta, f)$ of the function $f \in A_{\Omega,\theta}$ is

$$F(\theta, f; x) = [I(\theta, f; x), S(\theta, f; x)], \quad x \in \Omega.$$

The set $\mathbf{F}_{\Omega,\theta}$ will be defined by analogy with $\mathbf{F}_{\Omega}$ as the set of all functions $f \in \mathbf{A}_{\Omega,\theta}$ for which $F(\theta, f) = f$. We define **H-distance with weight θ and parameter α** in the set $\mathbf{F}_{\Omega,\theta}$ by

$$(2.61) \quad r(\theta; \alpha; f, g) = \max \{ \sup_{x\in\Omega} \sup_{y\in f(x)} \inf_{\xi\in\Omega} \inf_{\eta\in g(\xi)} \max \{\frac{|x-\xi|}{\alpha}, \frac{|y-\eta|}{\theta(x)}\},$$
$$\sup_{x\in\Omega} \sup_{y\in g(x)} \inf_{\xi\in\Omega} \inf_{\eta\in f(\xi)} \max \{\frac{|x-\xi|}{\alpha}, \frac{|y-\eta|}{\theta(x)}\}\}.$$

According to the definition, if $f, g \in \mathbf{A}_{\Omega,\theta}$, then

$$r(\theta, \alpha; f, g) = r(\theta, \alpha; F(\theta, f), F(\theta, g)).$$

It follows from (2.61) that for every $\alpha > 0$,

$$r(\theta, \alpha; f, g) \le \sup_{x\in\Omega} \frac{|f(x) - g(x)|}{\theta(x)} = R(\theta; f, g),$$

where $R(\theta; f, g)$ is the uniform distance with weight θ. Moreover, the equality

$$\lim_{\alpha\to+0} r(\theta, \alpha; f, g) = R(\theta; f, g)$$

holds. The **modulus of H-continuity with weight θ** is given by

$$\tau(\theta, \alpha, f; \delta) = r(\theta, \alpha; S(\theta, \delta/2, f), I(\theta, \delta/2, f)).$$

It follows from (2.58) - (2.60) that $\lim_{\alpha\to+0} \tau(\theta, \alpha, f; \delta) = \omega(\theta, f; \delta)$. The set of functions $f \in \mathbf{A}_{\Omega,\theta}$ for which $\lim_{\delta\to+0} \tau(\theta, \alpha, f; \delta) = 0$ will be denoted by $\mathbf{H}_{\Omega,\theta}$.

All statements for H-distance hold true for H-distance with weight as well, with uniform distance replaced by uniform distance with weight. Thus, for instance we obtain the following assertion, whose proof is similar to the proof of Theorem 2.3.

Theorem 2.10. If $f, g \in \mathbf{A}_{\Omega,\theta}$, $r(\theta, \alpha; f, g)$ is the Hausdorff distance with weight between f and g, and $R(\theta; f, g)$ is the uniform distance with weight between f and g, and Ω is compact, then

$$r(\theta, \alpha; f, g) \le R(\theta; f, g) \le r(\theta, \alpha; f, g) + \omega(\theta; \alpha r(\theta, \alpha; f, g),$$

where $\omega(\theta, \delta) = \min \{\omega(\theta, f; \delta), \omega(\theta, g; \delta)\}$.

Remarks . The relationship between uniform convergence of support functions and the Hausdorff distance between closed and bounded convex sets was considered by L. Hörmander [1] (see also C. Castaing and M. Valadier [1]). The Hausdorff distance is widely used in works on differential equations with multivalued right-hand side (control theory, etc.). For instance: V. Blagodadskih [1,2,3], M. Jacobs [1], C. Olech [1], H. Hermes [1], A. Cellina and J.-P. Aubin [1], V. Makarov and A. Rubinov [1], E. Barbasin and E. Alimov [1], T. Gicev {1,2,3,4,5,6,7], T. Gicev and N. Rozov [1], A. L. Donchev [1], A. L. Donchev and E. M. Farkhi [1], V. M. Veliov [1,2], A. L. Donchev and V. M. Veliov [1]. The Hausdorff distance is mentioned in standard books on analysis and topology, as for instance, J. Dieudonné [1, § 3.16], C. Berge [1, ch. 6, §6] and Kuratowski [1, § 21]. Hausdorff distance and its relation to other set topologies is discussed in E. Klein and A. Thompson [1] and in E. Michael [1]. Spaces of metric continua equipped with Hausdorff distance are considered in S. Nadler [1].

It should be noted that Hausdorff distance between functions, which was considered in the present chapter, is called the distance of Hausdorff-Pompieu in the paper of A. Lupas and M. Müller [1]. The Hausdorff distance between graphs of discontinuous functions, rather than their completed graphs, was considered by G. Beer [4]. For other forms of graph convergence, the reader may consult S. Naimpally [1], H. Poppe [1], and G. Beer [5]. A distance which is a generalization of uniform distance and is related to Hausdorff distance was considered in the papers of G.Iliev [1, 2], G. Iliev and V. Popov [1], G. Iliev and A. Andreev [1].

It is shown in G. Beer [3] in the context of arbitrary metric spaces that Hausdorff metric convergence coincides with uniform converge on C_Ω if and only if each continuous function on Ω is uniformly continuous, a weaker condition than compactness of Ω. Compactness and completeness criteria for C_Ω equipped with Hausdorff distance have also been considered by G. Beer [6,7].

There is an extensive literature on the approximation of convex multivalued mappings defined on abstract spaces by well-behaved multivalued mappings and/or continuous functions. For instance: A. Cellina [1,2], F. DeBlasi [1], F. DeBlasi and J. Myjak [1], M. Hukahara [1], F. Deutsch and P. Kenderov [1].

The Hausdorff distance between complex valued functions is defined in a note of B. Boyanov [5]. Hausdorff distance between hypographs of upper semicontinuous functions was considered in G. Beer [1,2].

The analogue of the modulus of H-continuity for integral distances is called the averaged modulus of smoothness. See Bl. Sendov and V. A. Popov [11].

Chapter 3

Linear Methods of Approximation

§ 3.1. Convergence of sequences of positive operators

An operator L, defined on $B_\Omega \subset A_\Omega$, is called **linear** if $L(\alpha f + \beta g; x) = \alpha L(f; x) + \beta L(g; x)$, for all $f, g \in B_\Omega$ and for all constants α, β, and it is called **positive** if $L(f; x) \geq 0$ for all $x \in \Omega$ whenever $f(x) \geq 0$ for all $x \in \Omega$.

Let $\psi(\delta)$ be a nondecreasing continuous function such that $\psi(0) = 0$ and $0 < \psi(\delta) \leq 1$ for all $\delta > 0$. We shall consider the convergence of the sequence of positive linear operators $\langle L_n \rangle_1^\infty$ on $\Delta \subset \Omega$, where the operators satisfy the conditions

$$\lim_{n\to\infty} r(\Delta, \alpha; L_n(1), 1) = 0, \quad \lim_{n\to\infty} r(\Delta, \alpha; L_n(\psi(|x - t|); x), 0) = 0,$$

which, according to Corollary 2.3, are equivalent to

$$(3.1) \qquad \lim_{n\to\infty} \sup_{x\in\Delta} |1 - L_n(1; x)| = 0, \quad \lim_{n\to\infty} \sup_{x\in\Delta} |L_n(\psi(|x - t|); x)| = 0.$$

The following basic theorem on the convergence of linear positive operators holds.

Theorem 3.1. In order that every sequence of linear positive operators, for which (3.1) holds, satisfies the relation

$$\lim_{n\to\infty} r(\Delta, \alpha; L_n(f), f) = 0,$$

it is necessary and sufficient that

(a) the function $f \in B_\Omega$ be H-continuous on Δ, i. e., $F(f) \in H_\Delta^\Omega$,

(b) $\sup_{x\in\Omega} |f(x)| = M_f < \infty$.

Proof. First we shall prove sufficiency. Let $F(f) \in H_\Delta^\Omega$ with $M_f < \infty$. The inequalities $I(\delta, f; x) \le f(t) \le S(\delta, f; x)$ hold for all $x, t \in \Omega$ with $|x - t| \le \delta$ (see §1.3). On the other hand, for $x, t \in \Omega$ with $|x - t| > \delta$

$$-M_f\psi(|x - t|)/\psi(\delta) \le f(t) \le M_f\psi(|x - t|)/\psi(\delta),$$

since ψ is a monotone nondecreasing function and $|f(x)| \le M_f$ for all $x \in \Omega$. Since $I(\delta, f; x) \ge -M_f$, and $M_f \ge S(\delta, f; x)$, we obtain

$$f(t) \le S(\delta, f; x) + 2M_f\psi(|x - t|)/\psi(\delta) = h_1(t) \tag{3.2}$$

$$f(t) \ge I(\delta, f; x) - 2M_f\psi(|x - t|)/\psi(\delta) = h_2(t), \tag{3.3}$$

for all $x, t \in \Omega$. According to the monotonicity of L_n, it follows from (3.2) and (3.3) that

$$L_n(h_2; x) \le L_n(f; x) \le L_n(h_1; x). \tag{3.4}$$

It follows from (3.1) that for every $\varepsilon > 0$ we can find an integer $n(\varepsilon) > 0$ such that for $n > n(\varepsilon)$, the inequalities

$$r(\Delta, \alpha; L_n(h_1), S(\delta, f)) < \varepsilon, \qquad r(\Delta, \alpha; L_n(h_2), I(\delta, f)) < \varepsilon \tag{3.5}$$

hold. Then, using Lemma 2.10, from (3.2) - (3.4) and (3.5), we obtain

$$\begin{aligned} r(\Delta, \alpha; L_n(f), f) &\le r(\Delta, \alpha; L_n(f), L_n(h_1)) + r(\Delta, \alpha; L_n(h_1), S(\delta, f)) + \\ &\quad r(\Delta, \alpha; S(\delta, f), f) \\ &\le r(\Delta, \alpha; L_n(h_2), L_n(h_1)) + \varepsilon + r(\Delta, \alpha; S(\delta, f), I(\delta, f)) \\ &\le r(\Delta, \alpha; L_n(h_2), I(\delta, f)) + r(\Delta, \alpha; S(\delta, f), L_n(h_1)) + \varepsilon + \\ &\quad 2r(\Delta, \alpha; S(\delta, f), I(\delta, f)) \\ &\le 3\varepsilon + 2\tau(\Delta, \alpha, f; 2\delta). \end{aligned} \tag{3.6}$$

Since f is H-continuous on Δ, we can choose $\delta > 0$ such that $\tau(\Delta, \alpha, f; 2\delta) < \varepsilon$. Inserting this into (3.6) gives $r(\Delta, \alpha; L_n(f), f) < 5\varepsilon$, and sufficiency is established.

Now we turn to the necessity of these two conditions. Let us begin with condition (a). For every function $f \in B_\Omega$ that does not satisfy condition (a) (but may satsify condition (b)), we shall construct a divergent (with respect to H-distance) sequence of positive linear operators that satisfies (3.1). Fix $f_1 \in B_\Omega$ with

(3.7) $$\tau(\Delta, \alpha, f_1; \delta) \geq r(\Delta, \alpha; I(f_1); S(f_1)) = d > 0.$$

Let $\delta > 0$ and $x \in \Omega$. Choose $\delta(x) \in (0, \delta]$ such that

(3.8) $$\psi(\delta(x)) \leq \delta.$$

This is possible to do, since $\lim_{\delta \to +0} \psi(\delta) = 0$. With every $x \in \Omega$ we associate $x(\delta) \in \Omega$ such that $|x - x(\delta)| < \delta(x)$ and

$$f_1(x(\delta)) + \delta \geq \sup \{f_1(t) : |x - t| \leq \delta(x),\ t \in \Omega\}$$

We shall define an operator L'_δ in the follwoing way:

(3.9) $$L'_\delta(f; x) = f(x(\delta)).$$

It follows from here that

(3.10) $$L'_\delta(f_1; x) \geq S(f_1; x) - \delta$$

for all $x \in \Omega$. The operator L'_δ is linear and positive. We obtain a sequence of linear positive operators $\langle L'_n \rangle_1^\infty$ by setting $\delta = 1, 1/2, 1/3, \ldots$. Now we shall show that for this sequence, the relation (3.1) holds. Indeed, if we set $x_n = x(\delta)$ for $\delta = 1/n$, then

(3.11) $$|1 - L'_n(1; x)| = 0, \qquad |L'_n(\psi(|x - t|); x)| = \psi(|x - x_n|).$$

According to (3.8) and the choice of x_n, we have $\psi(|x - x_n|) \leq 1/n$ for each n. Hence the conditions of (3.1) hold for the sequence $\langle L'_n \rangle_1^\infty$. Similarly, we may construct a sequence of positive linear operators $\langle L''_n \rangle_1^\infty$ satisfying the conditions of (3.1) and such that for all n,

(3.12) $$L''_n(f_1; x) \leq I(f_1; x) + 1/n.$$

Next, we shall consider the sequence of positive linear operators $<L_n>_1^\infty$ with $L_{2n-1} = L_n'$ and $L_{2n} = L_n''$. Clearly, the sequence $<L_n>_1^\infty$ satisfies the condition (3.1). But it follows from (3.7), (3.11) and (3.12) that

$$r(\Delta, \alpha; L_{2n-1}(f_1), L_{2n}(f_1)) \geq r(\Delta, \alpha; S(f_1) - 1/n , I(f_1) + 1/n)$$

$$\geq r(\Delta, \alpha; S(f_1), I(f_1)) - 2/n \geq d - 2/n.$$

Thus, the sequence $<L_n>_1^\infty$ diverges, and the necessicity of (a) is established.

The necessity of (b) is easier. Let $f_1 \in B_\Omega$ satsify $\sup \{|f_1(x)| : x \in \Omega\} = \infty$. Then for every positive integer n, we can find $x_n \in \Omega$ such that

(3.13) $$|f_1(x_n)| > n^2.$$

Now for each n, let L_n be this operator:

(3.14) $$L_n(f; x) = f(x) + f(x_n)/n.$$

For each n the operator L_n is linear and positive. We claim that the sequence $<L_n>_1^\infty$ satisfies the conditions of (3.1). From (3.14), we have

$$0 \leq \lim_{n\to\infty} \max_{x\in\Delta} |1 - L_n(1; x)| \leq \lim_{n\to\infty} 1/n = 0,$$

and noting that $0 < \psi(\delta) \leq 1$, we obtain

(3.15) $$0 \leq \lim_{n\to\infty} \max_{x\in\Delta} |L_n(\psi(|x - t|); x)| \leq \lim_{n\to\infty} 1/n = 0.$$

Thus, the conditions in (3.1) are verified.

Without loss of generality, we can assume that all the values $f_1(x_n)$ have the same sign. We consider the case $f_1(x_n) > 0$ for each n (the other is similar and is left to the reader). By (3.13) and (3.14) we have $L_n(f_1; x_n) = f_1(x_n) + n$, and as a result,

$$r(\Delta, \alpha; L_n(f_1), f_1) \geq r(\Delta, \alpha; f_1 + n, f_1) \geq n,$$

i.e., the sequence $<L_n(f_1)>_1^\infty$ fails to converge with respect to H-distance to f_1. The theorem is proved.

Since $H_\Omega^\Omega = H_\Omega$, an immediate corollary of Theorem 3.1 is

Corollary 3.1. Let $f \in B_\Omega$ have bounded range. Let $<L_n>_1^\infty$ be a sequence of positive linear operators on B_Ω such that

$$\lim_{n\to\infty} r(\Delta, \alpha; L_n(1), 1) = 0, \quad \lim_{n\to\infty} r(\Delta, \alpha; L_n(|x - t|; x), 0) = 0 \tag{3.16}$$

hold. Then a necessary and sufficient condition for

$$\lim_{n\to\infty} r(\Delta, \alpha; L_n(f), f) = 0$$

for every such sequence $<L_n>_1^\infty$ is the H-continuity of f on Ω.

Corollary 3.1 is a special case of Korovkin's theorem [3] for the H-distance. On the other hand if $\Omega = \Delta$, where Δ is a finite interval $[a, b]$, since uniform distance and the H-distance are topologically equivalent on the set C_Δ of functions continuous on Δ, (3.16) is equivalent to the uniform convergence of the operators in this context:

$$\lim_{n\to\infty} L_n(1; x) = 1, \qquad \lim_{n\to\infty} L_n(\psi(|x - t|); x), 0) = 0. \tag{3.17}$$

The following statement is a well-known classical theorem of Korovkin [1, p. 21].

Theorem 3.2. If for a sequence of positive linear operators $<L_n>_1^\infty$ the following three conditions

$$\begin{aligned} &L_n(1; x) = 1 + \alpha_n(x), \\ &L_n(t; x) = x + \beta_n(x), \\ &L_n(t^2; x) = x^2 + \gamma_n(x), \end{aligned} \tag{3.18}$$

all hold, where the functions $\alpha_n(x)$, $\beta_n(x)$, $\gamma_n(x)$ tend uniformly to zero on the closed interval $\Delta = [a, b]$, then for each continuous function f on $[a, b]$ that is also bounded and continuous to the right at b and to the left at a, we have the uniform convergence of $\langle L_n(f)\rangle_1^\infty$ to f.

Proof. This follows directly from Theorem 3.1 and Theorem 2.6, where we can set $\psi(\delta) = c\delta^2$. The conditions in (3.18) are equivalent to those in (3.17).

§ 3.2. The order of approximation of functions by positive linear operators

It follows from Theorem 3.1 that the natural class of functions in which to study the problem of approximation by positive linear operators with respect to the H-distance is the class of H-continuous functions. Of course, the estimates of approximation will be expressed in terms of the modulus of H-continuity. All our estimates in what follows will be, in a certain sense, generalizations of the corresponding results for uniform approximation, since the H-distance $r(\alpha; f, g)$ tends to the uniform distance $R(f, g) = \|f - g\|$, as $\alpha \to 0$, and the modulus of H-continuity $\tau(\alpha; f, \delta)$ tends to the modulus of continuity $\omega(f; \delta)$ as $\alpha \to 0$ as well.

We next prove a general theorem on the order of approximation of functions by positive linear operators.

Theorem 3.3. Let the positive linear operator L be defined on $B_\Omega \subset A_\Omega$. Suppose $\Delta \subset \Omega$ and $f \in B_\Omega$. Let $M = \sup\{|f(x)| : x \in \Delta\}$, and write

$$\omega(x, \delta, f; t) = \begin{cases} 0 & \text{if } t \in [x - \delta, x + \delta] \cap \Omega \\ \omega(f; |x - t| - \delta) & \text{if } t \in \Omega \backslash [x - \delta, x + \delta] \end{cases}.$$

Under the assumption $\omega(x, \delta, f; t) \in B_\Omega$, we have

$$r(\Delta, \alpha; L(f), f) \le \tau(\Delta, \alpha, f; 2\delta) + \sup_{x \in \Delta} L(\omega(x, \delta, f); x) + M \sup_{x \in \Delta} |1 - L(1; x)|. \tag{3.19}$$

Proof. Obviously, for every $x \in \Delta$ and $\delta > 0$, we have

$$I(\delta, f; x) \le f(x) \le S(\delta, f; x). \tag{3.20}$$

Let us show that

$$I(\delta, f; x) - \omega(x, \delta, f; t) \le f(t) \le S(\delta, f; x) + \omega(x, \delta, f; t) \tag{3.21}$$

for every $t \in \Omega$. Obviously, (3.21) is satisfied for every $t \in [x - \delta, x + \delta] \cap \Omega$. For $t > x + \delta$ and $t \in \Omega$, we have

$$\begin{aligned} f(t) &= f(x+\delta) + f(t) - f(x+\delta) \\ &\le S(\delta, f; x) + \omega(f; t - x - \delta) = S(\delta, f; x) + \omega(x, \delta, f; t). \end{aligned}$$

For $t < x - \delta$ and $t \in \Omega$, we have

$$f(t) = f(x - \delta) + f(t) - f(x - \delta) \le S(\delta, f; x) + \omega(x, \delta, f; t)$$

holds in the same way. It is proved similarly that for $t \in \Omega$,

$$f(t) \ge I(\delta, f; x) - \omega(x, \delta, f; t).$$

From (3.21) and the fact that L is linear and positive, we obtain

$$I(\delta, f; x)L(1; x) - L(\omega(x, \delta, f); x) \le L(f; x) \le S(\delta, f; x)L(1; x) + L(\omega(x, \delta, f); x),$$

or for $x \in \Delta$,

$$L(f; x) \le S(\delta, f; x) + M \sup_{x\in\Delta} |1 - L(1; x)| + \sup_{x\in\Delta} L(\omega(x, \delta, f); x), \tag{3.22}$$

$$L(f; x) \ge I(\delta, f; x) - M \sup_{x\in\Delta} |1 - L(1; x)| - \sup_{x\in\Delta} L(\omega(x, \delta, f); x). \tag{3.23}$$

The inequality (3.19) is obtained from (3.20), (3.22), (3.23) and Lemma 2.10. The theorem is proved.

A similar theorem, but without the appearance of the modulus of continuity of the approximated function, was proved by V. Veselinov [12].

Now we consider a series of applications of Theorem 3.3 to the estimation of the order of approximation realized by different operators that are encountered in analysis.

§ 3.3. Approximation of periodic functions by positive integral operators

Let K be an even positive kernel that is square integrable on $[-\pi, \pi]$, i.e.,

(a) $K(t) \geq 0,\ t \in [-\pi, \pi]$, (b) $K(t) = K(-t),\ t \in [-\pi, \pi]$,

and let K be normalized:

$$\int_{-\pi}^{\pi} K(t)\,dt = 1 \tag{3.24}$$

We consider the operator

$$L(f; x) = \int_{-\pi}^{\pi} f(x+t)K(t)\,dt, \tag{3.25}$$

defined for all 2π-periodic functions that are square integrable. The operator (3.25) is linear and positive. It follows from (3.24) that

$$L(1; x) \equiv 1. \tag{3.26}$$

The set of real and 2π-periodic functions will be denoted by $A_{2\pi} \subset A_{[-\infty,\infty]}$. From Theorem 3.3 we obtain

Theorem 3.4. If $f \in A_{2\pi}$ and f is square integrable, then for every $\delta > 0$, we have

$$r(\alpha; f, L(f)) \leq \tau(\alpha, f; 2\delta) + 2\int_{\delta}^{\pi} \omega(f; t-\delta)K(t)\,dt \tag{3.27}$$

Proof. According to the definition of $\omega(x, \delta, f; t)$ and (3.26) it remains only to compute

$$L(\omega(x, \delta, f; x); x) = \int_{-\pi}^{-\delta} \omega(f; -t-\delta)K(t)\,dt + \int_{\delta}^{\pi} \omega(f; t-\delta)K(t)\,dt$$

$$= 2\int_{\delta}^{\pi} \omega(f; t-\delta)K(t)\,dt,$$

using the evenness of K. The proof is thus complete.

Corollary 3.2. If $f \in A_{2\pi}$ and f is square integrable, then

$$r(\alpha; f, L(f)) \leq \tau(\alpha, f; 2\delta) + 2\delta^{-1}\,\omega(f; \delta)\int_{\delta}^{\pi} tK(t)\,dt \tag{3.28}$$

and

$$r(\alpha; f, L(f)) \leq \tau(\alpha, f; 2\delta) + 4M\int_{\delta}^{\pi} K(t)\, dt \ , \tag{3.29}$$

where $M = \sup_x |f(x)|$.

Inequality (3.28) is obtained with the help of the inequality

$$\omega(f; t - \delta) \leq \omega(f; t)\left(1 + \frac{t - \delta}{\delta}\right) = \frac{t\,\omega(f; t)}{\delta},$$

whereas (3.29) follows from the inequality $\omega(f; t - \delta) \leq 2M$.

Let us consider some applications of Theorem 3.4 and its corollaries for concrete operators.

3.3.1. The Fejer operator

The **Fejer operator** of the type (3.25) is defined by

$$\sigma_n(f; x) = \frac{1}{2\pi n}\int_{-\pi}^{\pi} f(x + t)\left(\frac{\sin (nt/2)}{\sin (t/2)}\right)^2 dt$$

where

$$\frac{1}{2\pi n}\int_{-\pi}^{\pi} \left(\frac{\sin (nt/2)}{\sin (t/2)}\right)^2 dt \ = 1.$$

It should be pointed out that for every 2π-periodic square integrable function, $\sigma_n(f; x)$ is a trigonometric polynomial of degree n.

Theorem 3.5. If $f \in A_{2\pi}$ and f is square integrable, then for $n \geq 2$, the inequalities

$$r(\alpha; f, \sigma_n(f)) \leq \tau(\alpha, f; 2n^{-1}\ln n) + 2\pi\, \omega(f; n^{-1}\ln n) \tag{3.30}$$

and

$$r(\alpha; f, \sigma_n(f)) \leq \tau(\alpha, f; 2n^{-1/2}) + 2\pi M n^{-1/2} \tag{3.31}$$

hold, where $M = \sup_x |f(x)|$.

Proof. We obtain (3.30) from (3.28) for $\delta = n^{-1}\ln n$ by using

$$(3.32)\qquad \frac{1}{2\pi n}\int_{\delta}^{\pi} t\left(\frac{\sin(nt/2)}{\sin(t/2)}\right)^2 dt \le \frac{\pi^2}{2\pi n}\int_{\delta}^{\pi} t^{-1}\,dt \le \frac{\pi}{n}\ln\delta^{-1}.$$

The inequality (3.31) is obtained from (3.29) for $\delta = n^{-1/2}$ by using

$$(3.33)\qquad \frac{1}{2\pi n}\int_{\delta}^{\pi}\left(\frac{\sin(nt/2)}{\sin(t/2)}\right)^2 dt \le \frac{\pi^2}{2\pi n}\int_{\delta}^{\pi} t^{-2}\,dt \le \frac{\pi}{2n\delta}.$$

From Theorem 3.5, letting $\alpha \to 0$, we obtain

$$\sup_x |f(x) - \sigma_n(f; x)| \le 9\omega(f; n^{-1}\ln n),$$

which is the well-known Bernstein 's Theorem(see I. Natanson [1, p. 205]).

It follows from Theorem 3.5 that the sequence $<\sigma_n(f)>_1^\infty$ converges, with respect to Hausdorff distance, to f, provided f is H-continuous. The well-known Fejer's Theorem (see I. Natanson [1, p. 199]) states that the sequence $<\sigma_n(f)>_1^\infty$ converges uniformly to f provided f is continuous. Fejer's theorem follows from Theorem 3.5 and Corollary 2.1 of Theorem 2.3. This is a typical example of how results for uniform approximation fall out as corollaries of more general results for Hausdorff approximation.

We shall denote by $A_{2\pi}^{\lambda}$, the set of functions $f \in A_{2\pi}$ for which there exists a constant $c = c(f)$ such that

$$(3.34)\qquad \tau(\alpha, f; \delta) \le c\alpha^{-1}\delta^{\lambda}, \qquad 0 < \lambda \le 1.$$

Theorem 3.6. If $f \in A_{2\pi}^{\lambda}$, where $0 < \lambda \le 1$, and f is bounded, then

$$(3.35)\qquad r(\alpha; f, \sigma_n(f)) = O(\alpha^{-1}n^{-\lambda/(1+\lambda)}),$$

and this estimate is sharp.

Proof . Insertion of (3.33) and (3.34) into (3.29) gives $r(\alpha; f, \sigma_n(f)) \le 2c\alpha^{-1}\delta^{\lambda} + 2M(\delta n)^{-1}$. Substituting $\delta = n^{-1/(1+\lambda)}$ yields (3.35).

Let us prove that (3.35) is sharp. To this end we shall consider for $0 < \lambda \le 1$ the 2π-periodic function

$$f_\lambda(x) = \begin{cases} 1 & \text{if } -\pi \le x < 0 \\ (x/\pi)^\lambda & \text{if } 0 \le x < \pi \end{cases} .$$

It is easy to verify that

$$\tau(\alpha, f; \delta) = \max\,\{\alpha^{-1}\delta, (\delta/\pi)^\lambda\} = O(\alpha^{-1}\delta^\lambda)$$

so that $f_\lambda \in A^\lambda_{2\pi}$.

Let us show that there exists a constant $c > 0$ such that for all $x \in [-\pi/2, \pi/2]$, the following inequality is valid:

$$\sigma_n(f_\lambda; x) \ge cn^{-\lambda/(1+\lambda)} . \tag{3.36}$$

But if (3.36) holds, then $f_\lambda(0) = 0$, and consequently $r(\alpha; f_\lambda, \sigma_n(f_\lambda)) \ge cn^{-\lambda/(1+\lambda)}$, i.e., the estimate (3.35) is sharp. Therefore it is sufficient to prove (3.36). For $\sigma_n(f_\lambda; x)$ we have

$$\sigma_n(f_\lambda; x) = \frac{1}{2n}\int_{-\pi}^{0} \left(\frac{\sin\,(n(t-x)/2)}{\sin\,((t-x)/2)}\right)^2 dt + \frac{1}{2\pi n}\int_0^{\pi} \left(\frac{t}{\pi}\right)^\lambda \left(\frac{\sin\,(n(t-x)/2)}{\sin\,((t-x)/2)}\right)^2 dt \tag{3.37}$$

It is easy to verify that for $x \in [-\pi/2, 0]$ we have $\sigma_n(f_\lambda; x) \ge 1/2$, hence it remains to verify (3.36) for $x \in [0, \pi/2]$. For such x, we have

$$\begin{aligned}
\frac{1}{2\pi n}\int_0^{\pi} \left(\frac{t}{\pi}\right)^\lambda \left(\frac{\sin\,(n(t-x)/2)}{\sin\,((t-x)/2)}\right)^2 dt &= \frac{1}{2\pi n}\int_{-x}^{\pi-x} \left(\frac{x+t}{\pi}\right)^\lambda \left(\frac{\sin\,(nt/2)}{\sin\,(t/2)}\right)^2 dt \\
&\ge \frac{1}{2\pi n}\int_0^{\pi/2} \left(\frac{x+t}{\pi}\right)^\lambda \left(\frac{\sin\,(nt/2)}{\sin\,(t/2)}\right)^2 dt \\
&\ge \frac{x^\lambda}{2\pi^2 n}\int_0^{\pi/2} \left(\frac{\sin\,(nt/2)}{\sin\,(t/2)}\right)^2 dt \\
&\ge \frac{x^\lambda}{4\pi}\,\frac{1}{2\pi n}\int_{-\pi}^{\pi} \left(\frac{\sin\,(nt/2)}{\sin\,(t/2)}\right)^2 dt
\end{aligned}$$

and hence,

(3.38) $$\frac{1}{2\pi n}\int_0^{\pi}\left(\frac{t}{\pi}\right)^{\lambda}\left(\frac{\sin(n(t-x)/2)}{\sin((t-x)/2)}\right)^2 dt \geq \frac{x^{\lambda}}{4\pi}.$$

On the other hand,

(3.39) $$\frac{1}{2\pi n}\int_{-\pi}^{0}\left(\frac{\sin(n(t-x)/2)}{\sin((t-x)/2)}\right)^2 dt = \frac{1}{2\pi n}\int_{x}^{\pi+x}\left(\frac{\sin(nt/2)}{\sin(t/2)}\right)^2 dt$$

$$\geq \frac{1}{2\pi n}\int_{x}^{\pi}\left(\frac{\sin(nt/2)}{\sin(t/2)}\right)^2 dt \geq \frac{1}{\pi}\int_{xn/2}^{\pi n/2} t^{-2}\sin^2 t\, dt = A$$

holds for $x \in [0, \pi/2]$. We introduce some notation:

(3.40) $$p = [nx/2\pi] + 1, \qquad q = [n/2].$$

Then

$$A = \frac{1}{\pi}\int_{xn/2}^{\pi n/2} t^{-2}\sin^2 t\, dt \geq \frac{1}{\pi}\sum_{k=p}^{q-1}\int_{k\pi}^{(k+1)\pi} t^{-2}\sin^2 t\, dt$$

$$= \frac{1}{\pi}\sum_{k=p}^{q-1}\int_{k\pi}^{(k+1)\pi} t^{-1}\sin 2t\, dt\,,$$

but

$$\int_{k\pi}^{(k+1)\pi} t^{-1}\sin 2t\, dt = \int_0^{\pi}\frac{\sin t}{t+2k\pi}dt + \int_{\pi}^{2\pi}\frac{\sin t}{t+2k\pi}dt$$

$$= \int_0^{\pi}\frac{\sin t}{t+2k\pi}dt - \int_0^{\pi}\frac{\sin t}{t+(2k+1)\pi}dt\,.$$

Therefore,

$$A \geq \frac{1}{\pi}\int_0^{\pi}\sum_{k=p}^{q-1}\frac{\pi}{(t+2k\pi)(t+(2k+1)\pi)}dt$$

$$\geq \int_0^{\pi} \sum_{k=p}^{q-1} \frac{1}{(t + 2k\pi)(t + 2(k+1)\pi)} dt$$

$$= \frac{1}{2\pi} \int_0^{\pi} \left(\frac{1}{t + 2p\pi} - \frac{1}{t + 2q\pi}\right) \sin t \, dt \geq \frac{q - p}{\pi^2(2p + 1)(2q + 1)} .$$

According to (3.40), keeping in mind that $x \in [0, \pi/2]$, we see that there exists a positive constant c_1 such that

$$\text{(3.41)} \qquad \frac{1}{2\pi n} \int_{-\pi}^{0} \left(\frac{\sin(n(t - x)/2)}{\sin((t - x)/2)}\right)^2 dt \geq \frac{c_1}{(1 + nx)} ,$$

From (3.37), according to (3.38) and (3.41), we have for $x \in [0, \pi/2]$ that

$$\text{(3.42)} \qquad \sigma_n(f_\lambda; x) \geq c_2(x^\lambda + 1/(1 + nx)),$$

where $c_2 = \min\{c_1, 1/4\pi\}$. If $x \in [0, n^{-1/(1+\lambda)}]$, then from (3.42) we obtain

$$\sigma_n(f_\lambda; x) \geq c_2/(1 + n^{\lambda/(1+\lambda)}) \geq \frac{1}{2} c_2 n^{-\lambda/(1+\lambda)} .$$

If $x \in [n^{-1/(1+\lambda)}, \pi/2]$, then again from (3.42) we obtain

$$\sigma_n(f_\lambda; x) \geq c_2 n^{-\lambda/(1+\lambda)}$$

and (3.36) is established. The theorem is proved.

3.3.2. THE JACKSON OPERATOR

The **Jackson operator** of the type (3.25) is given by

$$U_n(f; x) = \frac{3}{2\pi n(2n^2 + 1)} \int_{-\pi}^{\pi} f(x + t) \left(\frac{\sin(nt/2)}{\sin(t/2)}\right)^4 dt ,$$

where

$$\text{(3.43)} \qquad \frac{3}{2\pi n(2n^2 + 1)} \int_{-\pi}^{\pi} \left(\frac{\sin(nt/2)}{\sin(t/2)}\right)^4 dt = 1.$$

Here, $U_n(f; x)$ is a trigonometric polynomial of order $2n$. It is easy to verify that

$$(3.44)\qquad \frac{3}{2\pi n(2n^2+1)}\int_{\delta}^{\pi} t\left(\frac{\sin (nt/2)}{\sin (t/2)}\right)^4 dt \le \frac{3\pi^3}{8n^3\delta^2}$$

and

$$(3.45)\qquad \frac{3}{2\pi n(2n^2+1)}\int_{\delta}^{\pi} \left(\frac{\sin (nt/2)}{\sin (t/2)}\right)^4 dt \le \frac{\pi^3}{4n^3\delta^3}.$$

Theorem 3.7. If $f \in A_{2\pi}$ and f is square integrable, then

$$(3.46)\qquad r(\alpha; f, U_n(f)) \le \tau(\alpha, f; 2n^{-1}) + 48\omega(f; n^{-1}),$$

and

$$(3.47)\qquad r(\alpha; f, U_n(f)) \le \tau(\alpha, f; 2n^{-3/4}) + \pi^3 M n^{-3/4},$$

where $M = \sup_x |f(x)|$.

Proof. The inequality (3.46) is otained from (3.28) and (3.44), and the inequality (3.47) is obtained from (3.29) and (3.45).

From Theorem 3.7, letting $\alpha \to 0$, we obtain the classical Jackson Theorem:

$$\sup_x |f(x) - U_n(f; x)| = O(\omega(f; n^{-1})).$$

Let us consider the Hausdorff approximation of functions of the class $A_{2\pi}^{\lambda}$.

Theorem 3.8. If $f \in A_{2\pi}^{\lambda}$ where $0 < \lambda \le 1$ and f is bounded, then

$$(3.48)\qquad r(\alpha; f, U_n(f)) = O(\alpha^{-1} n^{-3\lambda/(3+\lambda)}),$$

and this estimate is sharp.

Proof. Insertion of (3.34) and (3.35) into (3.29) gives

(3.49) $$r(\alpha; f, U_n(f)) \le 2c\alpha^{-1}\delta^{\lambda} + \pi^3 M(n\delta)^{-3},$$

where $M = \sup_x |f(x)|$. Substituting $\delta = n^{-3/(3+\lambda)}$ into (3.49), we obtain (3.48).

We now prove (3.48) is sharp. As in Theorem 3.5, it is sufficient to prove that for $x \in [-\pi/2, \pi/2]$, the inequality

(3.50) $$U_n(f_\lambda; x) \ge cn^{-3\lambda/(3+\lambda)}$$

holds, where c is a positive constant. If we set $v_n = 3/2\pi n(2n^2+1)$, then

(3.51) $$U_n(f_\lambda; x) = v_n \int_{-\pi}^{0} \left(\frac{\sin(n(t-x)/2)}{\sin((t-x)/2)}\right)^4 dt + v_n \int_0^{\pi} \left(\frac{t}{\pi}\right)^{\lambda} \left(\frac{\sin(n(t-x)/2)}{\sin((t-x)/2)}\right)^4 dt$$

$$= g_1(x) + g_2(x).$$

Obviously, for $x \in [-\pi/2, 0]$, we have $U_n(f_\lambda; x) \ge 1/2$. Let $x \in [0, \pi/2]$; then, as in Theorem 3.5, one can show the existence of a constant c_1 such that

(3.52) $$g_2(x) \ge c_1 x^{\lambda}.$$

On the other hand, for $x \in [0, \pi/2]$, we have

(3.53) $$g_1(x) \ge \frac{1}{2\pi n^3} \int_x^{\pi + x} \left(\frac{\sin(nt/2)}{\sin(t/2)}\right)^4 dt \ge \frac{8}{\pi n^3} \int_x^{\pi} t^{-4} \sin^4(nt/2)\, dt.$$

$$= \frac{1}{\pi} \int_{xn/2}^{\pi n/2} t^{-4} \sin^4 t\, dt\ .$$

Let $p = [nx/2\pi]$, and $q = [n/2]$; then

(3.54) $$\frac{1}{\pi} \int_{xn/2}^{\pi n/2} t^{-4} \sin^4 t\, dt \ge \frac{1}{\pi} \sum_{k=p}^{q-1} \int_{k\pi}^{(k+1)\pi} t^{-4} \sin^4 t\, dt$$

$$= \frac{2}{3\pi} \sum_{k=p}^{q-1} \int_{k\pi}^{(k+1)\pi} t^{-3} \sin 2t \sin^2 t\, dt$$

$$= \frac{8}{3\pi} \sum_{k=p}^{q-1} \int_0^{2\pi} (t + 2k\pi)^{-3} \sin t \sin^2(t/2)\, dt$$

$$= \frac{8}{3\pi}\int_0^{\pi} \sum_{k=p}^{q-1} [(t + 2k\pi)^{-3} - (t + (2k + 1)\pi)^{-3}] \sin t \sin^2(t/2)\, dt.$$

Having in mind that

$$\sum_{k=p}^{q-1} [(t + 2k\pi)^{-3} - (t + (2k + 1)\pi)^{-3}] \ge$$
$$\ge \frac{1}{8}\sum_{k=p}^{q-1} [(t + 2k\pi)^{-3} - (t + 2(k + 1)\pi)^{-3}]$$
$$= \frac{1}{8}[(t + 2\pi p)^{-3} - (t + 2\pi q)^{-3}],$$

from (3.53) and (3.54) we obtain $g_1(x) \ge |(2p + 1)^{-3} - (2q + 1)^{-3}|/3\pi^4$. Then, obviously, it is possible to find a constant $c_2 > 0$ such that

$$g_1(x) \ge c_2(1 + nx)^{-3}, \qquad x \in [0, \pi/2]. \tag{3.55}$$

We obtain form (3.51), (3.52) and (3.55) that

$$U_n(f_\lambda; x) \ge c_3[x^2 + (1 + nx)^{-3}]$$

for $x \in [0, \pi/2]$ where $c_3 = \min\{c_1, c_2\}$. Considering, as in Theorem 3.5, separately the cases when $x \in [0, n^{-3/(3+\lambda)}]$ and when $x \in [n^{-3/(3+\lambda)}, \pi/2]$, we obtain $U_n(f_\lambda; x) \ge cn^{-3\lambda/(3+\lambda)}$. This completes the proof.

3.3.3. THE GENERALIZED JACKSON OPERATOR

Let us consider the positive linear operator

$$U_{m,r}(f; x) = c_{m,r}\int_{-\pi}^{\pi} f(x + t) \left(\frac{\sin (mt/2)}{m\sin (t/2)}\right)^{2r} dt ,$$

where m and r are positive integers and the constant $c_{m,r}$ is determined from the equality

$$c_{m,r}\int_{-\pi}^{\pi} \left(\frac{\sin (mt/2)}{m\sin (t/2)}\right)^{2r} dt = 1, \tag{3.56}$$

i.e., $U_{m,r}(f; x)$ is an operator of the type (3.25).

It is easy to see that $U_{m,r}(f; x)$ is a trigonometric polynomial of order mr. For $r = 1$, $U_{m,r}$ is the Fejer operator, and for $r = 2$, $U_{m,r}$ is the Jackson operator.

Lemma 3.1. For every $\delta > 0$, we have

$$c_{m,r}\int_{\delta}^{\pi}\left(\frac{\sin(mt/2)}{m\sin(t/2)}\right)^{2r} dt \le \pi\delta\left(\frac{\pi^2}{2m\delta}\right)^{2r-1}\left(\frac{1}{8(2r-1)}\right) \tag{3.57}$$

and

$$c_{m,r}\int_{\delta}^{\pi}\left(\frac{\sin(mt/2)}{m\sin(t/2)}\right)^{2r} dt \le \left(\frac{\pi^2}{2m\delta}\right)^{2r-1}\left(\frac{1}{4(2r-1)}\right) \tag{3.58}$$

Proof. From (3.56) we find that

$$1/c_{m,r} = \int_{-\pi}^{\pi}\left(\frac{\sin(mt/2)}{m\sin(t/2)}\right)^{2r} dt \ge 2m^{-2r}\int_{0}^{\pi/m}\left(\frac{\sin(mt/2)}{\sin(t/2)}\right)^{2r} dt$$

$$\ge 2m^{-2r}\int_{0}^{\pi/m}\left(\frac{2m}{\pi}\right)^{2r} dt = 2^{2r+1}\pi^{-2r+1}m^{-1}.$$

Hence,

$$c_{m,r} \le 2^{-2r-1}\pi^{2r-1}m. \tag{3.59}$$

It now follows from (3.59) that

$$c_{m,r}\int_{\delta}^{\pi}\left(\frac{\sin(mt/2)}{m\sin(t/2)}\right)^{2r} dt \le c_{m,r}\int_{\delta}^{\pi} t\,(m\sin(t/2))^{-2r}\, dt$$

$$\le \left(\frac{\pi}{m}\right)^{2r} c_{m,r}\int_{\delta}^{\infty} t^{-2r+1}\, dt = \pi\delta\left(\frac{\pi^2}{2m\delta}\right)^{2r-1}\left(\frac{1}{8(2r-1)}\right),$$

and (3.57) is verified.

From (3.59) we obtain (3.58) as well:

$$c_{m,r}\int_{\delta}^{\pi}\left(\frac{\sin(mt/2)}{m\sin(t/2)}\right)^{2r}dt \le \left(\frac{\pi}{m}\right)^{2r}c_{m,r}\int_{\delta}^{\infty}t^{-2r}\,dt$$

$$=\left(\frac{\pi^2}{2m\delta}\right)^{2r-1}\left(\frac{1}{4(2r-1)}\right).$$

The lemma is proved.

The following theorem is obtained from Corollary 3.1 and Lemma 3.1 for an optimal choice of δ.

Theorem 3.9. If $f \in A_{2\pi}$ and f is square integrable, then

(3.60) $$r(\alpha; f, U_{m,r}(f)) \le \tau(\alpha, f; \pi^2 m^{-1}) + \frac{\omega(f; m^{-1})}{r-1}$$

and

(3.61) $$r(\alpha; f, U_{m,r}(f)) \le \tau(\alpha, f; \pi^2 m^{-1+1/2r}) + \frac{Mm^{-1+1/2r}}{2r-1}$$

for $r \ge 2$, where $M = \sup_x |f(x)|$.

The inequality (3.60) shows that the generalized Jackson operator $U_{m,r}$, applied to continuous functions, does not give better results than the Jackson operator $U_n = U_{n,2}$ itself. On the other hand, for discontinuous functions, as (3.61) shows, the generalized Jackson operator gives Hausdorff approximation which improves substantially with increasing r.

Let us consider the special case of the **generalized Jackson operator**, when

(3.62) $$m = [2n/\ln n], \qquad r = [\ln n/2].$$

In this case the generalized Jackson operator $U_{m,r}(f)$ will be denoted by $A_n(f; x)$. Thus:

$$A_n(f; x) = c_{m,r}\int_{-\pi}^{\pi} f(x+t)\left(\frac{\sin(mt/2)}{m\sin(t/2)}\right)^{2r}dt,$$

where m and r are positive integers defined by (3.62) and $c_{m,r}$ is defined by (3.56). It follows from (3.62) that $A_n(f; x)$ is a trigonometric polynomial of order no higher than n.

Theorem 3.10. If $f \in A_{2\pi}$ and f is square integrable, then

(3.63) $r(\alpha; f, A_n(f)) \le \tau(\alpha; f, \pi^2 e^4 n^{-1}\ln n) + e^4 M n^{-1}$,

for $n \ge 20 > e^3$, where $M = \sup_x |f(x)|$

Proof. According (3.62), we have $m^{-1+1/2r} \le \varphi(n) n^{-1}\ln n$, where $\varphi(n) = (2n/\ln n)^{2/(\ln n - 2)}$. But

$$\ln \varphi(n) = \frac{2(\ln n + \ln 2 - \ln(\ln n))}{\ln n - 2} \le 2 - (2 + \ln 2 - \ln 3) \le 4$$

holds for $n > e^3$, so that

(3.64) $m^{-1+1/2r} \le e^4 n^{-1}\ln n.$

The desired estimate (3.63) follows from (3.64) and Theorem 3.9. The theorem is proved.

Later on we shall prove that for the function $\sigma(x) = \operatorname{sgn} \sin x$, for which $\tau(\alpha, \sigma; \delta) = \alpha^{-1}\delta$ for $\delta < \pi$, there exists a positive constant c such that for every trigonometric polynomial T of order no higher than n, the inequality

$$r(\alpha; \sigma, T) \ge c\alpha^{-1} n^{-1}\ln n$$

holds.

The operator A_n has the remarkable property that it achieves the best order of approximation out of all the positive linear operators that are trigonometric polynomials of order no higher than n.

3.3.4. THE VALLÉE-POUSSIN OPERATOR

The classical **Vallée-Poussin operator** is defined by

$$V_n(f; x) = \frac{(2n)!!}{2\pi(2n-1)!!}\int_{-\pi}^{\pi} f(x+t)\cos^{2n}(t/2)\,dt ,$$

where

$$\frac{(2n)!!}{2\pi(2n-1)!!}\int_{-\pi}^{\pi} \cos^{2n}(t/2)\,dt = 1.$$

Having in mind the inequality

$$(3.65)\qquad \frac{\sqrt{\pi/n}}{4} \le \frac{(2n)!!}{2\pi(2n-1)!!} \le \sqrt{\pi/n},$$

we find that

$$(3.66)\qquad \frac{(2n)!!}{2\pi(2n-1)!!}\int_{\delta}^{\pi} t\cos^{2n}(t/2)\,dt \;\le \frac{1}{2}\pi^{3/2}\, n^{-1/2}(1-(\delta/\pi)^2)^n,$$

and

$$(3.67)\qquad \frac{(2n)!!}{2\pi(2n-1)!!}\int_{\delta}^{\pi} \cos^{2n}(t/2)\,dt \;\le (\pi n)^{1/2}\,(1-(\delta/\pi)^2)^n .$$

Theorem 3.11. If $f \in A_{2\pi}$ and f is square integrable, then

$$(3.68)\qquad r(\alpha; f, V_n(f)) \le \tau(\alpha, f; 2\pi n^{-1/2}) + \pi\omega(f; \pi n^{-1/2})$$

and

$$(3.69)\qquad r(\alpha; f, V_n(f)) \le \tau(\alpha, f; 2\pi(n^{-1}\ln n)^{1/2}) + 8Mn^{-1/2},$$

where $M = \sup_x |f(x)|$.

Proof. Keeping in mind (3.28), (3.66) and that $(1-(\delta/\pi)^2)^n \le (1-1/n)^n < 1/e$ holds for $\delta = \pi n^{-1/2}$, we obtain (3.68). The inequality (3.69) is obtained analagously from (3.29) and (3.67), using the inequality

$$(1-(\delta/\pi)^2)^n \le (1-n^{-1}\ln n)^n \le e^{-\ln n} = n^{-1}$$

for $\delta = \pi(n^{-1}\ln n)^{1/2}$.

From (3.68), letting $\alpha \to 0$, we obtain this well-known inequality (I. Natanson [1, p. 257]) :

$$\sup_x |f(x) - V_n(f; x)| = O(\omega(f; n^{-1/2})).$$

Theorem 3.12. If $f \in A^{\lambda}_{2\pi}$, where $0 < \lambda \le 1$, and f is bounded, then

$$(3.70)\qquad r(\alpha; f, V_n(f)) = O((n^{-1}\ln n)^{\lambda/2}),$$

and this estimate is sharp for $\lambda = 1$.

Proof. The estimate (3.70) is obtained directly from (3.69) and (3.34). Now we shall show that (3.70) is sharp for $\lambda = 1$ (V. Veselinov [17]). We shall prove that for the function

$$\sigma(x) = \begin{cases} 0 & \text{if } -\pi < x < 0 \\ 1 & \text{if } 0 \le x \le \pi \end{cases},$$

there exists a positive constant c such that

$$\text{(3.71)} \qquad r(\sigma, V_n(\sigma)) \ge c\sqrt{n^{-1}\ln n}\,,$$

i.e., the order of the extimate (3.70) is sharp. It follows from the definition of V_n and (3.65) that if $x_n = \frac{1}{2}\sqrt{n^{-1}\ln n}$, where $n \ge 2$, then

$$V_n(\sigma; x) = \frac{(2n)!!}{2\pi(2n-1)!!}\int_{-x_n}^{\pi - x_n} \cos^{2n}(t/2)\,dt$$

$$\ge \frac{1}{4}\sqrt{\frac{n}{\pi}}\int_{-x_n}^{-2x_n} \cos^{2n}(t/2)\,dt \ge \frac{1}{4}\sqrt{\frac{n}{\pi}}\int_{-x_n}^{-2x_n} (1 - t/4)^n\,dt\,.$$

Having in mind that for $x \in [0, 1/2]$, the inequality $1 - x \ge e^{-2x}$ holds, and we obtain

$$V_n(\sigma; x) \ge \frac{1}{4}\sqrt{\frac{n}{\pi}}\int_{-x_n}^{-2x_n} e^{-nt^2/2}\,dt \ge \frac{-x_n}{4}\sqrt{\frac{n}{\pi}}\, e^{-2nx_n^2} = \frac{1}{8}\sqrt{\frac{\ln n}{\pi n}}\,.$$

But from here it follows that

$$r(\alpha; \sigma, V_n(\sigma)) \ge \max\{\frac{1}{2\alpha}, \frac{1}{8}\pi^{-1/2}\}\sqrt{n^{-1}\ln n}\,,$$

because $\sigma(x) = 0$ for $-\pi \le x \le 0$. Hence, (3.71) is proved.

§ 3.4. Approximation of functions by positive integral operators on a finite closed interval

Let us consider the problem of the approximation of functions, defined on a finite closed interval $\Delta = [0, 1]$, by positive integral operators of the type

$$L(f; x) = \int_0^1 f(t)K(t - x)\,dt, \qquad x \in [0, 1]\,, \tag{3.72}$$

where K is a nonnegative even kernel that is square integrable on $[0, 1]$.

Theorem 3.12. If $f \in A_\Delta$ and f is square integrable on $\Delta = [0,1]$, then for every $\delta > 0$, the inequality

$$r(\Delta, \alpha; f, L(f)) \le \tau(\alpha, f; 2\delta) + 2\int_\delta^1 \omega(f; t - \delta)K(t)\,dt + M \sup\left\{\left|1 - \int_\delta^1 K(t - x)\,dt\right| : x \in \Delta\right\} \tag{3.73}$$

holds, where $M = \sup\{|f(x)| : x \in \Delta\}$.

Proof. It is sufficient to prove, using Theorem 3.3, that for the operator (3.72), we have

$$\begin{aligned} L(\omega(x, \delta, f); x) &= \int_0^{x-\delta} \omega(f; x - t - \delta)K(t)\,dt + \int_{x+\delta}^1 \omega(f; t - x - \delta)K(t)\,dt \\ &= \int_\delta^x \omega(f; t - \delta)K(t)\,dt + \int_\delta^{1-x} \omega(f; t - \delta)K(t)\,dt \\ &\le 2\int_\delta^1 \omega(f; t - \delta)K(t)\,dt. \end{aligned}$$

3.4.1. The Landau operator

The **Landau operator** serves as a concrete example of an operator of the type (3.72). The algebraic polynomial

$$L_n(f; x) = \sqrt{\frac{n}{\pi}}\int_0^1 f(t)(1 - (t - x)^2)^n\,dt$$

of degree no higher than 2n is called the **Landau polynomial** (p. Korovkin [1, p.34]). The Landau operator is a nonperiodic analogue of the Vallée-Pousin operator and the estimate

$$(3.74) \qquad r([a, b], \alpha; f, L_n(f)) \leq \tau(\alpha, f; 2(n^{-1}\ln n)^{1/2}) + O((n^{-1}\ln n)^{1/2})$$

holds for the Landau operator on every segment $[a, b]$ with $0 < a < b < 1$.

3.4.2. THE GENERALIZED LANDAU OPERATOR

The **generalized Landau operator** defined by

$$L_{n,k}(f; x) = \frac{kn^{1/2k}}{\Gamma(1/2k)} \int_0^1 f(t)(1 - (t - x)^{2k})^n \, dt ,$$

yields polynomials of degree 2kn, for which the following estimate

$$(3.75) \qquad r([a, b], \alpha; f, L_{n,k}(f)) \leq \tau(\alpha, f; 2(n^{-1}\ln n)^{1/2k}) + O((n^{-1}\ln n)^{1/2k})$$

for $0 < a < b < 1$ holds, and this estimate is sharp (V. Veselinov and I. Kirkorov [1]).

The comparison of (3.74) and (3.75) shows that the generalized Landau operator approximates the corresponding function worse with respect to the Hausdorff distnace for $k > 1$, although the degree of the corresponding algebraic polynomial is higher.

§ 3.5. Approximation of functions by summation formulas on a finite closed interval

Let $x_0 < x_1 < x_2 < \cdots < x_n$ be $n + 1$ points in the closed interval $\Delta = [a, b]$, and let $\varphi_0(x), \varphi_1(x), \ldots, \varphi_n(x)$ be $n + 1$ nonnegative continuous functions, defined on the closed interval Δ. We assume that the equality

$$(3.76) \qquad \sum_{k=0}^{n} \varphi_k(x) = 1$$

holds for every $x \in \Delta$. Then

$$(3.77) \qquad \Phi_n(f; x) = \sum_{k=0}^{n} f(x_k)\varphi_k(x)$$

is a positive linear operator in B_Δ, and according to (3.76) we have

(3.78) $$\Phi_n(1; x) \equiv 1.$$

The operators of the form (3.77) are called **summation formulas.**

Theorem 3.13. If $f \in A_\Delta$, then for every $\delta > 0$, the inequality

(3.79) $$r(\Delta, \alpha; f, \Phi_n(f)) \le \tau(\alpha, f; 2\delta) + \sup_{x\in\Delta} \sum_{k\in\Delta(x,\delta)} \omega(f; |x - x_k| - \delta)\varphi_k(x)$$

holds, where $\Delta(x,\delta)$ is the set of those k for which $x_k \in \Delta \backslash [x - \delta, x + \delta]$.

Proof. The inequality (3.79) follows directly from Theorem 3.3 and (3.78).

Let us consider examples of concrete summation formulas.

3.5.1. Bernstein Polynomials

The well-known **Bernstein polynomials**

$$B_n(f; x) = \sum_{k=0}^{n} f\left(\frac{k}{n}\right)\binom{n}{k}x^k(1 - x)^{n-k}$$

are summation formulas on the closed interval [0, 1], since

$$\varphi_{n,k}(x) = \binom{n}{k}x^k(1 - x)^{n-k} \ge 0$$

for $x \in [0, 1]$ and

$$\sum_{k=0}^{n}\binom{n}{k}x^k(1 - x)^{n-k} = 1.$$

Theorem 3.14. (V. Veselinov [12, 19]). If $f \in A_{[0, 1]}$ then for sufficiently large n ($\ln n > 10$), we have

(3.80) $$r([0, 1], \alpha; f, B_n(f)) \le \tau(\alpha, f; 2n^{-1/2}) + \omega(f; n^{-1/2})$$

and

(3.81) $\quad r([0, 1], \alpha; f, B_n(f)) \leq \tau(\alpha, f; 2(n^{-1}\ln n)^{1/2}) + 2O(Mn^{-3/4}),$

where $M = \sup\{|f(x)| : x \in [0, 1]\}$.

Proof. According to Theorem 3.13, it is necessary to estimate

$$B(\delta, n; x) = \sum_{k\in\Delta(x,\delta)} \omega(f; |x - k/n| - \delta)\binom{n}{k}x^k(1 - x)^{n-k} .$$

In order to prove (3.80) and (3.81) it is necessary to consider the following two estimates:

$$B(\delta, n; x) \leq \delta^{-1}\omega(f; \delta) \sum_{k\in\Delta(x,\delta)} \binom{n}{k}x^k(1 - x)^{n-k} \tag{3.82}$$

and

$$B(\delta, n; x) \leq 2M \sum_{k\in\Delta(x,\delta)} \binom{n}{k}x^k(1 - x)^{n-k} . \tag{3.83}$$

Since the inequality $|x - k/n| \geq \delta$ holds for $k \in \Delta(x,\delta)$, then we have $\delta^{-m}(x - k/n)^m \geq 1$ for every even m. Therefore,

$$\sum_{k\in\Delta(x,\delta)} \binom{n}{k}x^k(1 - x)^{n-k} \leq \delta^{-m}\sum_{k=0}^{n}(x - k/n)^m \binom{n}{k}x^k(1 - x)^{n-k} . \tag{3.84}$$

Now we shall prove that for the polynomials

$$W_{m,n}(x) = \sum_{k=0}^{n} (x - k/n)^m x^k(1 - x)^{n-k} , \tag{3.85}$$

the inequality

$$\max_{0\leq x\leq 1} |W_{m,n}(x)| \leq \sqrt{2m\pi}\left(\frac{m}{e}\right)^m n^{1/5}(n\ln n)^{-m/2} \tag{3.86}$$

holds for $x \in [0, 1]$. For the function $\varphi_n(z,x) = e^{-zx}(1 - x + xe^{z/n})^n$, we have the following Maclaurin expansion (S. Bernstein [2]):

(3.87) $$\varphi_n(z,x) = e^{-zx}(1 - x + xe^{z/n})^n = \sum_{m=0}^{\infty} \frac{W_{m,n}(x)}{m!} z^m$$

Applying the Cauchy integral formula, we obtain form (3.87) that

(3.88) $$\frac{W_{m,n}(x)}{m!} = \frac{1}{2\pi i} \int_{|z| = R} z^{-m-1}\varphi_n(z,x)\, dz \, .$$

Let us denote by $A(R)$ the maximum modulus of $\varphi_n(z,x)$ for $|z| = R$ and $0 \le x \le 1$. Since

$$\varphi_n(z,x) = \left(1 + x(1 - x) \sum_{k=2}^{\infty} \left(\frac{z}{n}\right)^k \left(\frac{(1 - x)^{k-1} - (-x)^{k-1}}{k!}\right)\right)^n$$

then

$$A(\sqrt{n\ln n}\,) \le \left(1 + \frac{1}{4}\sum_{k=2}^{\infty} \frac{1}{k} \left(\frac{\ln n}{n}\right)^{k/2}\right)^n \le \left(1 + \frac{\ln n}{4n}\sum_{k=2}^{\infty} \frac{1}{k!}\right)^n ,$$

or

(3.89) $$A(\sqrt{n\ln n}\,) \le \left(1 + \frac{\ln n}{5n}\right)^n .$$

We obtain form (3.88) and (3.89) that

(3.90) $$\max_{0 \le x \le 1} |W_{m,n}(x)| \le m! \left(1 + \frac{\ln n}{5n}\right)^n (n\ln n)^{-m/2} \le m!\, n^{1/5}(n\ln n)^{-m/2} .$$

Recalling the Stirling inequality $m! \le \sqrt{3m\pi}\, m^m e^{-m}$, we obtain (3.86) from (3.90). Then, for $\delta = (n^{-1}\ln n)^{1/2}$ and $m = 2[\ln n/2]$, from (3.84) and (3.86) we have

(3.91) $$\sum_{k \in \Delta(x,\delta)} \binom{n}{k} x^k(1 - x)^{n-k} \le \sqrt{9m\pi} \left(\frac{m}{e\delta}\right)^m n^{1/5}(n\ln n)^{-m/2}$$

$$\le \sqrt{3\pi \ln n}\, (e^{-2}\, n\ln n)^{m/2}\, n^{1/5}(n\ln n)^{-m/2}$$

$$\le \sqrt{3\pi\ln n}\, e^2 e^{-\ln n}\, n^{1/5} = e^2\, n^{-4/5} \sqrt{3\pi\ln n} \, .$$

Using (3.82) and (3.83) and keeping in mind (3.91), we obtain (3.80) and (3.81), respectively. The theorem is proved.

V. Veselinov [19] has found the exact constant inside the modulus of H-continuity in (3.81). In (3.81), instead of $\tau(\alpha, f; 2(n^{-1}\ln n)^{1/2})$, we can insert $\tau(\alpha, f; (n^{-1}\ln n)^{1/2})$; this result cannot be improved.

From (3.80) letting $\alpha \to 0$, we obtain a classical result of T. Popoviciu [1] for continuous functions:

$$\|f - B_n(f)\| = \sup \{|f(x) - B_n(f; x)| : x \in \Delta\} \leq 3\omega(f; n^{-1/2}).$$

The following result is obtained directly from Theorem 3.14.

Corollary 3.3. If $f \in A_{[0,1]}^{\lambda}$ and f is bounded, then

$$r([0, 1], \alpha; f, B_n(f)) = O(\alpha^{-1} (n^{-1}\ln n)^{\lambda/2}). \tag{3.92}$$

Let us show that the order of the estimate (3.92) is exact (V. Veselinov [15]). Consider the function $f_\lambda \in A_{[0,1]}^{\lambda}$ with $\lambda \in (0, 1]$, and

$$f_\lambda(x) = \begin{cases} 2^\lambda(1/2 - x) & \text{if } 0 \leq x < 1/2 \\ 1 & \text{if } 1/2 \leq x \leq 1 \end{cases}.$$

Theorem 3.15. There exists a constant $c_1 > 0$ depending only on λ and such that for every positive integer $n \geq 2$, we have

$$r([0, 1], \alpha; f_\lambda, B_n(f_\lambda)) \geq c\alpha^{-1} (n^{-1}\ln n)^{\lambda/2}. \tag{3.93}$$

Proof. We shall show that for every $x \in [0, 1]$ and every positive integer $n \geq 2$, the inequaltiy

$$B_n(f_\lambda; x) \geq c(n^{-1}\ln n)^{\lambda/2} \tag{3.94}$$

holds, where $c > 0$ is a constant depending only on λ. The inequality (3.93) follows from (3.94) and the definition of Hausdorff distance.

Let $\varphi_{n,k}(x) = \binom{n}{k} x^k(1 - x)^{n-k}$. We next decompose $B_n(f_\lambda)$ into a sum:

(3.95) $$B_n(f_\lambda; x) = A(x) + B(x),$$

where

$$A(x) = \sum_{n/2 \le k \le n} \varphi_{n,k}(x) \quad \text{and} \quad B(x) = \sum_{0 \le k \le n/2} 2\lambda(1/2 - k/n)\varphi_{n,k}(x).$$

First we shall prove that for every $x \in [2/5, 1/2]$, and for every positive integer $k \in [n/2, 3n/5]$, the inequality

(3.96) $$\varphi_{n,k}(x) \ge (6n)^{-1/2}\, e^{-8n(k/n-x)^2}$$

holds. From Stirling's formula, it follows that

(3.97) $$\varphi_{n,k}(x) \ge \left(\frac{n}{2\pi e k(n-k)}\right)^{1/2} \left(\frac{nx}{k}\right)^k \left(\frac{n(1-x)}{n-k}\right)^{n-k}.$$

We write $Q = \left(\frac{nx}{k}\right)^k \left(\frac{n(1-x)}{n-k}\right)^{n-k}$. We have (G. Lorentz [1]) :

$$-\ln Q = \frac{n}{x(1-x)}\left(\frac{k}{n} - x\right)^2 \left(1 - \frac{\alpha k}{2nx} + 1 - \frac{\beta}{2}\frac{1 - k/n}{1 - x}\right),$$

where

$$\alpha = (1 + \omega_1(k/n - x)/x)^{-2}, \qquad \omega_1 > 0,$$

$$\beta = (1 - \omega_2(k/n - x)/(1 - x))^{-2}, \qquad \omega_2 < 1.$$

Consequently, $-\ln Q \le 8n(k/n - x)^2$, i.e.,

(3.98) $$Q \ge \exp(-8n(k/n - x)^2) \quad \text{for } x \in [2/5, 1/2].$$

Furthermore, we have for $k \in [n/2, 3n/5]$:

(3.99) $$\frac{n}{k(n-k)} \ge \frac{10}{3n}.$$

Combining (3.97), (3.98) , and (3.99) yields (3.96).

Let us write $x_n = 1/2 - (\lambda n^{-1}\ln n)^{1/2}/8$ and $\Delta_n = [1/2, 1 - x_n]$. From (3.96) we get

$$A(x_n) \geq \sum_{k/n\in\Delta_n} \varphi_{n,k}(x_n) \geq (6n)^{-1/2} \sum_{k/n\in\Delta_n} \exp(-8nk^2/(n - x_n)^2)$$

$$\geq (6n)^{-1/2} \sum_{k/n\in\Delta_n} \exp(-8n(1 - 2x_n)^2)$$

$$\geq (6n)^{-1/2}((\lambda n\ln n)^{1/2}/8 - 1)\exp(-\lambda\ln n/2),$$

or finally,

$$A(x_n) \geq c_1(n^{-1}\ln n)^{\lambda/2} \tag{3.100}$$

for $n > N_1$, where c_1 and N_1 are positive constants, depending only on λ.

It is known that if f is a monotone function, then $B_n(f; x)$ is also monotone (G. Lorentz [2]). Therefore it follows from (3.100) that

$$A(x) \geq c_3(n^{-1}\ln n)^{\lambda/2} \tag{3.101}$$

for $n > N_1$ and for every $x \in [x_n, 1]$. Furthermore, we have

$$B(x_n) \geq 2^\lambda \sum_{0\leq k\leq nx_n} (1/2 - k/n)^\alpha \varphi_{n,k}(x_n) \geq \lambda^{\lambda/2}4^{-\lambda}(n^{-1}\ln n)^{\lambda/2} \sum_{0\leq k\leq nx_n} \varphi_{n,k}(x_n). \tag{3.102}$$

Now let us recall the following asymptotic equality (G. Lorentz [2, p.15]):

$$\sum_{n(x - \delta_n)\leq k\leq nx_n} \varphi_{n,k}(x_n) \approx \pi^{-1/2} \int_0^{v_n} e^{-t^2}\,dt, \tag{3.103}$$

where $v_n = \delta_n n^{1/2}(2x(1 - x))^{-1/2}$. Using the method of proof of (3.103), we can prove

$$\sum_{0\leq k\leq nx_n} \varphi_{n,k}(x_n) \geq 1/5 \tag{3.104}$$

for $n \geq N_2$, where N_2 is an absolute constant. We obtain from (3.102) and (3.104) that $B(x_n) \geq (\lambda^{\lambda/2}/20)(n^{-1}\ln n)^{\lambda/2}$ and hence

$$B(x) \geq (\lambda^{\lambda/2}/20)(n^{-1}\ln n)^{\lambda/2} \tag{3.105}$$

for $n \geq N_2$ and every $x \in [0, x_n]$, since $B(x)$ is a Bernstein polynomial of a monotone decreasing function.

We obtain from (3.95), (3.101) and (3.105) that $B_n(f_\lambda; x) \geq c(n^{-1}\ln n)^{\lambda/2}$ for every $x \in [0, 1]$. The theorem is proved.

3.5.2. FEJER INTERPOLATIONAL POLYNOMIALS

Let $T_n(x) = \cos(n \arccos x)$ be the nth Chebyshev polynomial and denote its zeros by $x_{n,k} = \cos(\pi(2k - 1)/2n)$. Let

$$\Phi_n(f; x) = \sum_{k=1}^{n} f(x_{n,k}) \left(\frac{T_n(x)}{n(x - x_{n,k})}\right)^2 (1 - xx_{n,k}) \tag{3.106}$$

where f is a function defined on the segment $[-1, 1]$. But $\Phi_n(f)$ is the **Hermite-Fejer interpolational polynomial** of degree $2n - 1$ for the function f (I. Natanson [1, p.549]), and

$$\Phi_n(f; x_{n,k}) = f(x_{n,k}), \qquad \Phi_n'(f; x_{n,k}) = 0, \tag{3.107}$$

for $k = 1, 2, 3, \ldots, n$ and $n = 1, 2, 3, \ldots$.

Here, an important result is Fejer's Theorem: if the function f is continuous, then the sequence $\langle\Phi_n(f)\rangle_1^\infty$ uniformly tends to f on the closed interval $[-1, 1]$.

For Φ_n we have the equality

$$\Phi_n(1; x) = \sum_{k=1}^{n} \left(\frac{T_n(x)}{n(x - x_{n,k})}\right)^2 (1 - xx_{n,k}) \equiv 1,$$

which is clearly connected with the interpolational condition (3.107). Hence, Φ_n is a summation formula.

Theorem 3.16. If $f \in A_{[-1,1]}$ then for every positive integer n,

(3.108) $$r(\Delta,\ \alpha; f, \Phi_n(f)) \le \tau(\alpha, f; 2n^{-1/2}) + 2\omega(f; n^{-1/2})$$

and

(3.109) $$r(\Delta,\ \alpha; f, \Phi_n(f)) \le \tau(\alpha, f; 2n^{-1/2}) + cMn^{-1/2},$$

where $M = \sup\{|f(x)| : |x| \le 1\}$, $\Delta = [-1, 1]$, and c is an absolute constant.

Proof. Let $\Delta(x,\delta)$ be the set of those $k \in \{1,2,\ldots,n\}$ for which $|x - x_{n,k}| > \delta$ holds. It is necessary to estimate

$$\Phi_n(\delta, n; x) = \sum_{k\in\Delta(x,\delta)} \omega(f; |x - x_k| - \delta)\, \varphi_{n,k}(x)\ ,$$

where $x_k = x_{n,k} = \cos(\pi(2k - 1)/2n)$ and $\varphi_{n,k}(x) = \left(\frac{T_n(x)}{n(x - x_k)}\right)^2(1 - xx_k)$. In order to prove (3.108) and (3.109) it is necessary to employ the following two estimates:

(3.110) $$\Phi_n(\delta, n; x) \le \delta^{-1}\omega(f; \delta) \sum_{k\in\Delta(x,\delta)} |x - x_k|\varphi_{n,k}(x)$$

and

(3.111) $$\Phi_n(\delta, n; x) \le 2M \sum_{k\in\Delta(x,\delta)} \varphi_{n,k}(x)\ .$$

From the definition of $T_n(x)$ and x_k we obtain

(3.112) $$\sum_{k\in\Delta(x,\delta)} |x - x_k|\varphi_{n,k}(x) \le 2n^{-2} \sum_{k\in\Delta(x,\delta)} |x - x_k|^{-1}$$

$$\le 2n^{-2}\delta^{-1}\sum_{k=1}^{n} |x - x_k|^1\, |x - x_k|^{-1} = \frac{2}{\delta n}\ .$$

It can be proved that there exists an absolute constant c such that

(3.113) $$\sum_{k\in\Delta(x,\delta)} \varphi_{n,k}(x) \le \frac{c}{\delta n}\ .$$

We obtain (3.108) from (3.110) and (3.112) for $\delta = n^{-1/2}$, by using Theorem 3.13, and we obtain (3.109) from (3.111) and (3.113) in the same way. The theorem is proved.

From (3.108), letting $\alpha \to 0$, we get

(3.114) $$\|f - \Phi_n(f)\|_{[-1,1]} \le 4\omega(f; n^{1/2}).$$

The inequality (3.114) is a theorem of Fejer. If $f \in A^{\lambda}_{[-1,1]}$ then we obtain from (3.109) this relation:

(3.115) $$r([-1,1], \alpha; f, \Phi_n(f)) = O(\alpha^{-1} n^{-\lambda/2}), \qquad 0 < \lambda \le 1.$$

We shall show that the order ot the estimate (3.115) is sharp for $\lambda = 1$. Specifically, we shall show for the function

$$\sigma(x) = \begin{cases} 0 & \text{if } -1 \le x < 0 \\ 1 & \text{if } 0 \le x \le 1 \end{cases},$$

where $\sigma \in A^{\lambda}_{[-1,1]}$, there exists a constant $c > 0$ independent of n such that

(3.117) $$r([-1,1], \alpha; \sigma, \Phi_n(\sigma)) \ge cn^{-1/2}.$$

To this end it is sufficient to prove that there exists $\xi_n < -c\alpha n^{-1/2}$ such that

(3.118) $$\Phi_n(\sigma; \xi_n) \ge cn^{-1/2}.$$

The Fejer polynomial for the function described in (3.116) is

(3.119) $$\Phi_n(\sigma; x) = \sum_{k=1}^{s} \left(\frac{T_n(x)}{n(x - x_k)}\right)^2 (1 - xx_k),$$

where $s = [(n-1)/2]$.

Let $p = [n(1 - n^{-1/2})/2]$ and $\xi_n = \cos(\pi p/n)$ (note that $\xi_n < -n^{-1/2}$). Since $\xi_n x_k \le 0$ for $k = 1, 2, \ldots, s$, from (3.119) we have

(3.120) $$\Phi_n(\sigma; \xi_n) \geq n^{-2} \sum_{k=1}^{s} (1 - \xi_n x_k)(\xi_n - x_k)^{-1} \geq n^{-2} \sum_{k=1}^{s} (\xi_n - x_k)^{-2} .$$

It is easy to check that

(3.121) $$\sum_{k=1}^{s} (\xi_n - x_k)^{-2} \geq \frac{n}{2\pi} \sum_{k=1}^{s} \int_{x_k}^{x_{k-1}} (\xi_n - t)^{-2}\, dt$$

$$\geq \frac{n}{2\pi} \int_{x_s}^{1} (\xi_n - t)^{-2}\, dt \geq \frac{n}{2} \left(\frac{1}{x_s - \xi_n} - \frac{1}{1 - \xi_n} \right) \geq cn^{3/2}.$$

The inequality (3.118) follows from (3.121) and (3.200). The theorem is proved.

Other summation formulas can be analyzed in a similar manner. V. Spiridonov [1] considered the **interpolational trigonometric polynomials of Rappoport:**

$$R_n(f; x) = \frac{(2n)!!}{(2n-1)!!} \sum_{k=0}^{2n} f(x_k) \cos^{2n}\left(\frac{x_k - x}{2}\right), \quad x_k = \frac{2k\pi}{2n-1} ,$$

which are a summation analogue of the Vallée-Poussin integral operator (P. Natanson [1, p.574]). It can be proven that for every $f \in A_{2\pi}$ we have

(3.122) $$r(\alpha; f, R_n(f)) \leq \tau\,(\alpha, f; 2(n^{-1}\ln n)^{1/2}) + 4\pi M(n^{-1}\ln n)^{1/2} ,$$

where $M = \sup_x |f(x)|$.

§ 3.6. Approximation of nonperiodic functions by integral operators on the entire real axis

Let us consider within $A_{(-\infty, \infty)}$ the class A of functions that are bounded on the entire real axis. Let $K(t)$ be a nonnegative even kernel with an integrable square on $(-\infty, \infty)$ satisfying

$$\int_{-\infty}^{\infty} K(t)\, dt = 1.$$

The positive linear operator

$$(3.123) \qquad L(f; x) = \int_{-\infty}^{\infty} f(x + t)K(t)\, dt$$

is defined on the square integrable functions in A. The next theorem follows directly from Theorem 3.3.

Theorem 3.17. If $f \in A$ and f has integrable square, then for every $\delta > 0$ we have

$$(3.124) \qquad r(\alpha; f, L(f)) \le \tau\,(\alpha, f; 2\delta) + 2\int_{\delta}^{\infty} \omega(f; t - \delta)K(t)\, dt\ .$$

Here, as well as in the periodic case, it is possible to obtain the corresponding estimates for a sequence of classical operators of the form (3.123).

3.6.1. THE FEJER OPERATOR IN THE NONPERIODIC CASE

If $K(t) = (2/\nu\pi)\, t^{-2}\sin^2 (\nu t/2)$, we obtain the **Fejer integral operator**

$$\bar{\sigma}_\nu(f; x) = \frac{2}{\nu\pi}\int_{-\infty}^{\infty} f(x + t)\, t^{-2}\sin^2 (\nu t/2)\, dt$$

It should be noted that for every $f \in A$ where f is square integrable, the function $\bar{\sigma}_\nu(f)$ is of the exponential type ν.

From Theorem 3.17, in a manner similar to Theorem 3.5, we have

Theorem 3.18. If $f \in A$ and f has integrable square, then

$$r(\alpha; f, \bar{\sigma}_\nu(f)) \le \tau\,(\alpha, f; 2\nu^{-1}\ln \nu) + 2\pi\omega(f; \nu^{-1}\ln \nu)$$

and

$$r(\alpha; f, \bar{\sigma}_\nu(f)) \le \tau\,(\alpha, f; 2\nu^{-1/2}) + 2\pi M\nu^{-1/2}\ ,$$

where $M = \sup_x |f(x)|$. For $f \in A^\lambda$, we have

$$(3.125) \qquad r(\alpha; f, \bar{\sigma}_\nu(f)) = O(\nu^{-\lambda/(1+\lambda)})\ .$$

In the same way as in Theorem 3.6, one can show that the estimate (3.125) cannot be improved. A theorem for the uniform approximation by the Fejer operator (3.123) can be obtained as a corollary of Theorem 3.18.

The **Jackson integral operator** can be analyzed similarly (V. Veselinov [7]):

$$\bar{U}_\nu(f\,;x) = c_\nu \int_{-\infty}^{\infty} f(x+t)\, t^{-4} \sin^4(\nu t/2)\, dt$$

where

$$c_\nu \int_{-\infty}^{\infty} t^{-4} \sin^4(\nu t/2)\, dt = 1.$$

For $f \in A^\lambda$ it can be shown that $r(\alpha; f, \bar{U}_\nu(f)) = O(\nu^{-3\lambda/(3+\lambda)})$.

3.6.2. THE GENERALIZED JACKSON OPERATOR IN THE NONPERIODIC CASE

The **generalized Jackon operator** on A is defined analagously to the periodic case as

$$\bar{U}_{m,r}(f; x) = \bar{c}_{m,r} \int_{-\infty}^{\infty} f(x+t)\, t^{-2r} \sin^{2r}(mt/2)\, dt \tag{3.126}$$

where $\bar{c}_{m,r}$ is determined from the equality

$$\bar{c}_{m,r} \int_{-\infty}^{\infty} t^{-2r} \sin^{2r}(mt/2)\, dt = 1.$$

For every square integrable function $f \in A$, $\bar{U}_{m,r}(f)$ is a function of exponential type $\nu' = mr$. It is directly seen that

$$(\bar{c}_{m,r})^{-1} \geq 2 \int_0^{\pi/m} t^{-2r} \sin^{2r}(mt/2)\, dt \geq \frac{2\pi}{m}\left(\frac{m}{\pi}\right)^{2r},$$

so that

$$\bar{c}_{m,r} \leq \frac{1}{2}\left(\frac{\pi}{m}\right)^{2r-1}. \tag{3.127}$$

The following assertion can be proved in the same way as Theorem 3.9.

Theorem 3.19. For every square integrable function $f \in A$, and for $r \geq 2$, we have

$$r(\alpha; f, \bar{U}_{m,r}(f)) \leq \tau\,(\alpha, f; \pi^2 m^{-1}) + \frac{\omega(f; m^{-1})}{r-1} \tag{3.128}$$

and

$$r(\alpha; f, \bar{U}_{m,r}(f)) \leq \tau\,(\alpha, f; \pi^2 m^{-1+1/2r}) + \frac{M m^{-1+1/2r}}{2r-1}, \tag{3.129}$$

where $M = \sup_x |f(x)|$.

For a suitable choice of m and r we obtain from Theorem 3.17, as in the proof of Theorem 3.10, the following assertion:

Theorem 3.20. Suppose $f \in A$ and f has integrable square. Write $m = [2\nu/\ln \nu]$ and $r = [\ln \nu/2]$, then for sufficiently large ν, we have

$$r(\alpha; f, \bar{U}_{m,r}(f)) \leq \tau\,(\alpha, f; \pi^2 e^4 \nu^{-1} \ln \nu) + e^4 M \nu^{-1},$$

where $M = \sup_x |f(x)|$.

From (3.128) for $r = 2$, letting $\alpha \to 0$, we obtain Bernstein's theorem, stating that every bounded and uniformly continuous function f defined on $(-\infty, \infty)$ can be approximated uniformly by functions of exponential type ν with accuracy or order $\omega(f; \nu^{-1})$.

3.6.3. The Weierstrass operator

As another example, let us consider the **Weierstrass operator**

$$W_\nu(f; x) = \frac{\nu}{2} \int_{-\infty}^{\infty} f(t) e^{-\nu|t - x|}\, dt.$$

Obviously for every $f \in A$ where f is square integrable, the function $W_\nu(f)$ is of exponential type ν. Since

$$\frac{\nu}{2} \int_{\delta}^{\infty} t e^{-\nu t}\, dt = \left(\frac{\delta}{2} + \frac{1}{2\nu}\right) e^{-\nu\delta}, \qquad \frac{\nu}{2} \int_{\delta}^{\infty} e^{-\nu t}\, dt = \frac{1}{2} e^{-\nu\delta},$$

from Theorem 3.17 we obtain the following statement.

Theorem 3.21. If $f \in A$ and f has integrable square, then

$$r(\alpha; f, W_\nu(f)) \leq \tau\,(\alpha, f; 2\nu^{-1}) + \omega(f; \nu^{-1})$$

and

$$r(\alpha; f, W_\nu(f)) \leq \tau\,(\alpha, f; 2\nu^{-1} \ln \nu) + 2M\nu^{-1},$$

where $M = \sup_x |f(x)|$.

§ 3.7. Convergence of derivatives of linear operators

In §3.1 we proved a quite general theorem on the converence of sequences of positive linear operators. Usually, operators with a certain degree of smoothness are encountered, so that the problem of finding conditions which ensure the convergence of the derivatives of a given sequence of operators is a natural one. Considering the convergence with respect to Hausdorff distance, it is appropriate to extend the notion of derivative, as was done in §1.3.2, and to use the S-derivative (the segment derivative), which is defined for all functions.

Convexity conditions are natural in the study of the convergence of the derivatives of linear operators. In §1.3.2, we gave definitions for the m-th order convexity for segment functions. The function $f \in A_\Omega$ is called convex of m-th order on Ω if and only if for every $x \in \Omega$, the m-th segment derivative of the function f is positive, i.e.,

$$D^m(f; x) \geq 0, \qquad x \in \Omega. \tag{3.130}$$

Definition 3.1. By K^m_Ω we shall denote the set of all real valued functions $f \in A_\Omega$ for which (3.130) holds, i.e., K^m_Ω is the set of all functions that are convex of m-th order on Ω.

Let us recall (as was noted in §1.3.2) that K^0_Ω consists of all nonnegative functions, K^1_Ω consists of all the monotone functions, K^2_Ω consists of all functions that are usually called convex, etc.

Definition 3.2. Let L be an operator defined on a certain set of functions B_Ω such that $B_\Omega \cap K^m_\Omega \neq \emptyset$. The operator L will be called **convex of m-th order** if for every
$f \in B_\Omega \cap K^m_\Omega$ we have $L(f) \in K^m_\Omega$.

We note that convexity of order zero for an operator means that the operator is positive. From Definition 3.1 and Theorem 1.6, we have

Lemma 3.2. If $f \in K^m_\Omega$ and k is a positive integer with $k \leq m$, then $D^k(f) \in K^{m-k}_\Omega$

We shall consider a sequence of linear operators $\langle L_n\rangle_1^\infty$ defined on a certain set of functions $B_\Omega \subset A_\Omega$. Let us assume in advance that for every $f \in B_\Omega$, the function $L_n(f)$ has a derivative of m-th order. This means we shall need the existence of the usual derivative $L_n^{(m)}(f)$ (the S-derivative $D^m(L_n(f))$ always exists). This restriction is not too substantial, since the linear operators we usually consider are sufficiently smooth.

If the linear operators $\langle L_n\rangle_1^\infty$ are convex of m-th order, then the operators $\langle L_n^{(m)}\rangle_1^\infty$ are positive. Hence, keeping in mind Theorem 3.1, it is natural to expect that the H-continuity of $D^m(f)$ should be a necessary condition for the convergence of the sequence $\langle L_n^{(m)}(f)\rangle_1^\infty$ with respect to Hausdorff distance.

In what follows, our task will be to determine the necessary conditions for the convergence of $\langle L_n^{(m)}(f)\rangle_1^\infty$ with respect to the Hausdorff distance. It turns out that the notion of S-derivative (which always exists) can be used very fruitfully.

On the integrable functions defined on $\Omega = [a, b]$, we have the operator

(3.131) $$I_k(f; x) = \int_a^x \frac{(x-t)^{k-1}}{(k-1)!} f(t)\, dt, \qquad k = 1, 2, 3, \ldots$$

Obviously, the function $I_k(f; x)$ has derivatives up to the k-th order for every integrable function f and

(3.132) $$I_k^{(k)}(f; x) = f(x), \quad x \in \Omega.$$

Lemma 3.3. If $f \in A_\Omega$ has a bounded S-derivative of k-th order, then the functions

$$f_1(x) = f(x) - I_k(I(D^k(f)); x) \qquad f_2(x) = I_k(S(D^k(f)); x) - f(x)$$

are convex of k-th order, i.e., $f_1, f_2 \in K_\Omega^k$.

Proof. Since $D^k(f)$ is a bounded function, then its Baire functions

(3.133) $$\varphi(x) = I(D^k(f); x), \qquad \psi(x) = S(D^k(f); x)$$

are integrable. From the second part of Theorem 1.5 and (3.132), we obtain

$$D^k(f_1; x) = D^k(f; x) - I(D^k f; x)) \geq 0$$

$$D^k(f_2; x) = S(D^k f; x)) - D^k(f; x) \geq 0,$$

since $I(g; x) \leq g(x) \leq S(g; x)$, $x \in \Omega$, holds for every $g \in A_\Omega$. Hence, according to Definition 3.1, the lemma is proved.

Let L be a linear and convex operator of k-th order defined on $B_\Omega \subset A_\Omega$ and for every $f \in B_\Omega$ the function $L(f)$ has an ordinary k-th derivative $L^{(k)}(f)$. Keeping in mind the definitions of φ and ψ in (3.133), the functions $L(f; x) - L(\varphi; x)$ and $L(\psi; x) - L(f; x)$ belong to K_Ω^k by Lemma 3.3. Then according to the second part of Theorem 1.5 and the assumed existence of the k-th derivative $L^{(k)}(f)$, the inequality

$$Q(\varphi; x) \leq L^{(k)}(f; x) \leq Q(\psi; x) \tag{3.134}$$

holds for every $x \in \Omega$, where

$$Q(f) = L^{(k)}(I_k(f)). \tag{3.135}$$

It is easy to see that the operator defined by (3.135) is linear and positive.

Lemma 3.4. Let L be a linear and convex operator of k-th order, defined on $B_\Omega \subset A_\Omega$, where $\Omega = \Delta = [a, b]$, for which $L(f)$ has a k-th derivative for every $f \in B_\Omega$. Let $f \in B_\Omega$ have an H-continuous k-th S-derivative. Then the inequality

$$r(\Delta; D^k(f), L^{(k)}(f)) \leq 2r(\Delta; \psi, Q(\psi)) + r(\Delta; \varphi, Q(\varphi)) \tag{3.136}$$

holds, where φ, ψ and Q are described in (3.133) and (3.135).

Proof. Since $D^k(f)$ is H-continuous, we have $r(\Delta; D^k(f), \psi) = r(\Delta; \varphi, \psi) = 0$. Keeping this in mind, the inequality (3.134), and the properties of Hausdorff distance, we obtain directly

$$\begin{aligned} r(\Delta; D^k(f), L^{(k)}(f)) &\leq r(\Delta; D^k(f), \psi) + r(\Delta; \psi, Q(\psi)) + r(\Delta; Q(\psi), L^{(k)}(f)) \\ &\leq r(\Delta; \psi, Q(\psi)) + r(\Delta; Q(\psi), Q(\varphi)) \\ &\leq r(\Delta; \psi, Q(\psi)) + r(\Delta; Q(\psi), \psi) + r(\Delta; \psi, \varphi) + r(\Delta; \varphi, Q(\varphi)) \\ &= 2r(\Delta; \psi, Q(\psi)) + r(\Delta; \varphi, Q(\varphi)). \end{aligned}$$

The lemma is proved.

From Lemma 3.4 and Theorem 3.3 we obtain

Theorem 3.22. Let L be a linear and convex operator of k-th order, defined on $B_\Omega \subset A_\Omega$, where $\Omega = \Delta = [a, b]$, for which $L(f)$ has a k-th derivative for every $f \in B_\Omega$. Let $f \in B_\Omega$ have an H-continuous k-th S-derivative. Then the inequality

$$r(\Delta, \alpha; D^k(f), L^{(k)}(f)) \leq 3\tau(\Delta, \alpha, D^k(f); 2\delta) + 3 \sup_{x\in\Delta} Q((\omega(x, \delta, D^k(f); x) + 3M \sup_{x\in\Delta} |1 - \frac{1}{k!} L^{(k)}((t - a)^K; x)| \tag{3.137}$$

holds for every $\delta > 0$, where $M = \sup_{x\in\Delta} |f(x)|$, Q is obtained from (3.135), (3.131), and

$$\omega(x, \delta, f; t) = \begin{cases} 0 & \text{if } t \in [x - \delta, x + \delta] \cap \Omega \\ \omega(f; |x - t| - \delta) & \text{if } t \in \Omega\backslash[x - \delta, x + \delta] \end{cases}. \tag{3.138}$$

Proof. Since Q is a linear and positive operator, and for every H-continuous $g \in A_\Omega$ the equalities

$$\tau(\Delta, \alpha, g; \delta) = \tau(\Delta, \alpha, I(g); \delta) = \tau(\Delta, \alpha, S(g); \delta),$$

$$\omega(g; \delta) = \omega(I(g); \delta) = \omega(S(g); \delta)$$

hold, then from (3.19), (3.136), using (3.133), we obtain

$$r(\Delta, \alpha; \varphi, Q(\varphi)) \leq \tau(\Delta, \alpha, \varphi; 2\delta) + M \sup_{x\in\Delta} |1 - Q(1; x)| + \sup_{x\in\Delta} Q(\omega(x, \delta, \varphi); x),$$

$$r(\Delta, \alpha; \psi, Q(\psi)) \leq \tau(\Delta, \alpha, \psi; 2\delta) + M \sup_{x\in\Delta} |1 - Q(1; x)| + \sup_{x\in\Delta} Q(\omega(x, \delta, \psi); x),$$

or

$$r(\Delta, \alpha; D^k(f), L^{(k)}(f)) \leq 3\tau(\Delta, \alpha, D^k(f); 2\delta) + 3M \sup_{x\in\Delta} |1 - Q(1; x)| + 3 \sup_{x\in\Delta} Q(\omega(x, \delta, \psi); x).$$

This inequality and

$$I_k(1; x) = \int_a^x \frac{(x - t)^{k-1}}{(k - 1)!}\, dt = \frac{(x - a)^k}{k!}$$

yield (3.137). The theorem is proved.

Corollary 3.4. Under the conditions of Theorem 3.21, the following inequality

$$r(\Delta, \alpha; D^k(f), L^{(k)}(f)) \le 3\tau(\Delta, \alpha, D^k(f); 2\delta) + 3M \sup_{x\in\Delta} \left|1 - \frac{1}{k!}L^{(k)}((t - a)^k; x)\right|$$
$$+ 3\delta^{-2}\omega(D^k(f); \delta)\, \beta(L^{(k)})$$

holds, where

$$\beta(L^{(k)}) = \sup_{x\in\Delta} \left\{\frac{2}{(k + 2)!}L^{(k)}((t - a)^{k+2}; x) - \frac{2(x - a)}{(k + 1)!}L^{(k)}((t - a)^{k+1}; x)\right.$$
$$\left. + \frac{(x - a)^2}{k!}L^{(k)}((t - a)^k; x)\right\}.$$

Proof. It is sufficient to apply the inequality $\omega(f; |x - t| - \delta) \le \delta^2(x - t)\omega(f; \delta)$, and to calculate

$$I_k((\zeta - t)^2; x) = \int_a^x \frac{(x - t)^{k-1}}{(k - 1)!}(\zeta - t)^2\, dt$$
$$= \frac{2(x - a)^{k+2}}{(k + 2)!} - \frac{2(\zeta - a)(x - a)^{k+1}}{(k + 1)!} + \frac{(\zeta - a)^2(x - a)^k}{k!}.$$

Corollary 3.4 above can be considered as a generalization of an analagous theorem for the uniform distance of H. B. Knoop and P. Pottinger [1, Theorem 2.1]. But Knoop and Pottinger impose the additional assumption that L maps all the algebraic polynomials of degree $k - 1$ into algebraic polynomials of degree no higher than $k - 1$.

An interesting problem is to find for the operators L_n simple conditions which will guarantee the uniform convergence of $\langle L_n^{(k)}((t - a)^{k+i}; x)\rangle_1^\infty$ to the function

$(x - a)^i(k + i)!/i!$ for $i = 0, 1, 2$ on Ω. This problem was solved (Bl. Sendov and V. Popov [4]) in the case when the operators L_n are convex of order m for $m = 0, 1, 2, \ldots, k$, where the convergence takes place on every segment $\Delta = [c, d] \subset \Omega = [a, b]$, where $a < c < d < b$.

To establish this result, we require some preliminary facts.

Lemma 3.5. Let $\langle f_n \rangle_1^\infty$ be a sequence of monotone (with respect to the segment order relation) increasing functions where $f_n \in \mathbf{F}_\Delta$, on $\Delta = [a, b]$, and the inclusion

$$\operatorname*{Slim}_{n \to \infty} f_n(x) \subset f(x) \tag{3.139}$$

holds for every $x \in (a, b)$. If $f \in \mathbf{H}_\Delta$ and f is bounded, then

$$\lim_{n \to \infty} r(\Delta; f_n, f) = 0. \tag{3.140}$$

Proof. Write $g(x) = \operatorname{Slim}_{n \to \infty} f_n(x)$. According to the inclusion (3.139), we have $g \subset f$. But since $f \in \mathbf{H}_\Delta$ then the completed graph $F(g)$ of g must coincide with the completed graph of f, i.e., $F(g) = F(f) = f$. The monotonicity of f follows from the monotonicity of each f_n. By the monotonicity of f, and since $f \in \mathbf{H}_\Delta$, we see that f must be single valued at each endpoint of the interval, right continuous at a, and left continuous at b.

In order to prove (3.140) it is necessary to establish that for every $\varepsilon > 0$, there exists a positive integer $n(\varepsilon)$ such that for $n > n(\varepsilon)$ the inequality $r(\Delta; f_n, f) < \varepsilon$ holds. In terms of the definition of Hausdorff distance, this means

$$= \max \{ \max_{x \in \Delta} \min_{\xi \in \Delta} [\max [|x - \xi|, \max_{y \in f_n(x)} \min_{\eta \in f(\xi)} |y - \eta|],$$
$$\max_{x \in \Delta} \min_{\xi \in \Delta} [\max [|x - \xi|, \max_{y \in f(\xi)} \min_{\eta \in f_n(x)} |y - \eta|]\} < \varepsilon.$$

Let us assume the contrary. Then there exist $\varepsilon_0 > 0$ and an infinite sequence of indices $N_1 = \{n_1, n_2, \ldots\}$ such that for each $n \in N_1$, $r(\Delta; f_n, f) \geq \varepsilon_0$. There are two possibilities:

$$\text{(a)} \quad \max_{x \in \Delta} \min_{\xi \in \Delta} \max [|x - \xi|, \max_{y \in f_n(x)} \min_{\eta \in f(\xi)} |y - \eta|]\} \geq \varepsilon_0,$$

or

$$\text{(b)} \quad \max_{x \in \Delta} \min_{\xi \in \Delta} \max [|x - \xi|, \max_{y \in f(x)} \min_{\eta \in f_n(\xi)} |y - \eta|] \geq \varepsilon_0$$

for all $n \in N_1$. This means that for every positive integer $n \in N_1$ there exists a point $x_n \in \Delta$ for which either

(3.141) (a) $\min_{\xi\in\Delta} \max\,[\,|x_n - \xi|, \max_{y\in f_n(x_n)} \min_{\eta\in f(\xi)} |y-\eta|]\} \geq \varepsilon_0,$

or

(b) $\min_{\xi\in\Delta} \max\,[\,|x_n - \xi|, \max_{y\in f(x_n)} \min_{\eta\in f_n(\xi)} |y-\eta|] \geq \varepsilon_0.$

Since the sequence $\langle x_n\rangle_{n \in N_1}$ is bounded, it is possible to choose a convergent subsequence $\langle x_n\rangle_{n \in N_2}$ where $N_2 \subset N_1$ with limit $x_0 \in \Delta$. Without loss of generality, we may assume that

(3.143) $|x_n - x_0| < \varepsilon_0/2, \quad n \in N_2$

From (3.141), (3.142), and (3.143), it follows that for every

(3.144) $\xi \in [x_0 - \varepsilon_0/2, x_0 + \varepsilon_0/2] \cap \Delta$

we have either

(3.145) $\max_{y\in f_n(x_n)} \min_{\eta\in f(\xi)} |y-\eta| \geq \varepsilon_0,$

or

(3.146) $\max_{y\in f(x_n)} \min_{\eta\in f_n(\xi)} |y-\eta| \geq \varepsilon_0,$

since it follows from (3.143) and (3.144) that $|x_0 - \xi| < \varepsilon_0$.

We shall denote by N_3 (resp. N_4) the integers n in N_2 for which (3.145) (resp. 3.146) holds. At least one of these sets is infinite because $N_2 = N_3 \cup N_4$ and N_2 is infinite. Let us consider the case when N_3 is infinite. It follows from the boundedness of f and from (3.139), since the f_n are monotone functions, that the sequence $\langle f_n\rangle_1^\infty$ is uniformly bounded. Then it follows from (3.145) that for every $n \in N_3$ there exists $y_n \in f_n(x_n)$ such that

(3.147) $\min_{\eta\in f(\xi)} |y_n - \eta| \geq \varepsilon_0$

for $\xi \in [x_0 - \varepsilon_0/2, x_0 + \varepsilon_0/2]$. As $\{y_n : n \in N_3\}$ is a bounded set, we may choose a convergent subsequence $\langle y_n \rangle_{n \in N_5}$, where $N_5 \subset N_3$, with limit y_0. We can assume that for every $n \in N_5$, the inequality

(3.148) $|y_n - y_0| < \varepsilon_0/2$

holds. From (3.147) and (3.148) it follows that for every

(3.149) $\eta \in f(\xi), \quad \xi \in [x_0 - \varepsilon_0/2, x_0 + \varepsilon_0/2]$

the inequality

(3.150) $|\eta - y_0| \geq \varepsilon_0/2$

holds. From (3.146), we get the existence of $y_0 \in f(x_n)$ such that

$$\min_{\eta \in f_n(\xi)} |y_0 - \eta| \geq \varepsilon_0$$

holds for $\xi \in [x_0 - \varepsilon_0/2, x_0 + \varepsilon_0/2]$ so that the inequality

(3.151) $|\eta - y_0| \geq \varepsilon_0$

holds for every $\eta \in f_n(\xi)$ and $n \in N_4$.

It follows from (3.149) and (3.150) that the completed graph of f lies either above the straight line $y = y_0 + \varepsilon_0/2$ or below the line $y = y_0 - \varepsilon_0/2$, on the segment $[x_0 - \varepsilon_0/2, x_0 + \varepsilon_0/2]$. Let us assume that the completed graph of f lies below $y = y_0 - \varepsilon_0/2$ (the other case is considered similarly).

According to (3.139) there exists a positive integer n_0 such that

$$f_n(x_0 + \varepsilon_0/2) \subset f(x_0 + \varepsilon_0/2) + [-\varepsilon_0/2, \varepsilon_0/2]$$

for $n > n_0$, hence $f_n(x_0 + \varepsilon_0/2) \leq y_0 - \varepsilon_0/4$. Since f_n is a monotonically increasing function, it follows that

(3.152) $f_n(x) \leq y_0 - \varepsilon_0/4$

for all $x \le x_0 + \varepsilon_0/2$ and $n > n_0$. But according to the choice of (x_0, y_0), for every $\varepsilon > 0$ there exists a suficiently large n and a point (x_1, y_1) such that $|x_1 - x_0| < \varepsilon$, $|y_1 - y_0| < \varepsilon$ and $y_1 \in f_n(x_1)$. This contradicts (3.152) for $\varepsilon < \varepsilon_0/4$, which rules out case (a).

We now turn to case (b). According to (3.151) the graph of f_n for all n in N_4 lies either above the line $y = y_0 + \varepsilon_0$ or below the line $y = y_0 - \varepsilon_0$ on the segment $[x_0 - \varepsilon_0/2, x_0 + \varepsilon_0/2]$. We consider the "below" case, the "above" case being similar. Choose $N_6 \subset N_4$ such that for each $n \in N_6$ the graph of f_n lies below the line $y = y_0 - \varepsilon_0$ on the segment $[x_0 - \varepsilon_0/2, x_0 + \varepsilon_0/2]$. According to the choice of the point (x_0, y_0) and the monotonicity of f, the graph of the function f lies above the straight line $y = y_0$ for all $x \in [x_0 + \varepsilon_0/4, x_0 + \varepsilon_0/2]$. But this contradicts (3.139) for those x in $[x_0 + \varepsilon_0/4, x_0 + \varepsilon_0/2]$ since the graph of f_n for $n \in N_6$ lies below the line $y = y_0 - \varepsilon_0/2$ for $x \in [x_0 - \varepsilon_0/2, x_0 + \varepsilon_0/2]$. Hence the case (b) is also an impossible one. The obtained contradiction concludes the proof of the lemma.

Corollary 3.5. (Sendov and V. Popov [4]). Let $f \in C_\Delta$ where $\Delta = [a, b]$ and let $\langle f_n \rangle_1^\infty$ be a sequence of monotone nondecreasing real functions defined on Δ. If $\lim_{n \to \infty} f_n(x) = f(x)$ holds for every $x \in [a, b]$, then

$$\lim_{n \to \infty} \sup_{x \in \Delta} |f(x) - f_n(x)| = 0,$$

i. e., uniform convergence follows from pointwise convergence.

Lemma 3.6. Let $\langle f_n \rangle_1^\infty$ be a sequence of convex functions, $f_n \in K_\Delta^2$, $\Delta = [a, b]$, so that for every $x \in [a, b]$ we have

$$\text{(3.153)} \qquad \operatorname*{Slim}_{n \to \infty} f_n(x) \subset f(x).$$

If $D(f) \in \mathbf{H}_\Delta$ and $\Delta' = [a', b']$, where $a < a' < b' < b$, then

$$\lim_{n \to \infty} r(\Delta'; D(f_n), D(f)) = 0.$$

Proof. According to Lemma 3.5, it is sufficient to prove that for every interior point x_0 of the segment Δ, we have the inclusion

$$\text{(3.154)} \qquad \operatorname*{Slim}_{n \to \infty} D(f_n; x_0) \subset D(f; x_0)$$

because $D(f) \in H_\Delta$ and $D(f_n)$ are monotone, due to the convexity of the f_n. Also from the convexity of f_n on Δ follows the continuity of f_n on every segment $\Delta' = [a', b']$, where $a < a' < b' < b$.

Let $x_0 \in (a', b')$. Then all the functions f_n belong to $C_{\Delta'}$ and the sequence $<f_n>_1^\infty$ is uniformly convergent to f on Δ'. For each $\varepsilon > 0$ there exists $n(\varepsilon)$ such that for $n > n(\varepsilon)$ the inequalitites

$$(3.155)\qquad f(x) - \varepsilon \le f_n(x) \le f(x) + \varepsilon$$

hold for all $x \in [a', b']$. Let us assume that (3.154) fails, and let us write $g(x) = \underset{n \to \infty}{\text{Slim}} D_n(f_n; x)$. Then either (a) $S(g; x_0) > S(D(f); x_0)$ or (b) $I(g; x_0) < I(D(f); x_0)$. We shall consider only the case (a), since the case (b) is entirely analagous. Since $S(g; x_0) > S(D(f); x_0)$, there exists $\delta > 0$ for which $S(g; x_0) > S(D(f); x_0) + \delta$. For $x_1 > x_0$ and x_1 sufficiently close to x_0 with $x_1 \in [a, b]$, then

$$(3.156)\qquad \frac{f(x_1) - f(x_0)}{x_1 - x_0} + \frac{\delta}{2} < S(D(f); x_0) + \delta < S(g;x_0).$$

It follows from the convexity of f_n and $x_1 > x_0$ that the following inequality holds:

$$D(f_n; x_0) \le \frac{f_n(x_1) - f_n(x_0)}{x_1 - x_0}$$

Consequently, according to (3.155), we have

$$D(f_n; x_0) \le \lim_{n \to \infty} \frac{f_n(x_1) - f_n(x_0)}{x_1 - x_0} = \frac{f(x_1) - f(x_0)}{x_1 - x_0}.$$

or

$$(3.157)\qquad S(g; x_0) \le \frac{f(x_1) - f(x_0)}{x_1 - x_0}.$$

But (3.157) contradicts (3.156). The theorem is proved.

Corollary 3.6. (Bl. Sendov and V. Popov [4]). Let $<f_n>_1^\infty$ be a sequence of convex functions, $f_n \in K_\Delta^2$, $\Delta = [a, b]$, uniformly convergent to a function f on Δ. If $f' \in C_\Delta$ and f_n' exists in $\Delta' = [a', b']$ where $a < a' < b' < b$, then

$$\lim_{n\to\infty} \sup_{x\in\Delta'} |f'(x) - f_n'(x)| = 0.$$

Let us return to the remark made prior to the formulation of Lemma 3.5. We shall prove the following statement.

Lemma 3.7. Let $\langle L_n\rangle_1^\infty$ be a sequence of linear operators defined on $B_\Delta \subset A_\Delta$ where $L_n \in K_\Delta^m$ for $m = 0, 1, 2, \ldots, k$. Suppose $\langle L_n(f)\rangle_1^\infty$ converges uniformly to f on $\Delta = [a, b]$ for every continuous f. If $L_n^{(k)}((t-a)^l; x)$ exists for $n = 1, 2, 3, \ldots$, then the sequence $\langle L_n^{(k)}((t-a)^l; x)\rangle_1^\infty$ converges uniformly to $(x-a)^{l-k}\, l!/(l-k)!$ on all segments $\Delta' = [a', b']$ where $a < a' < b' < b$, where $l \geq k$ is a positive integer.

Proof. According to the hypotheses of the lemma,

$$\lim_{n\to\infty} \sup_{x\in\Delta} |L_n((t-a)^l; x) - (x-a)^l| = 0. \tag{3.158}$$

But $(x-a)^l \in K_\Delta^2$ and as a result $L_n((t-a)^l; x) \in K_\Delta^2$. According to (3.158) and Corollary 3.6,

$$\lim_{n\to\infty} \sup_{x\in\Delta_1} |L_n'((t-a)^l; x) - l(x-a)^{l-1}| = 0 \tag{3.159}$$

where $\Delta_1 = [a_1, b_1]$ with $a < a_1 < b_1 < b$.

On the other hand $(x-a)^l \in K_\Delta^3$; therefore, $L_n((t-a)^l; x) \in K_\Delta^3$, i.e, $L_n'((t-a)^l; x) \in K_\Delta^2$. From (3.159) and Corollary 3.6 we obtain

$$\lim_{n\to\infty} \sup_{x\in\Delta_2} |L_n''((t-a)^l; x) - l(l-1)(x-a)^{l-2}| = 0,$$

where $\Delta_2 = [a_2, b_2]$ with $a_1 < a_2 < b_2 < b_1$. Continuing in this way, we obtain the uniform convergence $\langle L_n^{(k)}((t-a)^l; x)\rangle_1^\infty$ to the function $l(l-1)\cdots(l-k+1)(x-a)^{l-k}$ on all segments $\Delta' = [a', b']$ where $a < a' < b' < b$, since we can choose each $[a_i, b_i]$ to contain $[a', b']$. The lemma is proved.

From Theorem 3.22, Theorem 3.1, and Lemma 3.7, we obtain the following theorem on the convergence of a sequence of derivatives of linear operators.

Theorem 3.23. Let $<L_n>_1^\infty$ be a sequence of linear operators for which

$$\lim_{n\to\infty} \sup_{x\in\Delta} |1 - L_n(1; x)| = 0, \qquad \lim_{n\to\infty} \sup_{x\in\Delta} L_n((x-t)^2; x) = 0,$$

and $L_n \in K_\Delta^m$ for $m = 0, 1, 2, \ldots, k$. Suppose $L_n^{(k)}(f)$ exists and $D^K(f) \in H_\Delta$. Then for every segment $\Delta' = [a', b']$ where $a < a' < b' < b$, we have

$$\lim_{n\to\infty} r(\Delta', \alpha; L_n^{(k)}(f), D^k(f)) = 0.$$

Theorem 3.23 generalizes a theorem of V. Popov and V. Veselinov [2], since the condition $D^K(f) \in H_\Delta$ is considerably weaker than the existence of a Lebesgue integrable k-th derivative of the function f.

By using Corollary 3.4, we can find estimates for the order of convergence of the derivatives of some concrete sequences of operators in the same way, as it is done by P. Pottinger [1] in the uniform case.

§ 3.8. A-distance

In order to cover the problem of convergence of a sequence of positive linear operators with respect to a number of classical distances, including the Hausdorff distance, in a unified manner, P. P. Korovkin [2,3] introduced a general notion of distance in function spaces. Let $\Delta \subset \Omega$ and let B_Ω be a certain subset of A_Ω.

Definition 3.3. We say that $r_A(\Delta; f, g) = r_A(f, g)$ where $f, g \in B_\Omega$ is an **A-distance** in B_Ω on Δ if:

(1) $r_A(f, g) = r_A(g, f) \geq 0$;

(2) $r_A(f, g) \leq r_A(f, h) + r_A(h, g)$;

(3) if $\varphi, \psi \in B_\Omega$ satisfy $\varphi(x) \leq f(x) \leq \psi(x)$ and $\varphi(x) - c \leq g(x) \leq \psi(x) + c$ for all $x \in \Delta$, then $r_A(f, g) \leq r_A(\varphi, \psi) + |c|$;

(4) if c is a constant, then $r_A(f, f + c) = 0 \Leftrightarrow c = 0$.

Definition 3.3 is more restrictive than Korovkin's definition in [2,3] of an A-metric. In his definition, the conditions (3) and (4) are replaced by

(3') if $f(x) \leq g(x) \leq \varphi(x)$ for all $x \in \Delta$, then $r_A(f, \varphi) \geq \max\{r_A(g, f), r_A(g, \varphi)\}$;

(4') there exist two bounded functions f and g bounded on Δ for which $r_A(f, g) > 0$;

(5') $r_A(f, g) \leq r_C(f, g) = \sup\limits_{x \in \Delta} |f(x) - g(x)|$.

Obviously, from (4) follows (4'), and from (3) follow both (3') and (5'), so that our definition is more restrictive than Korovkin's. Nevertheless, a number of classical distances satisfy (1) - (4). Some standard A-distances are these:

(a) the uniform distance:

$$r_C(f, g) = \sup_{x \in \Delta} |f(x) - g(x)|;$$

(b) the uniform distance with weight $\theta(x) \geq 1$:

$$r_{C,\theta}(f, g) = \sup_{x \in \Delta} |f(x) - g(x)|/\theta(x);$$

(c) the L_p distances for $p \geq 1$:

$$L_p(f,g) = \left(\frac{1}{b - a} \int_a^b |f(x) - g(x)|^p \, dx\right)^{1/p}$$

(d) the Hausdorff distance $r(\alpha; f, g)$ with parameter $\alpha > 0$ (see Lemma 2.10):

Definition 3.4. The **modulus of A-continuity** of the function $f \in B_\Omega$ on Δ is

$$\tau_A(f; \delta) = r_A(I(\delta/2, f), I(\delta/2, f)),$$

where $I(f)$ and $S(f)$ are the upper and lower Baire functions for f respectively.

Obviously, $\tau_A(f; \delta)$ is a monotone increasing function of the variable $\delta \geq 0$. The modulus as defined coincides with the usual modulus of continuity

$$\tau_C(f; \delta) = \omega(f; \delta) = \sup\{ |f(x_1) - f(x_2)| : |x_1 - x_2| \leq \delta, \ x_1, x_2 \in \Delta\}$$

with respect to the uniform distance. It follows from (3) above that for every A-distance, one has

$$\tau_A(f;\ \delta) \le \omega(f;\ \delta). \tag{3.160}$$

Definition 3.5. A function $f \in B_\Omega$ is called **A-continuous on** Δ if

$$\lim_{\delta \to +0} \tau_A(f;\ \delta) = 0.$$

From the definition of the modulus of A-continuity, we obtain at once

Lemma 3.8. A necessary and sufficient condition for the A-continuity of a function $f \in B_\Omega$ on Δ is $r_A(I(f), S(f)) = 0$.

For the uniform distance, A-continuity coincides with ordinary uniform continuity. From Lemma 3.8 and Theorem 1.7, we have

Lemma 3.9. A bounded function $f \in B_\Omega$ is A-continuous with respect to the integral distance L_1 if and only if f is Riemann integrable.

The properties of A-continuity in the case of Hausdorff distance, i.e., H-continuity, were discussed in §1.3.4.

Let $\psi(\delta)$ be a nondecresing continuous function for $\delta > 0$ satisfying $0 < \psi(\delta) \le 1$ and $\psi(0) = 0$. We shall examine sequences of positive linear operators defined on B_Ω satisfying the conditions

$$\lim_{n \to \infty} r_A(L_n(1), 1) = 0, \qquad \lim_{n \to \infty} r_A(L_n(\psi(|\xi - t|, 0) = 0. \tag{3.161}$$

The following statement is obtained by repeating the proof of Theorem 3.1.

Theorem 3.24. Let $\langle L_n \rangle_1^\infty$ be an arbitrary sequence of positive linear operators for which the conditions of (3.161) hold. Necessary and sufficient conditions for the convergence of the sequence $\langle L_n(f) \rangle_1^\infty$ to a function f with respect to A-distance on Δ are the A-continuity of f on Δ and $\sup\{|f(x)| : x \in \Omega\} = M_f < \infty$.

To determine the rate of convergence of sequence of positive operators, the following theorem can be used. The proof mimics the proof of Theorem 3.3.

Theorem 3.25. Let the positive linear operator L be defined on B_Ω and suppose $f \in B_\Omega$. Let $M = \sup\{|f(x)| : x \in \Delta\}$, and write

$$\omega(x, \delta, f; t) = \begin{cases} 0 & \text{if } t \in [x-\delta, x+\delta] \cap \Omega \\ \omega(f; |x-t|-\delta) & \text{if } t \in \Omega \setminus [x-\delta, x+\delta] \end{cases}.$$

Under the assumption $\omega(x, \delta, f; t) \in B_\Omega$, we have

$$r_A(L(f), f) \le \tau_A(f; 2\delta) + \sup_{x \in \Delta} L(\omega(x, \delta, f); x) + M \sup_{x \in \Delta} |1 - L(1; x)|$$

for each $\delta > 0$.

More generally, we can carry over to the A-distance almost everything that we proved about the Hausdorff distance, when approximations by positive linear operators are considered. The same is true for the convergence of the derivatives of linear operators (M. Müller [1], P. Pottinger [1], H. B. Knoop and P. Pottinger [1], G. Schmid [1], etc.).

§ 3.9. Approximation by partial sums of Fourier series

Up to now in this chapter we considered approximation of functions by positive linear operators. The positivity condition of operators ensures a natural extension and generalization of the results for uniform approximation not only to Hausdorff approximation, but for approximation with respect to an arbitrary A-distance as well.

The partial sums of a Fourier series are a very important and classical means for the approximation of periodic functions. If f is a 2π-periodic integrable function, then **the nth partial sum for its Fourier series** is

$$S_n(f; x) = \frac{a_0}{2} + \sum_{\nu=1}^{n} (a_\nu \cos \nu x + b_\nu \sin \nu x),$$

where

$$a_\nu = \frac{1}{\pi} \int_{-\pi}^{\pi} f(t) \cos \nu t \, dt \qquad \nu = 0, 1, 2, \ldots$$

$$b_\nu = \frac{1}{\pi}\int_{-\pi}^{\pi} f(t)\sin \nu t\, dt \qquad \nu = 1, 2, 3, \ldots$$

The integral representation of $S_n(f)$ by the singular Dirichlet integral is well-known:

(3.162) $$S_n(f; x) = \frac{1}{2\pi}\int_{-\pi}^{\pi} f(x + t)\,\frac{\sin((2n+1)t/2)}{\sin(t/2)}\,dt .$$

The kernel of the operator of (3.162) assumes negative values and therefore this operator is not a positive one. The continuity of f is necessary for the uniform convergence of $\langle S_n(f)\rangle_1^\infty$ to the initial function f, but it is far from sufficient. A classical sufficient condition for uniform convergence of the partial sums of the Fourier series is the Dini-Lipschitz condition: the modulus of continuity $\omega(\delta; f)$ satisfies

(3.163) $$\lim_{\delta\to+0} \omega(\delta; f)\ln\delta = 0.$$

We shall study the Hausdorff convergence of the partial sums of Fourier series, not only for continuous functions, but also for certain discontinuous functions as well. As a point of departure, let us look at a very simple 2π-periodic function, namely $\sigma(x) = \operatorname{sgn}\sin x$. The partial sums of its Fourier series are given by

(3.164) $$S_{2n}(\sigma; x) = S_{2n-1}(\sigma; x) = \frac{4}{\pi}\sum_{\nu=1}^{n}\frac{\sin(2\nu+1)x}{2\nu+1}$$

$$= \frac{2}{\pi}\int_0^x \frac{\sin(2nt)}{\sin t}\,dt = \frac{1}{n\pi}\int_0^{2nx}\frac{\sin t}{\sin(t/2n)}\,dt .$$

The last expression can be represented as a sum:

$$S_{2n}(\sigma; x) = \frac{1}{n\pi}\left\{\int_0^{\pi} + \int_{\pi}^{2\pi} + \cdots + \int_{(k-1)\pi}^{k\pi} + \int_{k\pi}^{2nx}\right\}\frac{\sin t}{\sin(t/2n)}\,dt ,$$

for $0 < x \le \pi/2$, where $k = [2nx/\pi]$. Setting

(3.165) $$v_\nu = \left|\int_{\nu\pi}^{(\nu+1)\pi}\frac{\sin t}{\sin(t/2n)}\,dt\right| ,$$

we obviously have for $\nu = 0, 1, 2, \ldots, n-1$,

(3.166) $\qquad v_\nu > 0,\ v_{\nu+1} < v_\nu,\ \nu = 0, 1, 2, \ldots, n-2.$

Finally, $S_{2n}(\sigma; x) = S_{2n-1}(\sigma; x) = v_0 - v_1 + \cdots + (-1)^{k-1}v_{k-1} + (-1)^k v_k'$ where by $(-1)^k v_k'$ we mean a remaining summand with a sign $(-1)^k$ and absolute value less than v_k. It follows from (3.164) that $S_{2n}'(\sigma; x) = \dfrac{2}{\pi}\dfrac{\sin(2nx)}{\sin x}$, and consequently, the function $S_{2n}(\sigma)$ attains its extrema at the points $x_i = \pi i/2n$, $i = \pm 1, \pm 2, \ldots$.

Now we shall turn to the study of the Hausdorff convergence of $S_{2n}(\sigma; x)$ when n tends to infinity. The equality (3.164) shows that for every x we have $\lim_{n\to\infty} S_{2n}(\sigma; x) = \sigma(x)$, and this convergence is uniform on every segment $[a, b]$ which does not contain the points $k\pi$, $k = 0, \pm 1, \pm 2, \ldots$. Since uniform convergence is stronger than Hausdorff convergence, it is necessary to examine the character of the convergence only in neighborhoods of the discontinuities. Since $\sigma(x) = \operatorname{sgn} \sin x$ is a 2π-periodic odd function as is $S_{2n}(\sigma; x)$, it is sufficient to examine the nature of the graph of $S_{2n}(\sigma)$ to the right in a neighborhood of the point $x = 0$.

The first extremum of $S_{2n}(\sigma)$ on the right of the point $x = 0$ is attained for $x = \pi/2n$ and its value is

$$M_1(n) = S_{2n}(\sigma; \pi/2n) = \frac{1}{n\pi}\int_0^\pi \frac{\sin t}{\sin(t/2n)}\,dt$$

It is easy to see that

(3.167) $$\mu_1 = \lim_{n\to\infty} M_1(n) = \frac{2}{\pi}\int_0^\pi \frac{1}{t}\sin t\,dt = 1.179\ldots .$$

On the other hand $\sigma(+0) = 1 < 1.179\ldots$, and consequently $S_{2n}(\sigma)$ cannot tend to the completed graph of σ with respect to Hausdorff distance. This is the so-called **Gibbs effect**: we have pointwise convergence of the partial sums of the Fourier series, but the Hausdorff convergence to the completed graph of the initial function does not take place. This effect was noticed at the end of the last century and has been thoroughly studied.

Explaining the Gibbs effect in his famous text on calculus, G. Fichtengholz (vol. III, p. 596) says:

"It can be said that the 'limiting geometric image' of the curves $y = S_{2n-1}(\sigma; x)$ as

$n \to \infty$ is not the polygonal line in Figure 3.1(a) (as it is natural to expect) but the polygonal line in Figure 3.1(b) with correspondingly elongated (approximately 18%) upright segments."

The polygonal line in Figure 3.1(a) is the completed graph of the function σ, and the polygonal line in Figure 3.1(b) is the point set to which $\langle S_{2n}(\sigma)\rangle_1^\infty$ converges with respect to Hausdorff distance. This motivates us to introduce the following notation.

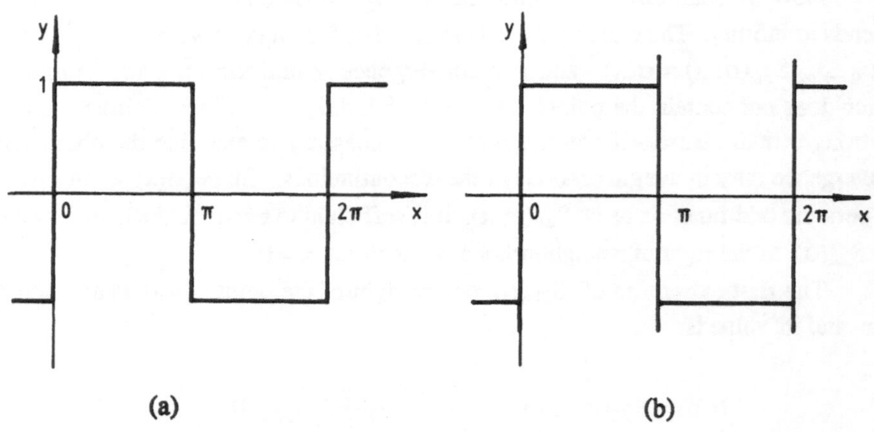

FIGURE 3.1

Definition 3.5. By $D_{2\pi}$ we shall denote the set of all 2π-periodic functions f without discontinuities of the second kind, such that at each point x we have the inequality

$$(f(x) - f(x - 0))(f(x) - f(x + 0)) \leq 0.$$

Obviously the set $D_{2\pi}$ consists of all the H-continuous 2π-periodic functions without discontinuities of the second kind. We denote by $H(f; x) = |f(x - 0) - f(x + 0)|$ the magnitude of the discontinuity of f at the point x. If $f \in D_{2\pi}$ then $H(f; x)$ is nonzero only at a countable set of values of x in $[0, 2\pi)$.

Defintion 3.6. Let $f \in D_{2\pi}$. The **Gibbs completed graph** of f is the segment function $\Phi(f) \in F_{2\pi}$ defined by

(3.168) $$\Phi(f; x) = [I(f; x) - \gamma H(f; x), S(f;x) + \gamma H(f; x)],$$

where $\gamma = \mu_1 - 1 = 0.179\ldots$, $I(f)$ and $S(f)$ are the upper and lower Baire functions for the function f, and μ_1 is determined from (3.167).

Of course, Figure 3.1(b) shows the Gibbs completed graph $\Phi(\sigma)$ of our function σ. From our above discussion with regards to the Hausdorff convergence of $\langle S_n(\sigma)\rangle_1^\infty$ we obtain

Lemma 3.9. Let $\sigma(x) = \operatorname{sgn} \sin x$. The relation $\lim_{n\to\infty} r(\alpha; S_n(\sigma), \Phi(\sigma)) = 0$ holds, i.e., the partial sums of the Fourier series of the function σ converge to the Gibbs completed graph of the function σ with respect to Hausdorff distance.

From the particular example we have analyzed, it is natural expect in a more general setting Hausdorff convergence of partial sums to the Gibbs completed graph and not to the completed graph of the initial function. It should be noted that the completed graph and the Gibbs completed graph of continuous functions coincide with the graph of the function itself. Before we consider more general functions, we shall indicate the order of approximation of the Gibbs completed graph $\Phi(\sigma)$ for the function σ above by partial sums $S_n(\sigma)$ of its Fourier series.

Lemma 3.10. The inequality

$$r(\alpha; S_n(\sigma), \Phi(\sigma)) \le c\alpha^{-1}n^{-1/2}$$

holds, where c is an absolute constant.

Proof. Having in mind Lemma 2.2 and the already established properties of $S_{2n}(\sigma)$, it is sufficient to consider the distance between the points $(x_i, S_{2n}(\sigma; x_i))$ and the graph of $\Phi(\sigma)$. It should be recalled that $S_{2n}(\sigma)$ is a 2π-periodic odd function, symmetric with respect to the line $x = \pi/2$. Hence, we can restrict our attention to $0 \le x_i \le \pi/2$.

Since the extrema of the function $S_{2n}(\sigma) = S_{2n-1}(\sigma)$ are at the points $x_i = \pi i/2k$, in view of (3.166) and (3.165), it is sufficient to estimate the distance between the point $A(0,1)$ and the point $B(x_i, S_{2n}(\sigma; x_i))$ nearest to it. Geometrically, it is clear that

$$r(\alpha; S_{2n}(\sigma), \Phi(\sigma)) \le \min\{\rho_\alpha(A(0,1),\ B(x_i, S_{2n}(\sigma; x_i))) : 0 \le i \le n\}. \tag{3.169}$$

It is not difficult to see, using (3.165) and (3.166), that

$$\min \{\rho_\alpha(A(0,1),\ B(x_i, S_{2n}(\sigma; x_i))) : 0 \le i \le n\} \le c\, \alpha^{-1} n^{-1/2},$$

where c is an absolute constant. The lemma is proved.

The exact value of the constant c in Lemma 3.10 has not been computed yet, but the order $n^{-1/2}$ is exact.

Definition 3.7. Let f be an arbitrary 2π-periodic function. The **modulus of nonmonotocity** of f is

$$\mu(f; \delta) = \frac{1}{2} \sup_x \{\sup \{|f(x_1) - f(x)| + |f(x_2) - f(x)| - |f(x_1) - f(x_2)| : x_2 - x_1 \le \delta\},$$

where $x_1 < x < x_2$. It is at once seen that

(3.170) $$\mu(f; \delta) \le \omega(f; \delta)$$

for every function f and $\delta > 0$. A function f, for which $\lim_{\delta \to +0} \mu(f; \delta) = 0$ is called **locally monotone.** With respect to the class $D_{2\pi}$, we have

Lemma 3.11. Let f be a 2π-periodic function. A necessary and sufficient condition for f to belong to $D_{2\pi}$ is the relation

(3.171) $$\lim_{\delta \to +0} \mu(f; \delta) = 0,$$

i.e., $D_{2\pi}$ coincides with the set of locally monotone 2π-periodic functions.

Proof. First we show that if (3.171) is valid, then $f(x - 0)$ and $f(x + 0)$ exist at each point x, i.e., f has no discontinuities of the second kind. We confine our attention to $f(x - 0)$. Let us assume to the contrary that there exists an increasing sequence $\langle x_n \rangle_1^\infty$ convergent to x_0 such that $\langle f(x_n) \rangle_1^\infty$ diverges. But since

$$\mu(f; x_0 - x_1) \ge |f(x_0) - f(x_n)| + |f(x_n) - f(x_1)| - |f(x_0) - f(x_1)|,$$

the sequence $\langle f(x_n) \rangle_1^\infty$ is bounded. Let A and B be two different limit points of $\langle f(x_n) \rangle_1^\infty$ with $A - B = q > 0$. It is possible to find two subsequences $\langle \xi_i \rangle_1^\infty$ and $\langle \eta_i \rangle_1^\infty$ of the sequence $\langle x_n \rangle_1^\infty$ such that

(3.172) $\lim_{i\to\infty} f(\xi_i) = A, \qquad \lim_{i\to\infty} f(\eta_i) = B$

and

(3.173) $\xi_i \le \eta_i \le \xi_{i+1}, \qquad i = 1, 2, 3, \ldots .$

As a result,

$$\mu(f; \xi_{i+1} - \xi_i) \ge |f(\xi_{i+1}) - f(\eta_i)| + |f(\xi_i) - f(\eta_i)| - |f(\xi_{i+1}) - f(\xi_i)|.$$

By (3.172) and (3.173), for every $\varepsilon > 0$ we have $\mu(f; \xi_{i+1} - \xi_i) \ge 2q - \varepsilon$ for sufficiently large i. The last inequality contradicts (3.171), so that $f(x - 0)$ and $f(x + 0)$ exist for each x.

We next show that the condition

$$(f(x) - f(x - 0))(f(x) - f(x + 0)) \le 0$$

holds at each x. Suppose this fails for some fixed x. Then both $(f(x) - f(x - 0))$ and $(f(x) - f(x + 0))$ have the same sign. We consider the case both are positive; the case both are negative is left to the reader. If both are positive, then for some $\varepsilon > 0$, and all $\delta > 0$ there exist x_1 and x_2 with $x_1 < x < x_2$, $|x_1 - x_2| < \delta$, $f(x) > f(x_1) + \varepsilon$ and $f(x) > f(x_2) + \varepsilon$. But then

$$|f(x_1) - f(x)| + |f(x_2) - f(x)| - |f(x_1) - f(x_2)| > 2\varepsilon,$$

and we would have $\mu(f; \delta) \ge \varepsilon$ for each positive δ. This also contradicts (3.171).

We have shown that a 2π-periodic function belongs to $D_{2\pi}$ if (3.171) holds for this function. The converse is proved more easily. If $f \in D_{2\pi}$, then

$$\lim_{\delta\to+0} \mu(f; \delta) = \max_x \{|f(x) - f(x + 0)| + |f(x) - f(x - 0)| - |f(x + 0) - f(x - 0)|\}$$

and since $(f(x) - f(x - 0))(f(x) - f(x + 0)) \le 0$ for all x, we get zero for this limit. The lemma is proved.

V. Hristov [2] and P. Petrusev and V. Hristov [1] studied the convergence of the partial sums $\langle S_n(f)\rangle_1^\infty$ of the Fourier seies of an arbitrary 2π-periodic function f to the Gibbs completed graph $\Phi(f)$ of the function f with respect to Hausdorff distance. Using the methods of the above two papers, the following result can be proved.

Theorem 3.25. If for a function $f \in D_{2\pi}$ the condition $\lim_{\delta \to +0} \mu(f; \delta) \ln \delta = 0$ holds, then $\lim_{n \to \infty} r(\alpha; S_n(f), \Phi(f)) = 0$.

Keeping in mind (3.170), the Dini-Lipschitz criterion (3.163) is obtained as a corollary of Theorem 3.25, since Hausdorff convergence to a continuous function f implies uniform convergence (see Corollary 2.1).

It has been shown that a criterion for uniform convergence of Fourier series that uses the modulus of nonmonotonicity, which is stonger than the Dini-Lipschitz criterion, can be obtained as a corollary of Theorem 3.25 (V Hristov [1], V. Hristov and P. Petrushev [1,2]). P. Petrusev and V. Hristov [1] obtained an estimate for the Hausdorff distance $r(\alpha; S_n(f), \Phi(f))$, but it is quite complicated.

Remarks. Initially, the analogue of Korovkin's theorem [1] (see Theorem 3.2) for Hausdorff convergence was proved only for the locally monotone functions, i.e., functions that are H-continuous and have no discontinuities of the second kind (Bl. Sendov [9,11], V. Popov and A. Andreev [1]). P. Korovkin [2] was the first to notice that the convergence holds for all the H-continuous functions. At firse, the modulus of nonmonotonicity was used for the estimation of the order of approximation by positive linear operators (Bl. Sendov [4,5], G. Freud and Bl. Sendov [1], B. Boyanov [2,3], V. Veselinov [10], Bl. Sendov and V. Popov [7,8], M. Müller and H. Walk [1]. See also G. Gasanov and V. Popov [1], S. Stoinski [2]).

Some remarks are in order on the convergence of positive linear operators. The Hausdorff approximation by the partial sums of a Haar series was considered in the paper of B. Boyanov and V. Veselinov [2]. The convergence of the Mirak'yan-Szaz operator and Baskakov's operator (see P. Korovkin [1]) with respect to the Hausdorff distance was studied by V. Veselinov [8], and the order of convergence of the Hermite-Fejer interpolation process was studied by G. Gasanov [1]. The papers of Bl. Sendov [3] and V. Popov and V. Veselinov [3] are devoted to the estimation of Hausdorff approximation by Bernstien polynomials, and Fejer's approximation process was considered by Bl. Sendov [6]. The Vallée-Poussin sums are not positive operators in general, so that for thier convergence, it is necessary to impose a specific condition on the choice of the indices, besides the H-continuity of the function. Such a condition was discovered by Bl. Sendov [23].

The Hausdorff convergence for unbounded functions and for functions defined on unbounded intervals was studied by B. Boyanov and V. Veselinov [1], V. Veselinov [5,6,13], S. Suleimanov [1], S. Suleimanov and G. Gasanov [1], and H. Walk [1]. Similar questions for the Hausdorff distance with weight were studied by V. Veselinov [14,18]. The papers of V. Veselinov [9,22] consider approximations by nonlinear operators.

V. Martynjuk [1,2] studied Hausdorff approximations by positive linear operators in the space of functions of two varaibles. He also considered in [3] the problem of the approximation of Steklov's function in Hausdorff distance.

In the papers of G. Gasanov [2], G. Gasanov and S. Suleimanov [1], A. Guseinov and G. Gasanov [1] can be found the utilization of the methods of approximation by linear operators with respect to Hausdorff distance to estimate the error in quadrature formulas.

The convergence of the sequence of derivatives of positive linear operators is discussed in the papers of H. Walk [2], and Bl. Sendov and V. Popov [2].

The problem of finding the best constants in the estimates for approximation by singular integrals with respect to Hausdorff distance was solved for special cases by V. Veselinov and F. Buong [1].

A number of other problems that are close to the contents of this chapter are considered in the papers of V. Veselinov [16], A. Dzafarov and G. Gasanov [1], A. Lupas [1], S. Markov [3], R. Minkova [1,2], V. Popov [12,14], Bl. Sendov and B. Boyanov [1], S. Stoinski [1], O. Shabazov [1], A. P. Petukhov [1], and H. N. Djidjev [1].

Chapter 4

Best Hausdorff Approximations

In this chapter we shall consider the problem for the best approximation of segment functions with regard to the Hausdorff distance. We shall confine ourselves to the consideration of functions defined on a finite or infinite interval.

Let $f \in A_\Delta$ where Δ is either a finite or infinite closed interval, and let W be a set of functions in A_Δ.

Definition 4.1. The **best approximation** of a function f by the elements of a set W with respect to Hausdorff distance with parameter $\alpha > 0$ on the closed interval Δ is the following number:

$$E(W, \Delta, \alpha; f) = \inf_{\varphi \in W} r(\Delta, \alpha; f, \varphi)$$

We shall abbreviate this by $E(W, \Delta; f)$ when $\alpha = 1$, and if the closed interval Δ is understood, we shall write $E(W, \alpha; f)$ or $E(W; f)$, respectively. For the class W, we will consider these classical means of approximation:

H_n = the set of algebraic polynomials of degree at most n;

T_n = the set of trigonometric polynomials of order at most n;

R_n = the set of rational functions that are quotients of two polynomials from H_n.

$S_{k,n}$ = the set of spline functions of order (k,n).

We recall that f is a **spline function** of order (k,n) if $f \in C^{k-1}_\Delta$ (the function has a continuous derivative of (k - 1)-st order on the interval $\Delta = [a, b]$), and there exist $n + 1$ points $a = x_0 < x_1 < x_2 < \cdots < x_n = b$ such that on every interval $[x_{i-1}, x_i]$, $i = 1, 2, 3, \ldots, n$, the function f coincides with an algebraic polynomial of degree at most k.

If $\varphi^* \in W$ and

$$E(W, \Delta, \alpha; f) = r(\Delta, \alpha; f, \varphi^*),$$

then φ^* is called the **element of best Hausdorff approximation** of f from W. Natural questions in the theory of best approximation, which we shall consider, are the following:

(1) The existence of an element of best Hausdorff approximation;

(2) The uniqueness of the element of best Hausdorff approximation;

(3) Characterizations of the elements of best Hausdorff approximation;

(4) Estimation of $E(W, \Delta, \alpha; f)$, subject to the set W and the properties of the approximated segment function f;

(5) The approximate determination of the element of best Hausdorff approximation.

§ 4.1. Best approximation by algebraic and trigonometric polynomials

The existence of an algebraic polynomial of best Hausdorff approximation for every closed and bounded point set in the plane is established, as expected, using the compactness of the set of uniformly bounded algebraic polynomials from H_n.

Theorem 4.1. Let $f \in A_\Delta$ where $\Delta = [a, b]$ is a finite closed interval. Suppose f is bounded. Then for each positive integer n, there exists an algebraic polynomial $P^* \in H_n$ such that

$$E(H_n, \Delta, \alpha; f) = r(\Delta, \alpha; f, P^*)$$

Proof. It follows from the definition of $E(H_n, \Delta, \alpha; f)$ that for each positive integer m, it is possible to find a polynomial $P_m \in H_n$ for which

$$r(\Delta, \alpha; f, P_m) \le E(H_n, \Delta, \alpha; f) + 1/m. \tag{4.1}$$

Let us write $M = \sup \{|y| : y \in f(x),\ x \in \Delta\}$. According to the definition of Hausdorff distance, for every $x \in \Delta$, there exists $\xi \in \Delta$ and $\eta \in F(f; \xi)$ so that

$$\max \{\alpha^{-1}|x - \xi|, |\eta - P_m(x)|\} \le r(\Delta, \alpha; f, P_m),$$

and, consequently, $|\eta - P_m(x)| \le E(H_n, \Delta, \alpha; f) + 1/m$, by (4.1). But since $|\eta| \le M$, we obtain from our calculations above that

$$|P_m(x)| \le M + 1 + E(H_n, \Delta, \alpha; f)$$

for every $x \in \Delta$. Hence, $\langle P_m \rangle_1^\infty$ is a uniformly bounded sequence defined on the closed interval Δ, and we can choose a convergent subsequence $\langle P_{m_k} \rangle_1^\infty$ such that

$$\lim_{k \to \infty} P_{m_k} = P^* \in H_n .$$

uniformly on Δ. From (4.1) and the last equality, we obtain the conclusion of the theorem.

The existence of a trigonometric polynomial of best approximation is established similarly.

Theorem 4.2. If $f \in A_{2\pi}$ and f is bounded, then there exists a trigonometric polynomial $Q^* \in T_n$ for which

$$E(T_n, \alpha; f) = r(\alpha; f, Q^*)$$

It turns out that the polynomial of best Hausdorff approximation is not always unique. Let us consider as an example the function $\varphi(x) = c|x|$ on $\Delta = [-1, 1]$, where c is a nonnegative constant. For simplicity, we shall use Hausdorff distance with parameter $\alpha = 1$. First we show that

$$E(H_1, [-1, 1]; c|x|) = c/2 \tag{4.2}$$

for $0 \le c \le 1$. To this end, it is sufficient to show that the inequality

$$E(H_1, [-1, 1]; c|x|) < c/2, \tag{4.3}$$

fails, since we have $r([-1,1]; P, \varphi) = c/2$ for the constant polynomial $P(x) = c/2$. Let us assume that (4.3) holds. Then there exists a polynomial $P(x) = ax + b$ for which $a \ge 0$ and the inequalities

$$c - (a + b) \le \varepsilon < c/2,$$

$$-\varepsilon a + b \leq \varepsilon < c/2,$$

$$c + (1 - \varepsilon)a - b \leq \varepsilon < c/2,$$

hold (see Figure 4.1). But it follows from the last two inequalities that $c > 1$, which contradicts the condition $0 \leq c \leq 1$. Thus (4.2) is proved.

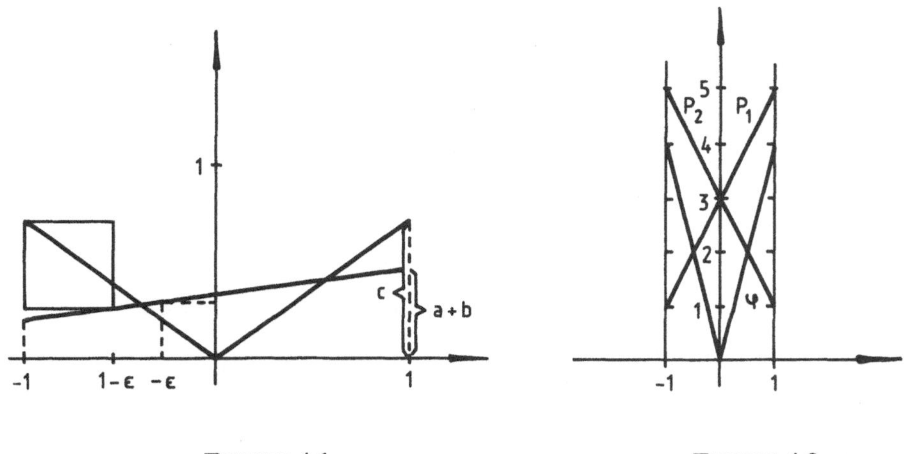

FIGURE 4.1 FIGURE 4.2

It is easily established that:

(1) For $0 \leq c < 1$, the function $\varphi(x) = c|x|$ has a unique algebraic polynomial of best Hausdorff approximation in H_1, and this polynomial is the constant one $P(x) = c/2$;

(2) For $c = 1$, the function $\varphi(x) = |x|$ has infinitely many algebraic polynomials of best approximation in H_1, and every such polynomial has the representation

$$P(x) = ax + \frac{1 + |a|}{2}, \qquad |a| \leq 2/3; \tag{4.4}$$

(3) For $c = 4$ the function $\varphi(x) = 4|x|$ has two algebraic polynomials of best approximation in H_1:

$$P_1(x) = 2x + 3 \qquad P_2(x) = -2x + 3,$$

and $E(H_1, [-1, 1]; 4|x|) = 1$ (see Figure 4.2).

This example gives some characteristic singularities of the best approximation with respect to the Hausdorff distance, besides the lack of uniqueness. If there is more than one polynomial of best approximation, then they are not obliged to form a convex set.

4.1.1. UNIQUENESS CONDITIONS FOR THE POLYNOMIAL OF BEST APPROXIMATION

The problem of finding necessary and sufficient conditions under which the polynomial of best approximation for the function $f \in \mathbf{A}_\Delta$ is unique has not yet been resolved. We shall consider only one sufficient condition for the uniqueness of the algebraic polynomial of best Hausdorff approximation, and this condition also holds for the generalized polynomials on an arbitrary Chebyshev system.

Since the Hausdorff distance between two functions is defined in terms of the Hausdorff distance between their completed graphs, it is sufficient to consider a function $f \in \mathbf{F}_\Delta$ only. For such an f, let us recall the definitions of the auxiliary functions $I(\delta, f; x)$ and $S(\delta, f; x)$ introduced in §1.3.1:

$$I(\delta, f; x) = I(\Delta, \delta, f; x) = \inf \{ y : y \in f(t),\ t \in [x - \delta, x + \delta] \cap \Delta\},$$

$$S(\delta, f; x) = S(\Delta, \delta, f; x) = \sup \{ y : y \in f(t),\ t \in [x - \delta, x + \delta] \cap \Delta\}.$$

From these, we introduce two more functions:

$$l(\delta, f; x) = l(\alpha, \delta, f; x) = I(\alpha\delta, f; x) - \delta,$$

$$u(\delta, f; x) = u(\alpha, \delta, f; x) = S(\alpha\delta, f; x) + \delta.$$

It follows at once from these definitions that $l(\delta, f)$ is lower semicontinuous and $u(\delta, f)$ is upper semicontinuous.

Lemma 4.1. Let $f \in \mathbf{F}_\Delta$, and let $\delta_1 > 0$, $\delta_2 > 0$. Then:

$$l(\delta_1, l(\delta_2, f); x) \geq l(\delta_1 + \delta_2, f; x) \tag{4.5}$$

and

$$u(\delta_1, u(\delta_2, f); x) \leq u(\delta_1 + \delta_2, f; x) \tag{4.6}$$

for all $x \in \Delta$.

Proof. We shall only establish (4.5), since (4.6) is obtained in the same way. For every $x \in \Delta$, we have

$$l(\delta_1, l(\delta_2, f); x) = I(\alpha\,\delta_1, I(\alpha\delta_2, f) - \delta_2; x) - \delta_1$$

$$= I(\alpha\delta_1, I(\alpha\delta_2, f); x) - \delta_1 - \delta_2$$

$$\geq I(\alpha\delta_1 + \alpha\delta_2, f; x) - \delta_1 - \delta_2 = l(\delta_1 + \delta_2, f; x).$$

The lemma is proved.

Definition 4.2. To each function $f \in \mathbf{F}_\Delta$ we associate a function χ of δ given by

(4.7) $$\chi(\alpha, f; \delta) = r(\Delta, \alpha; l(\delta/2, f), u(\delta/2, f)).$$

From the definitions of $l(\delta, f)$ and $u(\delta, f)$, it is clear that

(4.8) $$r(\Delta, \alpha; f, l(\delta, f)) \geq \delta,$$

(4.9) $$r(\Delta, \alpha; f, u(\delta, f)) \geq \delta.$$

It follows from (4.7)-(4.9) that for every $f \in \mathbf{F}_\Delta$, we have

(4.10) $$\chi(\alpha, f; \delta) \geq \delta.$$

From Definition 4.2 and the properties of Hausdorff distance, we see that

(4.11) $$\tau(\alpha, f; \alpha\delta) - \delta \leq \chi(\alpha, f; \delta) \leq \tau(\alpha, f; \alpha\delta) + \delta.$$

Lemma 4.2. Suppose $f \in \mathbf{F}_\Delta$, and that for some $\delta' > 0$, the equality

(4.12) $$\chi(\alpha, f; \delta') = \delta'$$

holds. Then for all $\delta \in (0, \delta')$ we have

(4.13) $$\chi(\alpha, f; \delta) = \delta.$$

Proof. Keeping in mind (4.8) and (4.9), it is sufficient to prove that in (4.8) and (4.9), strict inequality cannot hold under condition (4.12) for $\delta \in (0, \delta'/2)$.

We will only consider (4.8). Let us assume to the contrary, that we have strict inequality in (4.8), i.e., for a certain $\delta < \delta'/2$, we have

$$r(\Delta, \alpha; f, l(\delta, f)) = \delta + \varepsilon > \delta > 0. \tag{4.14}$$

The case of strict inequality in (4.9) is completely symmetric, and is left to the reader. It follows from the hypothesis of the lemma and inequalitites (4.8) and (4.9) that

$$r(\Delta, \alpha; f, l(\delta'/2, f)) = \delta'/2. \tag{4.15}$$

From the definition of $l(\delta, f)$ and (4.14), there exist $x_0, \xi_0 \in \Delta$, $y_0 \in f(x_0)$, and $\eta_0 = l(\delta, f; \xi_0)$ such that

$$\delta + \varepsilon = r(\Delta, \alpha; f, l(\delta, f)) = \max\{\alpha^{-1}|x_0 - \xi_0|, |y_0 - \eta_0|\}$$

and consequently,

$$\min\{l(\delta, f; \xi) : |x_0 - \xi| \le \alpha\delta\} \le y_0 - \delta - \varepsilon. \tag{4.16}$$

On the other hand, with $\delta_1 = \delta'/2 - \delta$, it follows again from the definition of $l(\delta, f)$ that

$$\min\{l(\delta_1, l(\delta, f); t) : |\xi - t| \le \alpha\delta_1\} \le l(\delta, f; \xi) - \delta_1.$$

In view of (4.16), we obtain from the last inequality that

$$\min\{l(\delta'/2 - \delta, l(\delta, f); t) : |t - x_0| \le \alpha\delta'/2\} \le y_0 - \delta'/2 - \varepsilon. \tag{4.17}$$

From Lemma 4.1 and (4.17) it follows that

$$\min\{l(\delta'/2, f; t) : |t - x_0| \le \alpha\delta'/2\} \le y_0 - \delta'/2 - \varepsilon,$$

and, consequently, $\max\{\alpha^{-1}|x_0 - t|, |y_0 - l(\delta'/2, f; t)|\} \ge \delta'/2 + \varepsilon$. By Lemma 2.3, we obtain from the last inequality

$$r(\Delta, \alpha; f, l(\delta'/2, f)) \ge \delta'/2 + \varepsilon > \delta'/2,$$

which contradicts (4.15). The lemma is proved.

Definition 4.3. The function $f \in \mathbf{F}_\Delta$ belongs to the class $\mathbf{H}_\Delta^{\alpha\nu}$ if $\chi(\alpha, f; \nu) = \nu$.

By Lemma 4.2, if $\nu_1 < \nu_2$, then

$$\mathbf{H}_\Delta^{\alpha\nu_2} \subset \mathbf{H}_\Delta^{\alpha\nu_1}. \tag{4.18}$$

From the left inequality in (4.11), it follows that for every $\alpha > 0$ and $\nu > 0$, we have

$$\mathbf{H}_\Delta^{\alpha\nu} \subset \mathbf{H}_\Delta. \tag{4.19}$$

Lemma 4.3. Let $f \in \mathbf{H}_\Delta^{\alpha\nu}$ and let $g \in \mathbf{F}_\Delta$. If $0 < \delta \leq \nu$ and $S(l(\delta, f); x) < g(x) < I(u(\delta, f); x)$ for all $x \in \Delta$, then $r(\Delta, \alpha; f, g) < \delta$.

Proof. Let us assume to the contrary that $r(\Delta, \alpha; f, g) \geq \delta$. Then at least one of the following statements holds:

(a) There exists $x_0 \in \Delta$ and $y_0 \in f(x_0)$ such that

$$\min\{\max\{\alpha^{-1}|x_0 - \xi|, |y_0 - \eta|\} : \xi \in \Delta, \eta \in g(\xi)\} \geq \delta;$$

(b) there exists $\xi_0 \in \Delta$ and $\eta_0 \in g(\xi_0)$ such that

$$\min\{\max\{\alpha^{-1}|x - \xi_0|, |y - \eta_0|\} : x \in \Delta, y \in f(x)\} \geq \delta.$$

Consider case (a). By (a) and Lemma 2.4, there exist points $(x_1, l(\delta, f; x_1))$ and $(x_2, u(\delta, f; x_2))$ such that

$$\max\{\alpha^{-1}|x_0 - x_1|, |y_0 - l(\delta, f; x_1)|\} \leq \delta,$$

$$\max\{\alpha^{-1}|x_0 - x_2|, |y_0 - u(\delta, f; x_2)|\} \leq \delta.$$

According to the hypotheses of the theorom, we have $I(g; x_1) > l(\delta, f; x_1)$ and $S(g; x_2) < u(\delta, f; x_2)$. By Lemma 1.7, there exist $\xi' \in \Delta$ and $\eta' \in g(\xi')$ such that $\max\{\alpha^{-1}|x_0 - \xi'|, |y_0 - \eta'|\} < \delta$. This contradicts assumption (a), so that this case is impossible.

Now consider case (b). Without loss of generality we can assume that $S(f; x_0) < \eta_0$; so, by upper semicontinuity,

$$S(f; x_0) < \eta_0 - \delta \qquad \text{for} \quad x \in (\xi_0 - \alpha\delta, \xi_0 + \alpha\delta) \cap \Delta$$

Then $S(l(\delta, f); x) \le \eta_0 - 2\delta$ for $x \in (\xi_0 - 2\alpha\delta, \xi_0 + 2\alpha\delta) \cap \Delta$. On the other hand, according to the hypotheses of the lemma, $I(u(\delta, f); \xi_0) > \eta_0$. By lower semicontinuity, there exists $\varepsilon > 0$ such that $I(u(\delta, f); x) > \eta_0$ for $x \in (\xi_0 - \varepsilon, \xi_0 + \varepsilon)$. But then obviously

$$r(\Delta, \alpha; I(l(\delta, f)), S(u(\delta, f))) > 2\delta.$$

This contradicts the hypotheses of the lemma. The lemma is proved.

Lemma 4.4. If $f \in \mathbf{H}_\Delta^{\alpha\nu}$, where $\Delta = [a, b]$, and $E(H_n, \Delta, \alpha; f) \le \nu$, then for each polynomial $P \in H_n$ which is a polynomial of best approximation for f, there exist $n + 2$ points in Δ

(4.20) $$a \le \xi_0 < \xi_1 < \xi_2 < \cdots \xi_n < \xi_{n+1} \le b$$

such that one of the two points sets

$$\{(\xi_{2i}, P(\xi_{2i})) : i = 0, 1, \ldots, [(n + 1)/2]\},$$

$$\{(\xi_{2i+1}, P(\xi_{2i+1})) : i = 0, 1, \ldots, [n/2]\},$$

belongs to the graph of $S(l(\delta, f))$ and the other belongs to the graph of $I(u(\delta, f))$.

Proof. First, we note that if $f \in \mathbf{H}_\Delta^{\alpha\nu}$, then for $\delta/2 = E(H_n, \Delta, \alpha; f) \le \nu$, we have

(4.21) $$r(\Delta, \alpha; f, l(\delta/2, f))) = r(\Delta, \alpha; f, u(\delta/2, f))) = E(H_n, \Delta, \alpha; f),$$

and if $P \in H_n$ is a polynomial of best approximation for f, then

(4.22) $$r(\Delta, \alpha; f, P) = E(H_n, \Delta, \alpha; f) = \delta/2.$$

From the above and the continuity of P, the inequalities

$$\varphi(x) = S(l(\delta/2, f); x) \leq P(x) \leq I(u(\delta/2, f); x) = \psi(x)$$

hold for all $x \in \Delta$.

In order to prove the lemma, we shall actually repeat the arguments employed in the proof of the existence of points of maximal deviation in the uniform case. If there does not exist such a system of $n + 2$ points as described in (4.20), then it is possible to find $m + 3$ points $\{t_0, t_1, \ldots, t_{m+2}\}$ (with $m < n$) such that

$$a_0 = t_0 < t_1 < t_2 < \cdots < t_{m+1} < t_{m+2} = b,$$

and such that $\varphi(t_i) < P(t_i) < \psi(t_i)$ for $i = 1, 2, \ldots, m + 1$, and in each of the adjacent intervals $\Delta_0 = [t_0, t_1)$, $\Delta_1 = (t_1, t_2)$, $\ldots$, $\Delta_m = (t_m, t_{m+1})$, $\Delta_{m+1} = (t_{m+1}, t_{m+2}]$, the graph of the polynomial P has consecutively either common points only with φ or common points only with ψ. Without loss of generality, we can assume that the polynomial P has common points with φ but not with ψ in Δ_{2i}, $i = 0, 1, 2, \ldots, [(m+1)/2]$, and that P has common points with ψ but not with φ in Δ_{2i+1}, $i = 0, 1, 2, \ldots, [(m+1)/2]$.

Denote by Q the following polynomial of degree $m + 1$ (where $m + 1 \leq n$):

$$Q(x) = (t_1 - x)(t_2 - x) \cdots (t_{m+1} - x).$$

It is not difficult to see that for sufficiently small positive λ, the polynomial $P^*(x) = P(x) + \lambda Q(x)$ will satisy the inequality $\varphi(x) < P^*(x) < \psi(x)$ on the interval Δ. Then by (4.21) and Lemma 4.3, we have $r(\alpha; f, P^*) < \delta/2 = E(H_n, \Delta, \alpha; f)$. This contradicts (4.22), and completes the proof.

Next we shall give a sufficient condition for uniqueness of the algebraic polynomial of best approximation. This theorem also holds for the generalized polynomials on an arbitrary Chebyshev system.

Theorem 4.3. If $f \in H_\Delta^{\alpha\nu}$ and $E(H_m, \Delta, \alpha; f) \leq \nu$, then for every $n \geq m$, the algebraic polynomial of best Hausdorff approximation of degree n for f is unique.

Proof. Let us assume to the contrary that there exist two polynomials $P, Q \in H_n$ which are polynomials of best approximation for f. First, we shall prove that the polynomial $R = (P + Q)/2$ is also a polynomial of best approximation for f.

Obviously the inequalities

$$l(\delta/2, f; x) \le P(x) \le u(\delta/2, f; x), \qquad l(\delta/2, f; x) \le Q(x) \le u(\delta/2, f; x)$$

hold for $x \in \Delta$ and $\delta/2 = E(H_n, \Delta, \alpha; f)$, since P and Q are polynomials of best approximation. But then

$$l(\delta/2, f; x) \le R(x) \le u(\delta/2, f; x)$$

so that $r(\Delta, \alpha; f, R) \le \delta/2 = E(H_n, \Delta, \alpha; f)$. This means that R is a polynomial of best approximation for f in H_n. Note that in the general case, as was shown by an earlier example, the polynomials of best Hausdorff approximation do not form a convex set.

Using the points of maximal deviation for R, we obtain by Lemma 4.4 that P and Q take equal values at $n + 2$ points, so that they are identically equal. The theorem is proved.

Now we shall give a characteristic property of the functions in the class $\mathbf{H}_\Delta^{\alpha\nu}$ from which it will follow that the class depends on the parameter $\lambda = \alpha\nu$.

Definition 4.4. A function $f \in \mathbf{H}_\Delta$ will be called **λ-monotonic** on $\Delta = [a, b]$, where $0 < \lambda \le b - a$, if f is monotonic on every segment $\Delta' \subset \Delta$ not longer than λ, and f is equal to a constant on each of the two segments $[a, a + \lambda]$ and $[b - \lambda, b]$.

Theorem 4.4. A necessary and sufficient condtion for a function f to belong to $\mathbf{H}_\Delta^{\alpha\delta}$ is the λ-monotonicity of f for $\lambda = \alpha\delta$.

Proof. First we prove sufficiency of the condition. Let f be a λ-monotone function on Δ for $\lambda = \alpha\delta$. Set $\overline{f}(x) = \max\{y : y \in f(x)\}$ and $\underline{f}(x) = \min\{y : y \in f(x)\}$. Consider

$$d_1 = \inf_{\xi \in \Delta} \max\{\alpha^{-1}|x - \xi|, |\overline{f}(x) - u(\delta/2, f; \xi)|\}$$

$$= \inf_{\xi \in \Delta} \max\{\alpha^{-1}|x - \xi|, |\overline{f}(x) - S(\alpha\delta/2, f; \xi) - \delta|\}$$

$$= \inf_{\xi \in \Delta} \max\{\alpha^{-1}|x - \xi|, \delta + \overline{f}(\xi + \varepsilon\alpha\delta/2) - \overline{f}(x)\},$$

where $\varepsilon = \pm 1$, since f is monotonic on the closed interval $[\xi - \alpha\delta/2, \xi + \alpha\delta/2]$. Assume that $f(x)$ is defined on $[a - \lambda/2, b + \lambda/2]$, with $f(x) = f(a)$ for $x \in [a - \lambda/2, a]$ and $f(x) = f(b)$ for for $x \in [b, b + \lambda/2]$. Then, if we set $\xi = x - \varepsilon\alpha\delta/2$, we obtain that $d_1 \leq \delta/2$. Analagously,

$$d_2 = \inf_{\xi \in \Delta} \max \{\alpha^{-1}|x - \xi|, |\overline{f}(\xi) - u(\delta/2, f; x)|\} \leq \delta/2.$$

It follows from the inequalities for d_1 and d_2 and (4.9) that

$$r(\Delta, \alpha; f, u(\delta/2, f)) = \delta/2,$$

and in the same way, that

$$r(\Delta, \alpha; f, l(\delta/2, f)) = \delta/2.$$

Consequently,

$$\chi(\alpha, f; \delta) = r(\Delta, \alpha; l(\delta/2, f)), u(\delta/2, f))) \leq \delta,$$

and according to (4.10), we have $\chi(\alpha, f; \delta) = \delta$, i.e., $f \in \mathbf{H}_\Delta^{\alpha\delta}$.

Now let us prove the necessity of the condition. Suppose to the contrary that $f \in \mathbf{H}_\Delta^{\alpha\delta}$ but that f is not a λ-monotone function for $\lambda = \alpha\delta$ on the segment Δ. This means there exists x_0 for which f is not monotonic on the interval $[x_0 - \lambda/2, x_0 + \lambda/2]$. Recall that f was defined on the entire segment $[a - \lambda/2, b + \lambda/2]$. Then either

$$S(\lambda/2, f; x_0) > \max \{\overline{f}(x_0 - \lambda/2), \overline{f}(x_0 + \lambda/2)\},$$

or

$$I(\lambda/2, f; x_0) < \min \{\underline{f}(x_0 - \lambda/2), \underline{f}(x_0 + \lambda/2)\}.$$

Consequently, either

$$r(\Delta, \alpha; f, S(\lambda/2, f) + \delta/2) = r(\Delta, \alpha; f, u(\delta/2, f)) > \delta/2,$$

or

$$r(\Delta, \alpha; f, I(\lambda/2, f) - \delta/2) = r(\Delta, \alpha; f, l(\delta/2, f)) > \delta/2.$$

Hence, $\chi(\alpha, f; \delta) > \delta/2$. But this contradicts our assumption $f \in \mathbf{H}_\Delta^{\alpha\delta}$. The proof is complete.

Using the same techniques as in the proof of Theorem 4.4, it is not difficult to see that the following statement is valid.

Corollary 4.1. If $f \in \mathbf{A}_\Delta$ and f is λ-monotonic on the segment Δ, then $\tau(\Delta, \alpha; f, \delta) \leq \alpha^{-1}\delta$ for all $\delta \leq \lambda$.

4.1.2. ESTIMATES FOR THE BEST APPROXIMATION

Initially, we shall consider certain examples for which it is possible to find the polynomial of best Hausdorff approximation, or to compute the best Hausdorff approximation, i.e., the minimal possible Hausdorff distance.

4.1.2.1 *Best Approximation of the Delta Function* . Denote by $\delta(x)$ the **delta function**

$$\delta(x) = \begin{cases} [0, 1] & \text{if } x = 0 \\ 0 & \text{if } x \neq 0 \end{cases},$$

and let us consider the problem of finding an algebraic polynomial of degree not higher than n, which approximates the function $M\delta(x)$ in the best way relative to the Hausdorff distance on [-1, 1], where M is a positive constant. The ε-neighborhood of the function $M\delta(x)$ is shown in Figure 4.3.

If for a certain polynomial $P \in H_n$ the inequality $r([-1, 1], \alpha; M\delta, P) \leq \varepsilon$ holds, then the graph of the polynomial P on [-1, 1] must be within the polygonal path $Aa_2b_2bBB_1A_1$ as shown in the figure, and at least one point of the graph of P must belong to the rectangle $a_1a_2b_2b_1$. The minimal ε for which there exists a polynomial $P \in H_n$ with the above properties will be the best approximation of $M\delta$ by algebraic polynomials from H_n in Hausdorff distance. Our task then reduces to finding the polynomial $P \in H_n$ satisfying these conditions for a minimal ε:

$$\text{(a)} \quad \max_{|x| \leq \alpha\varepsilon} P(x) \geq M - \varepsilon, \qquad \text{(b)} \quad \max_{\alpha\varepsilon \leq |x| \leq 1} |P(x)| \leq \varepsilon.$$

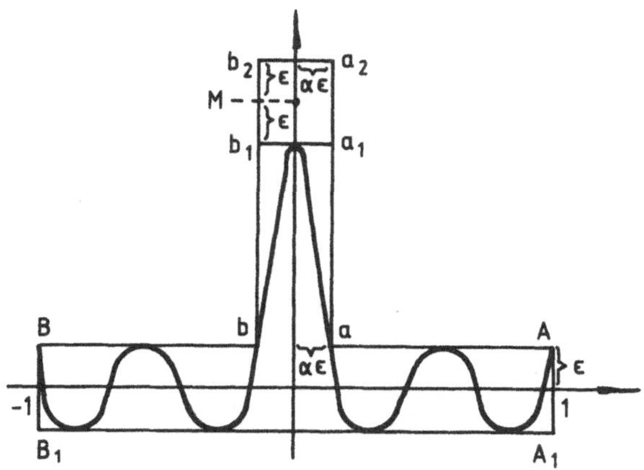

FIGURE 4.3

An analagous problem arises in finding the algebraic polynomial of least deviation from zero on two closed intervals (N. Ahiezer [1, p. 320, problem 36]).

Here it is natural to use Chebyshev polynomials of degree $k = [n/2]$:

$$T_k(x) = \cos(k \arc\cos x) = \frac{1}{2}\left[\left(x + \sqrt{x^2 - 1}\right)^k + \left(x - \sqrt{x^2 - 1}\right)^k \right], \tag{4.23}$$

along with the change of variable

$$x = \frac{(2t^2 - 1 - \alpha^2\varepsilon^2)}{(1 - \alpha^2\varepsilon^2)}. \tag{4.24}$$

The transformation (4.24) maps both of the intervals $[-1, -\alpha\varepsilon]$ and $[\alpha\varepsilon, 1]$ into $[-1, 1]$. Consequently, the polynomial

$$P(x) = (-1)^k \varepsilon \, T_k\left(\frac{2x^2 - 1 - \alpha^2\varepsilon^2}{1 - \alpha^2\varepsilon^2}\right)$$

satsifies condition (b) above, since $|T_k(x)| \le 1$ on the interval $[-1, 1]$. Moreover, $P \in H_n$ because $k = [n/2]$.

In order to satisfy the condition (a) for minimal ε, it is necessary to take

$$P(0) = (-1)^k \varepsilon T_k\left(\frac{-1 - \alpha^2\varepsilon^2}{1 - \alpha^2\varepsilon^2}\right) = M - \varepsilon.$$

The last equality, in view of (4.23), can be written in the following way:

$$\left(\frac{1 + \alpha\varepsilon}{1 - \alpha\varepsilon}\right)^k + \left(\frac{1 - \alpha\varepsilon}{1 + \alpha\varepsilon}\right)^k + 2 = \frac{2M}{\varepsilon}. \tag{4.25}$$

It is not difficult to check that for given α, M, and k, the equation (4.25) has only one positive root, since the left-hand side of (4.25) grows monotonically without bound as a function of ε, and the right-hand side decreases monotonically for $\varepsilon > 0$. Denote by ε_k the only positive root of (4.25), and let

$$P^*(x) = (-1)^k \varepsilon_k T_k\left(\frac{2x^2 - 1 - \alpha^2\varepsilon_k^2}{1 - \alpha^2\varepsilon_k^2}\right). \tag{4.26}$$

We shall show that P^* is the unique polynomial of best approximation of $M\delta$ among the polynomials in H_n, and that

$$E(H_n, [-1, 1], \alpha; M\delta) = \varepsilon_k. \tag{4.27}$$

The last assertion follows directly from P^* being a polynomial of best approximation.

Let us assume that there exists a polynomial $Q \in H_n$ for which

$$r([-1, 1], \alpha; Q, M\delta) = \varepsilon \le \varepsilon_k. \tag{4.28}$$

We must show $Q = P^*$. Then Q would satisfy the conditions (a) and (b). But the construction of the polynomial P^* shows that on both of the intervals $[-1, -\alpha\varepsilon_k]$ and $[\alpha\varepsilon_k, 1]$, P^* attains alternately $k + 1$ times the values $-\varepsilon_k$ and ε_k, and moreover, $P^*(-\alpha\varepsilon_k) = P^*(\alpha\varepsilon_k) = \varepsilon_k$. In addition, $P^*(0) = M - \varepsilon_k$. Thus, if the inequality (4.28) holds for the polynomial Q, then the equation

$$P^*(x) - Q(x) = 0 \tag{4.29}$$

will have at least k zeros on the interval $[-1,-\alpha\varepsilon_k)$, at least two zeros on the interval $[-\alpha\varepsilon_k, \alpha\varepsilon_k]$ and at least k zeros on the interval $(\alpha\varepsilon_k, 1]$. But the polynomial on the left-hand side of(4.29) is of degree no higher than n, and if it has $2k + 2 = 2[n/2] + 2 \geq n + 1$ zeros, then it is identically equal to zero, i.e., $P^* = Q$. Hence P^* is the polynomial of best Hausdorff approximation, and is unique in H_n.

In order to compute the best approximation ε_k in this case, it is necessarry to solve the algebraic equation (4.25). If we set $\lambda = \alpha\varepsilon_k$ in (4.25), then

$$\left(\frac{1+\lambda}{1-\lambda}\right)^k + \left(\frac{1-\lambda}{1+\lambda}\right)^k + 2 = \frac{2\alpha M}{\lambda}. \tag{4.30}$$

It is interesting to study the asymptotic behavior of ε_k when k tends to infinity. From (4.30) we see that when k tends to infinity, then the only positive root λ of (4.30) must tend to zero, and hence ε_k must also tend to zero.

The following expansion is valid:

$$\left(\frac{1+x}{1-x}\right)^k - e^{2kx} = 4kx^3 + a_4x^4 + a_5x^5 + \cdots,$$

and consequently

$$\left(\frac{1+x}{1-x}\right)^k = e^{2kx} + O(kx^3) \tag{4.31}$$

for small x. Using (4.31), from (4.30) we obtain $e^{2k\lambda} + e^{-2k\lambda} + 2 + O(k\lambda^3) = 2\alpha M/\lambda$, that is,

$$e^{n\lambda} + e^{-n\lambda} + 2 + O(n\lambda^3) = 2\alpha M/\lambda. \tag{4.32}$$

From the (4.32), we are lead to write

$$\lambda = \frac{\ln n}{n} + \frac{\varphi(n)}{n}, \tag{4.33}$$

where $\varphi(n)$ is to be estimated. Insertion of λ into (4.32) gives

$$ne^{\varphi(n)} + n^{-1}e^{-\varphi(n)} + 2 + O(n^{-2}\ln n^3) = \frac{2\alpha Mn}{\ln n + \varphi(n)},$$

and hence the function $\varphi(n)$ is bounded.

In this way we obtain from (4.33) that $\lambda = n^{-1}\ln n + O(n^{-1})$, or

$$\varepsilon_k = E(H_n, [-1, 1], \alpha; M\delta) = \alpha^{-1}n^{-1}\ln n + O(n^{-1}). \tag{4.34}$$

Note that the main term does not depend on M.

Let us consider a related example in which it is possible to find the polynomial of best approximation and to estimate the aymptotic behavior of the best approximation. Let $\chi(x) = \delta(x - 1)$, that is,

$$\chi(x) = \begin{cases} 0 & \text{if } -1 \le x < 1 \\ [0, 1] & \text{if } x = 1 \end{cases},$$

and let us consider the problem of finding the algebraic polynomial of degree no higher than n, which approximates the function $\chi(x)$ in the best way with respect to Hausdorff distance, on the interval $[-1, 1]$. The ε-neighborhood of the function $\chi(x)$ with respect to Hausdorff distance is shown in Figure 4.4.

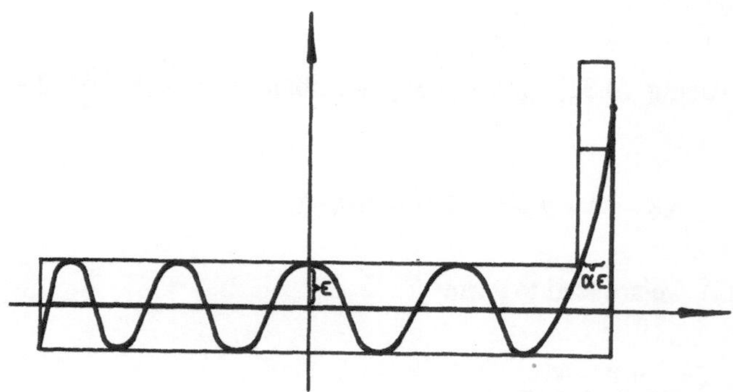

FIGURE 4.4

Proceeding in a way analagous to the delta function, we consider the polynomial

(4.35) $$P(x) = \varepsilon T_n\left(\frac{2x + \alpha\varepsilon}{2 - \alpha\varepsilon}\right),$$

where T_n is the Chebyshev polynomial. Obviously the polynomial of (4.35) satisfies the condition $|P(x)| \le \varepsilon$ for $x \in [-1, 1 - \alpha\varepsilon]$. To obtain $r([-1, 1], \alpha; \chi, P) = \varepsilon$, we require

(4.36) $$P(1) = \varepsilon T_n\left(\frac{2 + \alpha\varepsilon}{2 - \alpha\varepsilon}\right) = 1 - \varepsilon.$$

It is not difficult to see, using arguments similar to the ones of the previous example, that if we determine ε from (4.36), then the polynomial (4.35) is the unique polynomial of best Hausdorff approximation of χ among the polynomials in H_n.

Using (4.23), we may represent the equation (4.36) as follows:

$$\left(\frac{\sqrt{2} + \sqrt{\alpha\varepsilon}}{\sqrt{2} - \sqrt{\alpha\varepsilon}}\right)^n + \left(\frac{\sqrt{2} - \sqrt{\alpha\varepsilon}}{\sqrt{2} + \sqrt{\alpha\varepsilon}}\right)^n + 2 = 2\varepsilon^{-1},$$

which becomes, on setting $\alpha\varepsilon = 2\lambda^2$,

(4.37) $$\left(\frac{1 + \lambda}{1 - \lambda}\right)^n + \left(\frac{1 - \lambda}{1 + \lambda}\right)^n + 2 = \frac{\alpha}{\lambda^2}.$$

From (4.37), it is obvious that the only positive root λ tends to zero when n tends to infinity. Also from (4.37), we find that $\lambda = n^{-1}\ln n + O(n^{-1})$, and consequently

$$\varepsilon = E(H_n, [-1, 1], \alpha; \chi) = 2\alpha^{-1}\left(\frac{\ln n}{n}\right)^2 + O\left(\frac{\ln n}{n^2}\right).$$

4.1.2.2. *Universal Estimates* . Let us consider first the periodic case. Denote by $\mathbf{F}_{2\pi}$ the set of 2π-periodic segment functions in $\mathbf{F}_{(-\infty, \infty)}$. We have this immediate consequence of Theorem 3.10:

Corollary 4.2. If $g \in A_{2\pi}$ and g is bounded with integrable square, then for $n > 20$ there exists a trigonometric polynomial $P \in T_n$ of order no higher than n for which

$$r(\alpha; g, P) = r((-\infty, \infty), \alpha; g, P) \le \tau(\alpha, g; 600n^{-1}\ln n) + 600Mn^{-1},$$

where $M = \max_x |g(x)|$. Moreover if g is an even function, then P is an even polynomial containing only cosines.

Lemma 4.5. Let $f \in F_{2\pi}$ with finite segment values, and let q be an arbitrary positive integer. Then there exists a 2π-periodic continuous function g for which $\max_x \{|y| : y \in f(x)\} = \max_x |g(x)|$ that satisfies these conditions:

(1) $r(\alpha; f, g) \le \pi/\alpha q$; (2) $\tau(\alpha; g, \delta) \le \alpha^{-1}\delta$ for $\delta \le \pi/4q$;

(3) $\omega(g; \delta) \le 10\omega(f; \delta)$ for $\delta \le \pi/2q$.

Proof. We write $x_k = \pi k/4q$, for $k = 0, 1, 2, \ldots, 8q$, and

$$m_k = \min \{y \in f(x) : |x_k - x| \le \pi/2q\}$$

$$M_k = \max \{y \in f(x) : |x_k - x| \le \pi/2q\}.$$

Let us consider a 2π-periodic and continuous function g defined in the following way over one period: for $k = 0, 1, 2, \ldots, 2q - 1$, let $g(x) \equiv m_{4k+2}$ for x in $[x_{4k}, x_{4k+1}]$, let $g(x) \equiv M_{4k+2}$ for x in $[x_{4k+2}, x_{4k+3}]$, and let g be linear on the intervals $[x_{4k+1}, x_{4k+2}]$ and $[x_{4k+3}, x_{4k+4}]$. Since g is monotonic on every closed interval of length $\pi/4q$, i.e., g is $\pi/4q$ monotonic, then by Corollary 4.2, condition (2) above holds. Also, condition (1) follows immediately form the definition of g.

To prove (3), let us write

$$d = \max \{ \max \{M_{4k+2} - m_{4k+2}, M_{4k+2} - m_{4k+6}\} : k = 0,1, \ldots, 2q - 1\}.$$

Then obviously we have

$$\omega(f; \pi/2q) \ge d. \tag{4.38}$$

On the other hand, it is seen from the definition of g that for every $\delta \in (0, \pi/2q)$, we have the inequaltiy

$$\omega(g; \delta) \le \frac{4\delta q d}{\pi}. \tag{4.39}$$

According to the properties of the modulus of continuity, as well as (4.38) and (4.39), we have

$$\omega(g;\delta) \le \frac{4\delta q}{\pi}\omega(f; 2\pi/q) \le \frac{4\delta q}{\pi}(1 + 2\pi/\delta q)\omega(f;\delta),$$

so that for $\delta \le \pi/2q$, $\omega(g;\delta) \le 10\omega(f;\delta)$. This completes the proof.

From Corollary 4.2 and Lemma 4.5, we obtain a universal estimate for the best Hausdorff approximation of all the functions in $\mathbf{F}_{2\pi}$ by trigonometric polynomials.

Theorem 4.5. There exist absolute constants c_1 and c_2 such that for every $f \in \mathbf{F}_{2\pi}$, the inequality

$$E(T_n, \alpha; f) \le c_1\alpha^{-1}n^{-1}\ln n + c_2 M n^{-1},$$

holds, where $M = \max_x \{|y| : y \in f(x)\}$.

Proof. We set $q = 1 + (\pi n/2400\ln n)$ in order to determine the function g in Lemma 4.5. This means that $600n^{-1}\ln n \le \pi/4q$, and consequently,

$$\tau(\alpha, g; 600n^{-1}\ln n) \le 600\alpha^{-1}n^{-1}\ln n. \tag{4.40}$$

According to Lemma 4.5 and the choice of q, we have

$$r(\alpha; f, g) \le \pi/\alpha q \le 2400\alpha^{-1}n^{-1}\ln n. \tag{4.41}$$

On the other hand, by Corollary 4.2 there exists a trigonometric polynomial $P \in T_n$ for which, according to (4.40),

$$r(\alpha; g, P) \le 600\alpha^{-1}n^{-1}\ln n + 60Mn^{-1}, \tag{4.42}$$

where $M = \max_x \{|y| : y \in f(x)\} = \max_x |g(x)|$. Then

$$E(T_n, \alpha; f) \le r(\alpha; f, P) \le r(\alpha; f, g) + r(\alpha; g, P),$$

so that by (4.41) and (4.42), we obtain

$$E(T_n, \alpha; f) \le 3000\alpha^{-1}n^{-1}\ln n + 60Mn^{-1}.$$

The theorem is proved.

Now we move on to the the nonperiodic case. Let $f \in F_\Delta$ where $\Delta = [a, b]$ and a and b are nonnegative real numbers. By a linear change of variable, we may think of the domain as $[0, 2\pi]$. We extend f to an even periodic function on the entire real axis. Then by Theorem 4.5 and Corollary 4.2, for each n there exists an even polynomial $Q \in T_n$ containing only cosines, and such that

$$r(\alpha; f, Q) \leq 3000\alpha^{-1}n^{-1}\ln n + 60Mn^{-1},$$

where $M = \max_x \{|y| : y \in f(x)\}$. We now make the substitution $x = \arccos t$ for the independent variable of the function f. Then $Q(\arccos t) = P(t)$ will be some polynomial of degree at most n, i.e., $P \in H_n$. Keeping in mind Lemma 2.7 and Lemma 2.9, we obtain the following theorem.

Theorem 4.6. There exists absolute constants c_3 and c_4 such that the inequality

$$E(H_n, \Delta, \alpha; f) \leq (b - a)\,\alpha^{-1}(c_3 n^{-1}\ln n + c_4 M n^{-1}) \tag{4.43}$$

holds for every $f \in F_\Delta$, $\Delta = [a, b]$, where $M = \max\{|y| : y \in f(x), x \in \Delta\}$.

According to (4.34) the estimate (4.43) is exact in order. The same is true for the estimate in Theorem 4.5. From (4.43) it follows that $E(H_n, \Delta, \alpha; f)\alpha n/(b - a)\ln n$ is bounded asymptotically for n by an absolute constant for all finite segment functions $f \in F_\Delta$. It is natural to search for the least possible constant.

4.1.2.3. *Exact Asymptotic Behavior of the Best Approximation*. The set of all functions $f \in F_\Delta$ (resp. $f \in F_{2\pi}$) such that $\max\{|y| : y \in f(x), x \in \Delta\} \leq M$ (resp. $\max_x \{|y| : y \in f(x)\} \leq M$) will be denoted by F^M_Δ (resp. $F^M_{2\pi}$). The determination of the exact asymptotic behavior of best Hausdorff approximation on such classes will be our goal in this section.

Theorem 4.7. (Bl. Sendov and V. Popov [9]). The following equalities

$$\lim_{n\to\infty} \frac{\alpha n}{\ln n} E(H_n, \Delta, \alpha; F^M_\Delta) = \frac{b - a}{2},$$

$$\lim_{n\to\infty} \frac{\alpha n}{\ln n} E(T_n, \alpha; F^M_{2\pi}) = 1,$$

hold, where $E(H_n, \Delta, \alpha; F_\Delta^M) = \sup\{E(H_n, \Delta, \alpha; f) : f \in F_\Delta^M\}$ and $E(T_n, \alpha; F_{2\pi}^M) = \sup\{E(T_n, \alpha; f) : f \in F_{2\pi}^M\}$.

Keeping in mind (4.34), the above theorem can be obtained as a corollary of the following statement.

Theorem 4.8. For every $\delta > 0$ and every $M > 0$ there exists a positive integer $n_0 = n(\delta, M)$ such that for every function $f \in F_\Delta^M$ with $\Delta = [a, b]$ and for every positive integer $n \geq n_0$, there exists an algebraic polynomial $P_n \in H_n$ such that

$$r(\Delta, \alpha; f, P_n) \leq \frac{1}{2}(1 + \delta)(b - a)\alpha^{-1} n^{-1} \ln n,$$

and for every function $f \in F_{2\pi}^M$ and for every positive integer $n \geq n_0$ there exists a trigonometric polynomial $\pi_n \in T_n$ such that

$$r(\alpha; f, \pi_n) \leq (1 + \delta)\alpha^{-1} n^{-1} \ln n,$$

where $\alpha \in (0,1]$.

The proof of Theorem 4.8 is quite complicated and is executed via a series of lemmas. It should be noted that by Lemma 2.8, it is sufficient to prove the estimate just for $\alpha = 1$, since we are assuming $\alpha \in (0,1]$.

Lemma 4.6. For every $\lambda \in (0,\pi/4)$ there exists an even trigonometric polynomial $A_n(\lambda; t)$ of nth order such that

$$A_n(\lambda; t) \geq \frac{1}{2}\exp(n\lambda(1 - 2|t|/\lambda) \quad \text{if } |t| \leq \lambda/2; \tag{4.44}$$

$$0 \leq A_n(\lambda; t) \leq \exp(n\lambda(1 + \lambda - t^2/2\lambda^2)) \quad \text{if } |t| \leq \lambda; \tag{4.45}$$

$$|A_n(\lambda; t)| \leq 1 \quad \text{if } \lambda \leq |t| \leq \pi. \tag{4.46}$$

Proof. Let us consider the even trigonometric polynomial

$$A_n(\lambda; t) = T_n\left(\frac{2\cos t + 1 - \cos\lambda}{1 + \cos\lambda}\right), \qquad \lambda \in (0,\pi/4), \tag{4.47}$$

where $T_n(x)$ is the Chebyshev polynomial (4.23) of order n. For $\lambda \le |t| \le \pi$, we have $-1 \le (2\cos t + + 1 - \cos \lambda)/(1 + \cos \lambda) \le 1$, and hence (4.46) holds.

Obviously, $0 \le A_n(\lambda; t)$ for $|t| \le \lambda$;, i.e., the left inequality in (4.45) holds. It follows from (4.47) and (4.23) that

$$2A_n(\lambda; t) = \left(\frac{(\sqrt{1 + \cos t} + \sqrt{\cos t - \cos \lambda}\,)^2}{1 + \cos \lambda}\right)^n - \left(\frac{(\sqrt{1 + \cos t} - \sqrt{\cos t - \cos \lambda}\,)^2}{1 + \cos \lambda}\right)^n ,$$

and for $|t| \le \lambda$, we have

$$A_n(\lambda; t) \le \left(\frac{\cos t/2 + \sqrt{\sin((\lambda - t)/2)\sin((\lambda + t)/2)}}{\cos(\lambda/2)}\right)^{2n}$$

$$\le \left(\frac{1 + \sqrt{(\lambda^2 - t^2)/4}}{1 - \lambda^2/8}\right)^{2n} \le \left(\frac{1 + \lambda(1 - t^2/2\lambda^2)/2}{1 - \lambda^2/8}\right)^{2n}$$

$$\le (1 + \lambda(1 + \lambda - t^2/2\lambda^2)/2)^{2n} \le \exp(n\lambda(1 + \lambda - t^2/2\lambda^2)),$$

since we have $1 - x \le e^x$ for $x \ge 0$. Thus, the right inequality in (4.45) holds as well.

Similarly, for $|t| \le \lambda/2$ we have

$$2A_n(\lambda; t) \ge \left(\frac{\cos(t/2) + \sqrt{\sin((\lambda - t)/2)\sin((\lambda + t)/2)}}{\cos(\lambda/2)}\right)^{2n},$$

and then

$$2\left(\cos\frac{\lambda}{2}\right)^{2n} A_n(\lambda; t) \ge \left(\sin\frac{\pi - |t|}{2} + \sin\frac{\lambda - |t|}{2}\right)^{2n}$$

$$= \left(2\sin\left(\frac{\pi + \lambda}{4} - \frac{|t|}{2}\right) + \cos\frac{\pi - \lambda}{4}\right)^{2n}$$

$$= \left(2\cos\left(\frac{\pi - \lambda}{4} + \frac{|t|}{2}\right)\cos\frac{\pi - \lambda}{4}\right)^{2n}$$

$$\geq \left(\sqrt{2}\cos\left(\frac{\pi - \lambda}{4} + \frac{|t|}{2}\right)\right)^{4n} = (1 + \sin(\lambda/2 - |t|)^{2n}.$$

As a result,

$$2A_n(\lambda; t) \geq \left(\frac{1 + \sin(\lambda/2 - |t|)}{\cos(\lambda/2 - |t|)}\right)^{2n}$$

$$= \left(\frac{1 + \tan(\lambda - 2|t|)/4}{1 - \tan(\lambda - 2|t|)/4}\right)^{2n}$$

$$\geq \left(\frac{1 + (\lambda - 2|t|)/4}{1 - (\lambda - 2|t|)/4}\right)^{2n} \geq \exp(n(\lambda - 2|t|)),$$

since $(1 + x)/(1 - x) \geq e^{2x}$ for $x \in [0,1)$. Hence the inequality (4.44) is verified, and the lemma is proved.

Lemma 4.7. For every $\varepsilon \in (0, 1/16)$ and for every $M \geq 1$, there exists an even trigonometric polynomial $D_n(\varepsilon, M; t) = D_n(t)$ of order no higher than n and a number $n(M)$ depending only on M such that the following relations

(4.48) $\qquad D_n(t) \geq M \quad$ if $|t| \leq 1/2n$,

(4.49) $\qquad 0 \leq D_n(t) \leq (8M)^{1+4\varepsilon} \exp(n(\lambda_\varepsilon^2 - t^2/2\lambda_\varepsilon)) \quad$ if $|t| \leq \lambda_\varepsilon$,

(4.50) $\qquad |D_n(t)| \leq 4n^{-1-4\varepsilon} \quad$ if $\lambda_\varepsilon \leq |t| \leq \mu_\varepsilon$,

(4.51) $\qquad |D_n(t)| \leq n^{-4-4\varepsilon} \quad$ if $\mu_\varepsilon \leq |t| \leq \pi$,

hold for $n \geq n(M)$, where $\lambda_\varepsilon = \lambda(\varepsilon,n,M) = (1 - 4\varepsilon)n^{-1}\ln 8Mn$ and $\mu_\varepsilon = \mu(\varepsilon,n,M) = 3n^\varepsilon\lambda_\varepsilon$.

Proof. We write $p = [n(1 - n^{-\varepsilon})]$, $q = [n^{1-\varepsilon}]$, and we consider the even trigonometric polynomial

(4.52) $$D_n(\varepsilon, M; t) = D_n(t) = n^{-1-4\varepsilon} A_p(\lambda_\varepsilon; t) A_q(\mu_\varepsilon; t)/A_q(\mu_\varepsilon; 0),$$

where $A_n(\lambda; t)$ is the polynomial in Lemma 4.6. Obviously, $D_n(t)$ is of order no higher than $p + q = [n(1 - n^{-\varepsilon})] + [n^{1-\varepsilon}] \le n$. We shall show that the polynomial of (4.52) satisfies the inequalities (4.48) - (4.51).

From (4.44), (4.45) and (4.52), we have for $|t| \le 1/2n$ that

(4.53) $$D_n(t) \ge \frac{1}{4} n^{-1-4\varepsilon} \exp(p\lambda_\varepsilon(1 - \lambda_\varepsilon/n)) \exp(q\mu_\varepsilon(1 - \mu_\varepsilon/n)) / \exp(q\mu_\varepsilon(1 + \mu_\varepsilon))$$

$$\ge \frac{1}{4}(8M)^{1+4\varepsilon} \exp(-n^{1-\varepsilon}\lambda_\varepsilon - \lambda_\varepsilon^2 - 2n\mu_\varepsilon^2)$$

$$\ge 2M \exp(-n^{1-\varepsilon}\lambda_\varepsilon - \lambda_\varepsilon^2 - 18n^{1+2\varepsilon}\lambda_\varepsilon^2).$$

But $\lambda_\varepsilon^2 \le 2n^{-1}\ln 8Mn$, and consequently, we can choose $n(M)$ such that for $n \ge n(M)$, the following inequality holds:

$$n^{1-\varepsilon}\lambda_\varepsilon + \lambda_\varepsilon^2 + 18n^{1+2\varepsilon}\lambda_\varepsilon^2 \le \ln 2.$$

From this inequality and (4.53), we obtain (4.48).

The left inequality in (4.49) follows directly from (4.52) and the left inequality in (4.45). For $|t| \le \lambda_\varepsilon$, we have from (4.44), (4.45), and (4.52) that

$$D_n(t) \le n^{-1-4\varepsilon} A_p(\lambda_\varepsilon; t) \le n^{-1-4\varepsilon} \exp(p\lambda_\varepsilon(1 + \lambda_\varepsilon - t^2/2\lambda_\varepsilon^2))$$

$$\le (8M)^{1+4\varepsilon} \exp(n(\lambda_\varepsilon^2 - t^2/2\lambda_\varepsilon)),$$

i.e., the inequality (4.49) is established.

For $\lambda_\varepsilon \le |t| \le \mu_\varepsilon$, we obtain from (4.44), (4.45), (4.46), and (4.52) that

(4.54) $$|D_n(t)| \le 2n^{-1-4\varepsilon} \exp(q\mu_\varepsilon(1 + \mu_\varepsilon - \lambda_\varepsilon^2/2\mu_\varepsilon^2))/\exp(q\mu_\varepsilon) \le 2n^{-1-4\varepsilon} \exp(q\mu_\varepsilon^2).$$

Since $q\mu_\varepsilon^2 = 9n^{1+\varepsilon}\lambda_\varepsilon^2 \le 36n^{-1/2}(\ln 8Mn)^2$, we can choose $n(M)$ in such a way that for $n \ge n(M)$, the inequality $q\mu_\varepsilon^2 \le \ln 2$ holds. We now obtain (4.50) from (4.54).

For $\mu_\varepsilon \le |t| \le \pi$, we obtain from (4.44), (4.45), (4.46) and (4.52) that

(4.55) $$|D_n(t)| \le 2n^{-1-4\varepsilon} \exp(-q\mu_\varepsilon).$$

Since

$$q\mu_\varepsilon = 3[n^{1-\varepsilon}]n^\varepsilon\lambda_\varepsilon \geq 3(n^{1-\varepsilon} - 1)n^{\varepsilon-1} \ln 8Mn \geq 3\ln 8Mn - 3n^{-1/2} \ln 8Mn,$$

we can choose $n(M)$ such that for $n > n(M)$, the inequality $q\mu_\varepsilon \geq 3\ln 8Mn - \ln 2$ holds. Then from (4.55), we obtain (4.51), and we are done.

By τ_n we shall denote this operator on the space $C_{2\pi}$ of continuous real 2π-periodic functions:

$$\tau_n(f; x) = v_n \int_{-\pi}^{\pi} D_n(\varepsilon, M; x - t)f(t)\, dt,$$

where the constant v_n is determined from

$$v_n \int_{-\lambda_\varepsilon}^{\lambda_\varepsilon} D_n(\varepsilon, M; t)\, dt = 1.$$

Obviously, $\tau_n(f)$ is a trigonometric polynomial of order no higher than n.

We estimate v_n by using Lemma 4.7:

$$v_n^{-1} = \int_{-\lambda_\varepsilon}^{\lambda_\varepsilon} D_n(\varepsilon, M; t)\, dt \geq \int_{-1/2n}^{1/2n} D_n(\varepsilon, M; t)\, dt \geq Mn^{-1},$$

and hence,

(4.56) $\qquad v_n \leq n/M.$

If f, g are segment functions with domain Ω we define the **one-sided Hausdorff distance** from f to g by

$$h(f,g) = \sup_{x\in\Omega}\ \sup_{y\in f(x)}\ \inf_{t\in\Omega}\ \inf_{\eta\in g(y)} \max\{|x - t|, |y - \eta|\}.$$

For general segment functions, we have

(4.57) $\qquad r(f,g) = \max(h(f,g), h(g,f)).$

Obviously, if $f, g \in C_{2\pi}$, we have $h(f,g) = \max_x \min_t \{|x - t|, |f(x) - g(t)|\}$,

Lemma 4.8. For every $\varepsilon \in (0, 1/16)$ and for every $M \geq 0$, there exists $n_0 = n(\varepsilon, M)$ such that for every function $\varphi \in F_{2\pi}^M$ and for every positive integer $n > n_0$, there exists a trigonometric polynomial $\pi_n \in T_n$ for which

$$h(\pi_n, \varphi) \leq (1 + 4\varepsilon)n^{-1} \ln 8Mn = \lambda_\varepsilon .$$

Proof. Since every function in $F_{2\pi}^M$ can be approximated arbitrarily closely by a continuous function with respect to the Hausdorff distance, it is sufficient to consider only continuous functions in $F_{2\pi}^M$, i.e., in $C_{2\pi}^M$.

Let $\varphi \in C_{2\pi}^M$. We shall consider the trigonometric polynomial

$$\tau_n(\varphi; x) = \nu_n \int_{-\pi}^{\pi} D_n(\varepsilon, M; x - t)\varphi(t)\, dt$$

of order no higher than n. Then

$$\tau_n(\varphi; x) = \nu_n\left\{\int_{-\pi}^{-\lambda_\varepsilon} \varphi(x - t)\, D_n(t)\, dt + \int_{\lambda_\varepsilon}^{\pi} \varphi(x - t)\, D_n(t)\, dt\right\} \tag{4.58}$$

$$+ \nu_n \int_{-\lambda_\varepsilon}^{\lambda_\varepsilon} \varphi(x - t)\, D_n(t)\, dt - \varphi(x).$$

It follows from Lemma 4.7 and (4.56) that

$$\nu_n \left|\int_{-\pi}^{-\lambda_\varepsilon} \varphi(x - t)\, D_n(t)\, dt + \int_{\lambda_\varepsilon}^{\pi} \varphi(x - t)\, D_n(t)\, dt\right| \tag{4.59}$$

$$\leq n(\pi n^{-4-4\varepsilon} + 8n^{-1-4\varepsilon}\mu_\varepsilon) \leq Kn^{-1-2\varepsilon},$$

where K is an absolute constant.

Since φ is a continuous function, then from the mean value theorem it follows that there exists a point $\xi(x)$ with $|x - \xi(x)| \leq \lambda_\varepsilon$ such that $(D_n(t) \geq 0$ for $|t| \leq \lambda_\varepsilon)$:

$$\varphi(\xi(x)) - \tau_n(\varphi; x) = \nu_n \int_{-\lambda_\varepsilon}^{\lambda_\varepsilon} \varphi(x - t)\, D_n(t)\, dt - \tau_n(\varphi; x)). \tag{4.60}$$

From (4.58) - (4.60), we obtain

(4.61) $$h(\tau_n(\varphi), \varphi) \leq \max_x \max \{|x - \xi(x)|, |\tau_n(x) - \varphi(\xi(x))| \leq \max \{\lambda_\varepsilon, Kn^{-1-2\varepsilon}\}$$

We can choose $n(\varepsilon,M)$ such that the inequality

(4.62) $$Kn^{-2\varepsilon} \leq 1$$

holds for $n > n(\varepsilon,M)$. The conclusion of the lemma follows from (4.61) and (4.62).

Lemma 4.9. If $m = [n/2\varepsilon \ln 8Mn]$, $t_i = \pi i/m$, $i = 0, 1,2, \ldots, 2m -1$, then for every $l \in \{0,1,2, \ldots, 2m - 1\}$, the inequalities

(4.63) $$\sum_{\substack{i=0 \\ i \neq l}}^{2m-1} |D_n(\varepsilon, M; t_l - t_i)| \leq c_1(M)n^{-\varepsilon^2},$$

and

(4.64) $$\max_{|t - t_l| \geq \lambda_\varepsilon} \sum_{i=0}^{2m-1} |D_n(\varepsilon, M; t_l - t)| \leq c_2(M)n^{-1},$$

hold, where $c_1(M)$ and $c_2(M)$ are constants depending only on M.

Proof. Let us decompose the sum (4.63) into two parts S_1 and S_2. In S_1 we sum over i for $i \neq l$ and $|t_l - t_i| \leq \lambda_\varepsilon$. Hence, $|l - i| \leq 1 + [1/\varepsilon]$. Then according to Lemma 4.7 we obtain

(4.65) $$S_1 \leq (8m)^{1+4\varepsilon} \exp(n\lambda_\varepsilon^2) \sum_{k=1}^{1+[1/\varepsilon]} \exp(-\varepsilon^2 k^2 \ln 8Mn)$$

$$\leq c(M) \sum_{k=1}^{\infty} (8Mn)^{-k\varepsilon^2} = c_1(M)n^{-\varepsilon^2},$$

where $c_1(M)$ is a constant depending only on M.

In the second sum S_2 we shall sum over those i for which $|t_l - t_i| > \lambda_\varepsilon$ and consequently $|D(t_l - t_i)| \leq 4n^{-1-4\varepsilon}$. Since the number of terms in this sum will be less than n, then $S_2 = O(n^{-4\varepsilon})$. The inequality (4.63) follows from this and from (4.65).

In exactly the same way we divide the sum (4.64) into two parts S_1' and S_2'. In S_1' we sum over those i for which $\lambda_\varepsilon \le |t_i - t_l| < \mu_\varepsilon$. Then according to Lemma 4.7, we have

$$S_1' \le 4n^{-1-4\varepsilon} m/\pi \le 8\varepsilon^{-1} n^{-1-3\varepsilon} = O(n^{-1}). \tag{4.66}$$

In the second part S_2', we sum over those i for which $\mu_\varepsilon \le |t_i - t_l|$ and consequently $|D(t_i - t_l)| \le n^{-4-4\varepsilon}$. Since the number of terms in this sum is less than n, we have $S_2' = O(n^{-3})$. Hence (4.64) follows from this remark and (4.66). The lemma is proved.

Lemma 4.10. Let $f, g \in C_{2\pi}$ with

$$h(f,g) \le \theta. \tag{4.67}$$

If $\psi \in C_{2\pi}$ with $\psi(x) \ge 0$ for $x \in [x_0 - \alpha\theta, x_0 + \alpha\theta] = \Delta_0$, and

$$\max\{f(x) + \psi(x) : x \in \Delta_0\} \le g(x_0) - \theta, \tag{4.68}$$

then

$$\max_{x \in \Delta_0} \min_{\xi} \max\{|x - \xi|, |f(x) + \psi(x) - g(\xi)|\} \le \theta. \tag{4.69}$$

Proof. Let $x \in \Delta_0 = [x_0 - \alpha\theta, x_0 + \alpha\theta]$. It follows from (4.67) that there exists $\xi(x)$ such that $|x - \xi(x)| \le \theta$ and $|f(x) - g(\xi(x))| \le \theta$. On the other hand, it follows from (4.68) that the inequality $f(x) + \psi(x) \le g(x_0) - \theta$ holds for the chosen $x \in \Delta_0$. From this fact and the continuity of the function g, it follows that there exists a point $\bar{\xi}(x)$ between the points x_0 and $\xi(x)$, that is, $\bar{\xi}(x) \in x_0 \vee \xi(x)$, for which $|f(x) + \psi(x) - g(\bar{\xi}(x))| \le \theta$. Moreover, we have $|x - \bar{\xi}(x)| \le \theta$, so that (4.69) holds, and we are done.

Let $f \in \mathbf{F}_{2\pi}$ and let $m = [n/2\varepsilon \ln 8Mn]$. For $t_i = \pi i/m$, $i = 0, \pm 1, \pm 2, \ldots$, denote

$$m_i = \inf\{y : y \in f(x),\ x \in [t_{i-1}, t_{i+1}]\} = I(\pi/m, f; t_i),$$

$$M_i = \sup\{y : y \in f(x),\ x \in [t_{i-1}, t_{i+1}]\} = S(\pi/m, f; t_i).$$

We define the value of the operator $\varphi_n(\varepsilon, f) \in C_{2\pi}^M$ for every $f \in \mathbf{F}_{2\pi}^M$ in the following way :

$$\varphi_n(\varepsilon, f; t_{2i}) = m_{2i}, \qquad \varphi_n(\varepsilon, f; t_{2i+1}) = M_{2i+1}, \qquad i = 0, \pm 1, \pm 2, \ldots,$$

and the function $\varphi_n(\varepsilon, f)$ is linear on each segment $[t_i, t_{i+1}]$ for $i = 0, \pm 1, \pm 2, \ldots$. It follows directly from the definition of the operator $\varphi_n(\varepsilon, f)$ that

(4.70) $\qquad r(f, \varphi_n(\varepsilon, f)) \le \varepsilon n^{-1} \ln 8Mn.$

Lemma 4.11. Let $f \in \mathbf{F}_{2\pi}^M$ and $\varphi_n(\varepsilon, f)$ be as defined above. If for the function $\pi_n^* \in C_{2\pi}$ we have the inequalities

$$I(\theta, \pi_n^*; t_{2i}) \le \varphi_n(\varepsilon, f; t_{2i}) + \theta = m_{2i} + \theta$$

$$S(\theta, \pi_n^*; t_{2i+1}) \ge \varphi_n(\varepsilon, f; t_{2i+1}) - \theta = M_{2i+1} - \theta,$$

when $i = 0, 1, 2, \ldots, m - 1$, then

$$h(\varphi_n(\varepsilon, f), \pi_n^*) \le \theta + \pi/m.$$

Proof. Let $x \in [0, 2\pi)$ be arbitrary. Either $x \in [t_{2i}, t_{2i+1}]$ or $x \in [t_{2i-1}, t_{2i}]$; we just consider the first case. Since $m_{2i} \le \varphi_n(\varepsilon, f; x) \le M_{2i-1}$, then by virtue of the hypotheses of the lemma , there exist $\xi', \xi'' \in [t_{2i} - \theta, t_{2i+1} + \theta]$ such that $\pi_n^*(\xi') \ge M_{2i+1} - \theta$, and $\pi_n^*(\xi'') \le m_{2i} + \theta$. Consequently, $\pi_n^*(\xi') \ge \varphi_n(\varepsilon, f; x) - \theta$ and $\pi_n^*(\xi'') \le \varphi_n(\varepsilon, f; x) + \theta$. By virtue of the continuity of π_n^*, there exists $\xi \in [t_{2i} - \theta, t_{2i+1} + \theta]$ such that $|\varphi_n(\varepsilon, f; x) - \pi_n^*(\xi)| \le \theta$. The result now follows, since $|\xi - x| \le t_{2i+1} - t_{2i} + \theta = \theta + \pi/m$.

We return now to the proof of Theorem 4.8. Let $f \in \mathbf{F}_{2\pi}^M$ and $\varphi_n(\varepsilon, f)$ be as defined above, where $0 < \varepsilon < 1/16$. According to Lemma 4.8 for the chosen n and $\varepsilon > 0$, if $n > n_0 = n_0(\varepsilon, M)$, there exists a trigonometric polynomial $\pi_n \in T_n$ for which

(4.71) $\qquad h(\pi_n, \varphi_n(\varepsilon, f)) \le (1 + 4\varepsilon)n^{-1} \ln 8Mn.$

Keeping in mind (4.70), if the estimate (4.71) holds for $h(\varphi_n(\varepsilon, f), \pi_n)$ too, then the second part of Theorem 4.8 would be proved. The polynomial π_n will be replaced by another trigonometric polynomial π_n^* in T_n in such a way that the estimate (4.71) is not altered substantially, and at the same time, an analagous estimate will hold for $h(\varphi_n(\varepsilon, f),\pi_n^*)$ as well. We obtained this perturbed function by an iterative process. We shall use Lemma 4.10 to preserve in essence the estimate of (4.71), and we shall use Lemma 4.11 so that an analagous estimate will hold for $h(\varphi_n(\varepsilon, f),\pi_n^*)$, too.

We define a sequence of trigonometric polynomials $\langle p_j\rangle_0^\infty \subset T_n$ as follows:

(1) $p_0 = \pi_n$;

(2) if p_{j-1} is already defined, in order to produce p_j we do the following:

(2.1) If $j = 2km + 2i$ where k and i are positive integers with $i < m$, and $2m$ is the number of knots $t_0, t_1, \ldots, t_{2m-1}$ of $\varphi_n(\varepsilon, f)$, then we verify whether the inequality

$$I(\lambda_\varepsilon, p_{j-1}; t_{2i}) \le \varphi_n(\varepsilon, f; t_{2i}) + \lambda_\varepsilon = m_{2i} + \lambda_\varepsilon \tag{4.72}$$

holds, where $\lambda_\varepsilon = (1 + 4\varepsilon)n^{-1} \ln 8Mn$. If the inequality (4.72) holds, then we set $p_j = p_{j-1}$ and $a_{k,2i} = 0$. If the inequality (4.72) fails, we determine the positive numbers $a_{k,2i}$ in such a way that

$$\min\{p_{j-1}(x) - a_{k,2i}\, D_n(\varepsilon, M; x - t_{2i}) : x \in [t_{2i} - \lambda_\varepsilon, t_{2i} + \lambda_\varepsilon]\} = m_{2i} + \lambda_\varepsilon,$$

and we set $p_j(x) = p_{j-1}(x) - a_{k,2i}\, D_n(\varepsilon, M; x - t_{2i})$. According to (4.48), such $a_{k,2i}$ exist.

(2.2) If $j = 2km + 2i + 1$, where k and i are positive integers with $i < m$, then we verify whether the inequality

$$S(\lambda_\varepsilon, p_{j-1}; t_{2i+1}) \ge \varphi_n(\varepsilon, f; t_{2i+1}) - \lambda_\varepsilon = M_{2i+1} - \lambda_\varepsilon \tag{4.73}$$

holds. If it does, we set $p_j = p_{j-1}$ and $a_{k,2i+1} = 0$. If the inequality (4.73) fails, we determine the positive numbers $a_{k,2i+1}$ in such a way that

$$\max\{p_{j-1}(x) + a_{k,2i+1}\, D_n(\varepsilon, M; x - t_{2i+1}) : x \in [t_{2i+1} - \lambda_\varepsilon, t_{2i+1} + \lambda_\varepsilon]\} = M_{2i+1} + \lambda_\varepsilon.$$

The sequence of trigonometric polynomials $\langle p_j\rangle_0^\infty$ converges to the trigonometric polynomial

(4.74) $\pi_n^*(x) = \pi_n(x) + \sum_{i=0}^{2m-1} (-1)^{i-1} a_i D_n(\varepsilon, M; x - t_i),$

where $a_i = a_{0,i} + a_{1,i} + \cdots \geq 0$. If n is chosen sufficiently large, so that we have $c_1(M)n^{-\varepsilon^2} < M/2$ in (4.63), then, according to (4.48), (4.63), and the consecutive choice of the numbers $a_{k,i}$, we see that

(4.75) $0 \leq a_i \leq 2, \quad i = 0, 1, 2, \ldots, 2m - 1.$

It is directly seen that the polynomial π_n^* satisfies the hypotheses of Lemma 4.11 for $\theta = \lambda_\varepsilon$, hence

(4.76) $h(\varphi_n(\varepsilon, f), \pi_n^*) \leq \lambda_\varepsilon + \pi/m.$

On the other hand, from the definition of sequence of polynomials $\langle p_j \rangle_0^\infty$, (4.76) and Lemma 4.10, it follows that for every t_i and $\Delta_i = [t_i - \lambda_\varepsilon, t_i + \lambda_\varepsilon]$, we have

$$\max_{x \in \Delta_i} \min_{\xi} \max \{|x - \xi|, |p_j(x) - \varphi_n(\varepsilon, f; \xi)|\} \leq \lambda_\varepsilon,$$

and hence

$$\max_{x \in \Delta_i} \min_{\xi} \max \{|x - \xi|, |\pi_n^*(x) - \varphi_n(\varepsilon, f; \xi)|\} \leq \lambda_\varepsilon.$$

Then from (4.64), (4.74), (4.75), and the last inequality we obtain

$$h(\pi_n^*, \varphi_n(\varepsilon, f)) = \max_{x} \min_{\xi} \max \{|x - \xi|, |\pi_n^*(x) - \varphi_n(\varepsilon, f; \xi)|\}$$

$$\leq (1 + 4\varepsilon)n^{-1} \ln 8Mn + c_2(M)n^{-1}.$$

From the last inequality, (4.76), and (4.70), it follows that for sufficiently large n, the inequality

$$r(f, \pi_n^*) \leq (1 - 6\varepsilon)\, n^{-1} \ln n$$

holds. Keeping in mind Lemma 2.8, the last inequality proves the second half of Theorem 4.8, concerning the periodic case.

The proof of the first half of the theorem, which concerns the nonperiodic case, is obtained, as in the proof of Theorem 4.5, by the transformation $x = \arccos t$, and applying Lemmas 2.7 and 2.9. This completes the proof of Theorem 4.8.

The proof of Theorem 4.7 follows directly from Theorem 4.8, using (4.34), Lemma 2.9, and Lemma 2.8.

It should be noted that in the case of Hausdorff approximation there are functions for which the asymptotic behavior of best approximation is attained for the entire class of bounded functions, e.g., the delta-function $\delta(x)$ as considered in §4.1.2.1, which is zero at all points of $[-1, 1]$ except at $x = 0$, where $\delta(0) = [0, 1]$. It can be proved (N. Kjurkciev and Bl. Sendov [1]) that for the function $\delta(\lambda; x) = \delta(x - \lambda)$, with $\lambda \in (0,1)$, we have the equality

$$E(H_n, [-1, 1], \alpha; \delta(\lambda)) = (1 - \lambda^2)^{1/2}\alpha^{-1} n^{-1}\ln n + O(n^{-1}).$$

4.1.2.4. *Generalization of Jackson's Theorem*. The classical Jackson's theorem gives the estimate of best uniform approximation of continuous 2π-periodic real functions by trigonometric polynomials through the modulus of continuity of the approximated function. In §4.1.2.2, we obtained the estimate for best Hausdorff approximation of bounded 2π-periodic functions by trigonometric polynomials. Since the Hausdorff distance with parameter α tends to uniform distance when α tends to zero, then it is natural to expect that Jackson's theorem will fall out as a limiting case of the best Hausdorff approximation with parameter α when α tends to zero.

Theorem 4.9. (Bl. Sendov and V. Popov [10]). There exists an absolute constant c such that for every bounded function $f \in A_{2\pi}$ and for every $\alpha > 0$, we have

$$E(T_n, \alpha; f) \leq c\omega(f; n^{-1})\frac{\ln(e + \alpha n\omega(f; n^{-1}))}{1 + \alpha n\omega(f; n^{-1})}. \tag{4.77}$$

Proof. Since every bounded function on a closed interval can be approximated with arbitrary accuracy by a continuous function with respect to the Hausdorff distance, without loss of generality, we may assume that $f \in C_{2\pi}$. We hold fixed the positive integer n, and we introduce the following constants relative to f, α, and n:

$$\beta = \alpha n\omega(f; n^{-1}), \qquad m = [n/\ln \beta], \qquad p = [\ln \beta], \tag{4.78}$$

$$\delta = e^{1/2}\pi^2/2m, \qquad q = [n/4e^{1/2}\pi\ln \beta] + 1.$$

By Lemma 4.5, we may choose a function $g \in C_{2\pi}$ for which

(4.79) $$r(\alpha; f, g) \le \pi/\alpha q,$$

$$\tau(\alpha, g; \delta) \le \delta/\alpha \quad \text{for} \quad \delta \le \pi/4q,$$

$$\omega(g; \delta) \le 10\omega(f; \delta) \quad \text{for} \quad \delta \le \pi/2q .$$

As in Theorem 3.10, we shall consider the trigonometric polynomial of order no higher than n, $n = mp$,

$$\pi_n(x) = c_{m,p} \int_{-\pi}^{\pi} g(x + t) \left(\frac{\sin (mt/2)}{m \sin (t/2)}\right)^{2p} dt,$$

where $c_{m,p}$ is determined from the equality

$$c_{m,p} \int_{-\pi}^{\pi} \left(\frac{\sin (mt/2)}{m \sin (t/2)}\right)^{2p} dt = 1.$$

We obtain form Corollary 3.2 and (3.57) that

(4.80) $$r(\alpha; g, \pi_n) \le \tau(\alpha, g; 2\delta) + 2\delta^{-1}\omega(g; \delta)\, c_{m,p} \int_{\delta}^{\pi} t \left(\frac{\sin (mt/2)}{m \sin (t/2)}\right)^{2p} dt$$

$$\le \tau(\alpha, g; 2\delta) + \pi(\pi^2/2m\delta)^{2p-1}\, \omega(g; \delta)/4(p - 1).$$

According to (4.80) and (4.79), keeping in mind that $\delta \le \pi/4q$ with respect to the notation of (4.78), we obtain

(4.81) $$r(\alpha; f, \pi_n) \le r(\alpha; f, g) + r(\alpha; g, \pi_n)$$

$$\le \pi/\alpha q + 2\delta/\alpha + 10\pi(\pi^2/2m\delta)^{2p-1}\, \omega(g; \delta)/4(p - 1)$$

$$\le 4\pi^2 e^{1/2}\alpha^{-1} n^{-1} \ln \beta + \pi^2 e^{1/2}/\alpha[n/\ln \beta]$$
$$+ \ 5e^{1/2}\pi e^{-[\ln \beta]}\, \omega(f; e^{1/2}\pi^2/2[n/\ln \beta])/\ 2([\ln b] - 1).$$

For sufficiently large n, we will have $n/\ln \beta > 2$. Moreover, we may assume

$[\ln b] > 2$, for otherwise, $\omega(f; n^{-1}) = O(n^{-1})$, i.e., the function f satisfies a Lipschitz condition, in which case the theorem is trivial. With these comments in mind, from (4.81) we obtain

$$r(\alpha; f, \pi_n) \leq 6\pi^2 e^{1/2}\alpha^{-1}n^{-1} \ln \beta + 10\pi^3 e^2 \alpha^{-1} n^{-1}$$

$$\leq 1500\alpha^{-1}n^{-1} \ln (\alpha n\omega(f; n^{-1})) \leq 3000\omega(f; n^{-1})\frac{\ln(e + \alpha n\omega(f; n^{-1}))}{1 + \alpha n\omega(f; n^{-1})}.$$

This completes the proof.

If in (4.77) we set $\alpha n\omega(f; n^{-1})) = \beta = 3$, then we see that the best uniform approximation of a 2π-periodic continuous function f by trigonometric polynomials is of order $\omega(f; n^{-1})$, which is the content of the classical Jackson's theorem. If $f \in \mathbf{F}_{2\pi}$ is not a continuous function, then $\omega(f; n^{-1})$ does not tend to zero as n increases, and the inequality (4.77) is equivalent to Theorem 4.5.

As with Theorem 4.5, from the periodic case we can proceed to the nonperiodic case. In this way, we obtain the following result.

Theorem 4.10. There exists an absolute constant c such that for every bounded $f \in \mathbf{A}_\Delta$ and for every $\alpha > 0$, we have

$$E(H_n, \Delta, \alpha; f) \leq c\omega(f; n^{-1})\frac{\ln(e + \alpha n\omega(f; n^{-1}))}{1 + \alpha n\omega(f; n^{-1})}.$$

It is possible to obtain a similar theorem for the approximation by functions of a given exponential type (Bl. Sendov and V. Popov [10]).

4.1.2.5. *Approximation of Certain Concrete Functions* . We shall consider the best Hausdorff approximation by algebraic polynomials of the following three functions:

$$\varphi(\lambda; x) = |x|^\lambda,$$

$$\psi(\lambda; x) = |x|^\lambda \operatorname{sgn} x,$$

$$\sigma(\lambda; x) = (1 - x^2)^\lambda,$$

for $\lambda \in (0, 1)$ on the segment $[-1, 1]$. These functions are interesting because the order of their uniform approximation by algebraic polynomials depends essentially on λ, whereas the corresponding order of their Hausdorff approximation does not depend on λ.

We shall obtain estimates for the best Hausdorff approximation of these three functions by making use of a procedure of S. Bernstein for estimating the best approximation of the function $|x|$ with the help of so-called polynomial oscillators.

Definition 4.5. (S. Bernstein [1, p. 158]). We say that the polynomial

$$P(x) = A_0x^{\alpha_0} + A_1x^{\alpha_1} + A_2x^{\alpha_2} + \cdots + A_nx^{\alpha_n}$$

is a **polynomial oscillator** on the segment $[0, 1]$ corresponding to the sequence of nonnegative exponents $\alpha_0 < \alpha_1 < \alpha_2 < \cdots < \alpha_n$, if it achieves its maximum absolute value at $n+1$ points of the segment. The number n is called the order of the polynomial oscillator.

As usual, the uniform norm $\|P\|$ of P on the segment $[0, 1]$ will be

$$\|P\| = \max \{|P(x)| : x \in [0, 1]\}$$

In what follows, we shall need the following two oscillatory polynomials:

$$T_{2k}(x) = \cos (2k \arccos x) = (-1)^k(1 - 2k^2x^2 + \cdots + (-1)^k 2^{2k-1}x^{2k}), \tag{4.82}$$

$$T_{2k+1}(x) = \cos ((2k+1) \arccos x) = (-1)^k(x - \tfrac{2}{3}k(k+1)x^3 + \cdots + (-1)^k 2^{2k}x^{2k+1},. \tag{4.83}$$

for which we obviously have $\|T_{2k}\| = \|T_{2k+1}\| = 1$.

It is not hard to prove, using Descartes' theorem, that the coefficients of polynomial oscillators alternate in sign, and that two consecutive extrema on the segment $[0, 1]$ have opposite signs. The following two basic theorems of Bernstein hold.

Theorem 4.11. (S. Bernstein [1, p.160]). Let

$$P(x) = \sum_{i=0}^{n} A_i x^{\alpha_i}, \qquad Q(x) = \sum_{i=0}^{n} B_i x^{\alpha_i},$$

be two algebraic polynomials with the same degree of the variable x and with one common coefficient, for example $A_{i_0} = B_{i_0}$. If P is a polynomial oscillator, then $\|P\| \leq \|Q\|$, and equality takes place if and only if $Q = P$.

Proof. Indeed, if $\|P\| > \|Q\|$, then at $n+1$ consecutive points x_k, $k = 0, 1, 2, \ldots, n$ at which $|P(x)|$ attains its maximum value, we shall have

$$\eta(-1)^k[P(x_k) - Q(x_k)] \geq 0,$$

where $\eta = \pm 1$. As a result, the equation $P(x) - Q(x) = 0$ will have at least n positive roots. But this is impossible, since it contains only n terms, because of the condition $A_{i_0} = B_{i_0}$. This completes the proof.

Theorem 4.12. (S. Bernstein [1, p. 162]). If for two polynomial oscillators

$$P(x) = x^{\alpha_0} + A_1x^{\alpha_1} + A_2x^{\alpha_2} + \cdots + A_nx^{\alpha_n},$$

$$Q(x) = x^{\alpha_0} + B_1x^{\beta_1} + B_2x^{\beta_2} + \cdots + B_nx^{\beta_n}$$

the inequalities $0 < \alpha_0 < \beta_1 < \alpha_1 < \beta_2 < \alpha_2 < \cdots < \beta_n < \alpha_n$ hold, then $\|P\| > \|Q\|$.

Proof. Indeed, we know that the coefficients of the polynomial $P(x)$ alternate in sign, so that the coefficients of polynomial

$$Q(x) - P(x) = B_1x^{\beta_1} - A_1x^{\alpha_1} + B_2x^{\beta_2} - \cdots + B_nx^{\beta_n} - A_nx^{\alpha_n}$$

cannot have more than n alternations of sign. Then the equation $Q(x) - P(x) = 0$ will have at most n positive roots. If $\|Q\| \geq \|P\|$, then the difference $Q(x_k) - P(x_k)$ would have the same sign as $Q(x_k)$ (or would be equal to zero) at all the points x_k, where $|Q(x_k)| = \|Q\|$. Hence the equation $Q(x) - P(x) = 0$ would have necessarilly n positive zeros $\xi_1, \xi_2, \ldots, \xi_n$ satisfying the inequalities

$$x_1 \leq \xi_1 \leq x_2 \leq \cdots \leq \xi_n \leq x_{n+1}.$$

It follows from here that the difference $Q(x) - P(x)$ would have the same sign as B_1 in the interval $(0, \xi_1)$, i.e., the minus sign, and in general, the difference $Q(x) - P(x)$ would have the sign $(-1)^{i+1}$ in the interval (ξ_i, ξ_{i+1}). Moreover, since the number of points x_i is greater than the number of points ξ_i, there would exists at least one such point x_i for which $\xi_{i-1} \leq x_i \leq \xi_i$, if we agree to assume that $\xi_0 = 0$ and $\xi_{n+1} = \infty$ in this inequality. At this point x_i, we would have $(-1)^i [Q(x_i) - P(x_i)] > 0$,and consequently

$(-1)^i Q(x_i) > 0$. But $Q(x_1) > 0$, because for small positive values of x, the polynomial $Q(x)$ has the sign of its first term, hence $Q(x_2) < 0$, and in general $(-1)^i Q(x_i) < 0$. We have reached a contradiction, and the theorem is proved.

We have to establish two more assertions for polynomial oscillators.

Lemma 4.12. Suppose $P(x) = x^{\alpha_0} - \sum_{i=1}^{n} a_i x^{\alpha_i}$ is a polynomial oscillator with exponents $\alpha_0 < \alpha_1 < \alpha_2 < \cdots < \alpha_n$, then

$$v(x) = \sum_{i=1}^{n} a_i x^{\alpha_i}$$

is nonnegative on $[0, \xi_1]$, where ξ_1 is the leftmost local extremum of P in $(0, 1]$.

Proof. We denote by $0 < \xi_1 < \xi_2 < \cdots < \xi_n < \xi_{n+1} = 1$ the points at which $|P(x)| = \|P\|$. Obviously, the equality

$$P(\xi_k) = (-1)^{k-1}\|P\|, \quad k = 1, 2, 3, \ldots, n+1, \tag{4.84}$$

holds and

$$P'(\xi_k) = 0 \quad \text{for} \quad k = 1, 2, 3, \ldots, n. \tag{4.85}$$

We denote by u the polynomial

$$u(x) = xv'(x) - \alpha_0 v(x) = \sum_{i=1}^{n} b_i x^{\alpha_i}. \tag{4.86}$$

According to Descartes' law, $u(x)$ has at most $n - 1$ positive zeros. On the other hand, from (4.84) and (4.85), we obtain

$$(-1)^{k-1} u(\xi_k) > 0, \qquad k = 1, 2, 3, \ldots, n,$$

and consequently $u(x)$ has at least $n - 1$ zeros in the interval (ξ_1, ξ_n). Then $u(x) > 0$ throughout the interval $(0, \xi_1)$, since $u(\xi_1) > 0$. But then $v(x) > 0$ for $x \in (0, \xi_1)$. Indeed, if for an arbitrary $\eta \in (0, \xi_1)$ we had $v(\eta) > 0$, then there would exist $\eta_1 \in (0, \xi_1)$ for which we have simultaneously $v(\eta_1) > 0$ and $v'(\eta_1) < 0$. But from these inequalities and (4.86), we obtain $v(\eta_1) < 0$, which is impossible. The lemma is proved.

Lemma 4.13. If $s \neq m$ and P and Q are the polynomial oscillators given by

$$P(x) = a_0 x^s + x^m + \sum_{i=1}^{n} a_i x^{\alpha_i} ,$$

$$Q(x) = x^m + \sum_{i=1}^{n} b_i x^{\alpha_i} ,$$

then $\|P\| \geq (1 + 2^{s/|m-s|})^{-1}\|Q\|$.

Proof. We use Bernstein's method [1]. It is clear that for each $\mu \in [0, 1]$ and every $x \in [0, 1]$ we have

$$\left| a_0\left(\frac{x}{1+\mu}\right)^s + \left(\frac{x}{1+\mu}\right)^m + \sum_{i=1}^{n} a_i \left(\frac{x}{1+\mu}\right)^{\alpha_i} \right| \leq \|P\| ,$$

or

$$|a_0 x^s + (1+\mu)^{s-m} x^m + \cdots + a_n (1+\mu)^{s-\alpha_n} x^{\alpha_n}| \leq (1+\mu)^s \|P\|.$$

Taking into account the form of P, we obtain from the last inequality

$$|(1+\mu)^{s-m} - 1| \cdot |x^m + a_1' x^{\alpha_1} + \cdots + a_n' x^{\alpha_n}| \leq (1 + (1+\mu)^s)\|P\|,$$

and according to Theorem 4.11,

$$|(1+\mu)^{m-s} - 1| \cdot \|Q\| \leq (1 + (1+\mu)^s)\|P\|.$$

To complete the proof, we need only set $\mu = 2^{1/|m-s|} - 1$.

Using Theorem 4.11 and Theorem 4.12, we can prove the following assertion.

Lemma 4.14. The polynomial oscillator

$$R_k(x) = x + a_1 x^2 + a_2 x^4 + \cdots + a_k x^{2k}$$

satisfies the inequalities

$$1/2(1 + \sqrt{2})(2k + 1) \leq \|R_k\| \leq 1/(2k + 1) .$$

Proof. From Theorem 4.12 and the polynomial oscillator (4.83), we obtain that $\|R_k\| \leq 1/(2k+1)$. As in the proof of Lemma 4.13, for every $\mu \in [0, 1]$, we have

$$\left| \frac{x}{1+\mu} + \sum_{i=1}^{k} a_i \left(\frac{x}{1+\mu}\right)^{2i} \right| \leq \|R_k\|,$$

which we may rewrite as

$$|(1+\mu)x + a_1x^2 + a_2x^4(1+\mu)^{-2} + a_kx^{2k}(1+\mu)^{2k-2}| \leq (1+\mu)^2\|R_k\|,$$

or

$$|\mu x + a_2'x^4 + a_3'x^6 + \cdots + a_k'x^{2k}| \leq ((1+\mu)^2 + 1)\|R_k\|.$$

By Theorems 4.11 and 4.12, and the polynomial oscillator (4.83), we obtain from the last inequality that

$$\mu/(2k-1) \leq ((1+\mu)^2 + 1)\|R_k\|.$$

Setting $\mu = 1/\sqrt{2}$ in the last inequality, we obtain the other inequality.

Lemma 4.15. The polynomial oscillator

$$U_k(x) = x + b_1x^2 + b_2x^3 + \cdots + b_kx^{k+1}$$

satisfies the inequalities

$$1/2(k+1)^2 < \|U_k\| < 1/(k+1)^2.$$

Proof. From (4.82), we obtain

$$v_k(x) = T_{2k+2}(\sqrt{x}) = (-1)^{k+1}(1 - 2(k+1)^2x + \ldots + (-1)^{k+1}2^{2k+1}x^{k+1}),$$

and consequently

$$\|U_k\| < \frac{1}{2}(k+1)^{-2}\|v_k(x) - (-1)^{k+1}\| \leq (k+1)^{-2}.$$

On the other hand, we claim that

$$\|U_k\| > \frac{1}{2}(k+1)^{-2}\|v_k\| = \frac{1}{2}(k+1)^{-2}.$$

Indeed, let us assume to the contrary that the polynomial

$$U_k(x) - (-1)^k \frac{1}{2}(k+1)^{-2} v_k(x) = c_1 + c_2x^2 + c_3x^3 + \ldots + c_{k+1}x^{k+1} \tag{4.87}$$

has at least $k+1$ positive zeros, since $v_k(x)$ achieves its maximal absolute value at $k+2$ points, and these values of $v_k(x)$ alternate in sign. But the polynomial of (4.87) has $k+1$ terms, and according to Descartes' law, it cannot have more than k positive roots. The obtained contradiction proves the lemma.

We now introduce two continuous auxiliary functions g_n and h_n:

$$g_n(\lambda; x) = \max\{0, x^\lambda - x^{\lambda-1}/n\},$$

$$h_n(\lambda; x) = \min\{2x^\lambda, x^\lambda + x^{\lambda-1}/n\}.$$

Evidently, g_n and h_n have derivatives for all $x > 0$ except at $x = 1/n$.

Lemma 4.16. For each positive integer n and for each $\lambda \in (0, 1)$, there exists an even algebraic polynomial $P_n(\lambda; x) \in H_n$ such that

$$g_n(\lambda; x) \le P_n(\lambda; x) \le h_n(\lambda; x)$$

for all $x \in [0, 1]$.

Proof. Without loss of generality we may assume that λ is a rational number and $\lambda = 1 - p/q$ for positive integers p and q with $p < q$. Consider the polynomial oscillators

$$S(x) = x^q + a_1x^{2q+p} + a_2x^{4q+p} + \cdots + a_kx^{2kq+p},$$

$$Q(x) = x^q + b_1x^{3q} + a_2x^{5q} + \cdots + b_kx^{(2k+1)q}.$$

According to Theorem 4.12, we have $\|S\| < \|Q\|$, and by Theorem 4.11 and (4.83) we have

$$Q(x) = (-1)^k(2k+1)^{-1}\, T_{2k+1}(x^q),$$

i.e., $\|Q\| = (2k-1)^{-1}$. As a result,

$$\|S\| < (2k-1)^{-1}. \tag{4.88}$$

If we let ξ_1 denote the leftmost extremum of $S(x)$ in the interval $(0, 1]$, then from Lemma 4.12 we obtain

$$0 \le x^p v_k(x^{2q}) = -a_1x^{2q+p} - a_2x^{4q+p} - \cdots - a_kx^{2kq+p} \le x^q \tag{4.89}$$

for all $x \in [0, \xi_1]$. From (4.88) and (4.89) it follows that

$$x^p\, |x^{q-p} - v_k(x^{2q})| \le \|S\|,$$

or

$$|x^\lambda - v_k(x^2)| \le x^{\lambda-1}\|S\| \quad \text{for } x \in (0, 1), \tag{4.90}$$

and

$$0 \le v_k(x^2) \le x^\lambda \quad \text{for } x \in [0, x_1], \quad \text{with } x_1 = \xi_1. \tag{4.91}$$

Since $x_1^\lambda - v_k(x_1^2) = x_1^{\lambda-1}\|S\|$, or $v_k(x_1^2) = x_1^\lambda - x_1^{\lambda-1}\|S\| = (x_1 - \|S\|)x_1^{\lambda-1} > 0$, we have

$$x_1 > \|S\|. \tag{4.92}$$

Let n be an arbitrary positive integer. We denote $P_n(\lambda; x) = v_k(x^2)$ with $k = [n/2]$. From (4.90), (4.91), and (4.92) it follows that $P_n(\lambda; x) \ge 0$ for $x \in [0, \|S\|]$, and $P_n(\lambda; x) \le x^\lambda - x^{\lambda-1}\|S\| \le 2x^\lambda$ for $x \in [\|S\|, 1]$, or according to (4.88),

$$P_n(\lambda; x) \ge \max\{0, x^\lambda - x^{\lambda-1}\|S\|\} = g_n(\lambda; x)$$

for all $x \in [0, 1]$. On the other hand, again from (4.90), (4.91), and (4.92), we obtain $P_n(\lambda; x) \le 2x^\lambda$ for $x \in [0, \|S\|]$, and $P_n(\lambda; x) \le x^\lambda - x^{\lambda-1}\|S\| \le 2x^\lambda$, for $x \in [\|S\|, 1]$. Thus, using (4.88), we have

$$P_n(\lambda; x) \le \min\{2x^\lambda, x^\lambda - x^{\lambda-1}\|S\|\} \le h_n(\lambda; x).$$

This finishes the proof of the lemma.

Lemma 4.17. For each positive integer n and for each $\lambda \in (0, 1)$ there exists an odd algebraic polynomial $P_n^*(\lambda; x) \in H_n$ for which

$$g_n(\lambda; x) \leq P_n^*(\lambda; x) \leq h_n(\lambda; x).$$

for all $x \in [0, 1]$.

The proof of this lemma is similar to the one of Lemma 4.16, where this time we use the polynomial oscillator

$$S(x) = x^q + a_1x^{q+p} + a_2x^{3q+p} + \cdots + a_kx^{(2k-1)q+p} .$$

Lemma 4.18. For each positive integer n and for each $\lambda \in (0, 1)$, there exists an algebraic polynomial $U_n(\lambda; x) \in H_n$ such that

$$g_{n2}(\lambda; x) \leq U_n(\lambda; x) \leq h_{n2}(\lambda; x)$$

for all $x \in [0, 1]$.

Proof . The proof is similar to the one of Lemma 4.16, where we use the polynomial oscillator

$$S(x) = x^q + a_1x^{q+p} + a_2x^{2q+p} + \cdots + a_{n-1}x^{(n-1)q+p} .$$

By Lemma 4.12, if we denote by Q the polynomial oscillator

$$Q(x) = x^q + b_1x^{2q} + b_2x^{3q} + \cdots + b_{n-1}x^{nq},$$

then $\|S\| < \|Q\|$. But on the other hand, according to the lemma, $\|Q\| < n^{-2}$ and consequently, $\|S\| < n^{-2}$. The other arguments go on as in the proof of Lemma 4.16.

Lemma 4.19. For each positive integer n and for $\lambda \in (0, 1)$, it is impossible to find an even polynomial $P \in H_n$ for which the inequality

(4.93) $$|x^\lambda - P(x)| \leq 2^{-1-1/\lambda} x^{\lambda-1} n^{-1}$$

holds for all $x \in (0, 1]$.

Proof. Let us assume that such a polynomial exists. Without loss of generality, we can assume that λ is rational with $\lambda = 1 - p/q$ where p and q are positive integers with $p < q$. Replacing x by x^q and keeping in mind that P is an even polynomial, we get from (4.93) that

$$M = \max_{0\le x\le 1} \left| x^q - \sum_{k=0}^{[n/2]} c_k x^{2kq+p} \right| \le 2^{-1-1/\lambda} n^{-1}. \tag{4.94}$$

Denote by $S(x)$ the polynomial oscillator

$$S(x) = a_0 x^p + x^q + a_1 x^{2q+p} + a_2 x^{3q+p} + \cdots + a_m x^{2mq+p}, \qquad m = [n/2].$$

Then according to Theorems 4.11 and 4.12, Lemma 4.13 and (4.83), we obtain

$$M \ge \|S\| > \frac{\|T_{2m+1}\|}{(1 + 2^{p/(q-p)})(2m+1)},$$

or

$$M > \frac{1}{(2^{-1-1/\lambda})(n+1)} \ge \frac{1}{(2 + 2^{1/\lambda})n} > 2^{-1-1/\lambda} n^{-1}.$$

The last inequality contradicts (4.94), and completes the proof.

Lemma 4.20. For each positive integer n and for each $\lambda \in (0, 1)$, it is impossible to find an odd algebraic polynomial $P \in H_n$ for which the inequality

$$|x^\lambda - P(x)| \le 2^{-4-1/\lambda(1-\lambda)} x^{\lambda-1} n^{-1}$$

holds for all $x \in (0, 1]$.

Proof. Let us assume that such an odd polynomial exists. We again denote $\lambda = 1 - p/q$ where p and q are positive integers with $p < q$, as again the assumption that λ is rational does not affect the generality of the proof. According to our assumptions, we have

$$M = \max_{0\le x\le 1} \left| x^q - \sum_{k=0}^{[n/2]} c_k x^{(2k+1)q+p} \right| \le 2^{-4-1/\lambda(1-\lambda)} n^{-1}. \tag{4.95}$$

Denote by $S(x)$ the polynomial oscillator

$$S(x) = a_0x^p + x^q + a_1x^{q+p} + a_2x^{3q+p} + \cdots + a_mx^{(2m+1)q+p}, \qquad m = [n/2].$$

Applying Theorems 4.11 and 4.12 with the reiterated application of Lemma 4.13, we obtain

$$M \geq \|S\| > \frac{\|T_{2m+1}\|}{2(1 + 2^{p(q-p)})(1 + 2^{(q+p)p})(2m + 1)}.$$

Now applying (4.83) we obtain

$$M > \frac{1}{4n(1 + 2^{-1+1/\lambda})(1 + 2^{1+1/(1-\lambda)})},$$

which gives $M > 2^{-4-1/\lambda(1-\lambda)}n^{-1}$. The last inequality contradicts (4.95), finishing the proof.

Lemma 4.21. For each positive integer n and for each $\lambda \in (0, 1)$ it is impossible to find an algebraic polynomial $P \in H_n$ for which the inequality

$$|x^\lambda - P(x)| \leq 2^{-5-1/\lambda(1-\lambda)}x^{\lambda-1}n^{-2}$$

holds for all $x \in (0, 1]$.

Proof. As in the preceeding lemmas, let us assume the contrary, again writing $\lambda = 1 - p/q$ where p and q are positive integers with $p < q$. We have

$$M = \max_{0\leq x\leq 1}\left|x^q - \sum_{k=0}^{n} c_kx^{kq+p}\right| \leq 2^{-5-1/\lambda(1-\lambda)}n^{-2}. \tag{4.96}$$

If we let $S(x)$ denote the polynomial oscillator

$$S(x) = a_0x^p + x^q + a_1x^{q+p} + a_2x^{2q+p} + \cdots + a_nx^{nq+p},$$

then by Theorems 4.11 and 4.12, and Lemma 4.13, we have

$$M \geq \|S\| > \frac{\|U_n\|}{2(1 + 2^{p/(q+p)})(1 + 2^{(q+p)/p})},$$

where U_n is the polynomial oscillator $U_n(x) = x^1 + a_2x^{2q} + \cdots + a_nx^{nq}$. Then from Lemma 4.15 and the last inequalities, we obtain

$$M > \frac{1}{8n^2(1 + 2^{-1+1/\lambda})(1 + 2^{1+1/(1-\lambda)})},$$

or $M > 2^{-5-1/\lambda(1-\lambda)} n^{-2}$, which contradicts (4.96). The lemma is proved.

Lemma 4.22. If for an arbitrary positive integer n and for $\lambda \in (0, 1)$ there does not exists an algebraic polynomial $P \in H_n$ for which the inequality

$$|x^\lambda - P(x)| \leq 2\nu x^{\lambda-1}$$

holds for $x \in (0, 1]$, then the best Hausdorff approximation $E(H_n, [0, 1], \alpha; x^\lambda)$ of the function x^λ by algebraic polynomials on the segment $[0, 1]$ for $0 < \alpha \leq 1$ satisfies the inequality $E(H_n, [0, 1], \alpha; x^\lambda) > \nu$.

Proof. Let us assume to the contrary that $E(H_n, [0, 1], \alpha; x^\lambda) \leq \nu$. Then there exists $P \in H_n$ such that

$$P(x) \geq (x - \alpha\nu)^\lambda - \nu \qquad \text{for } x \in [\alpha\nu, 1], \tag{4.97}$$

$$P(x) \geq -\nu \qquad \text{for } x \in [0, \alpha\nu], \tag{4.98}$$

and

$$P(x) \leq (x + \alpha\nu)^\lambda + \nu \qquad \text{for } x \in [0, 1 - \alpha\nu], \tag{4.99}$$

$$P(x) \leq 1 + \nu \qquad \text{for } x \in [1 - \alpha\nu, 1]. \tag{4.100}$$

We first claim that

$$\varphi(x) = 2\nu x^{\lambda-1} - x^\lambda + (x - \alpha\nu)^\lambda - \nu \geq 0 \quad \text{for } x \in [\alpha\nu, 1]. \tag{4.101}$$

Indeed, since $0 < \alpha \leq 1$,

$$\varphi(\alpha\nu) = 2\nu(\alpha\nu)^{\lambda-1} - (\alpha\nu)^\lambda - \nu \geq \alpha^{-1}(\alpha\nu)^\lambda - \nu \geq 0, \tag{4.102}$$

(4.103) $\qquad \varphi(1) = 2\nu - 1 + (1 - \alpha\nu)^{\lambda} - \nu \geq \nu - 1 + 1 - \alpha\nu \geq 0,$

and

$$\varphi'(x) = -2\nu(1-\lambda)x^{\lambda-2} - \lambda x^{\lambda-1} + \lambda(x - \alpha\nu)^{\lambda-1}$$

$$= -2\nu(1-\lambda)x^{\lambda-2} + \lambda(1-\lambda)\alpha\nu(x - \theta\alpha\nu)^{\lambda-2},$$

where θ is some number in $(0, 1)$. It is seen from the last equality that $\varphi'(\alpha\nu) = +\infty$, i.e., for $\varepsilon > 0$ sufficiently small, the inequality $\varphi'(\alpha\nu + \varepsilon) > 0$ holds, and also that $\varphi'(x)$ has at most one zero on the segment $[\alpha\nu, 1]$. As a result, (4.101) follows from (4.102) and (4.103).

From (4.97) and (4.101) we have for $x \in [\alpha\nu, 1]$:

$$x^{\lambda} - P(x) \leq x^{\lambda} - (x - \alpha\nu)^{\lambda} + \nu \leq 2\nu\, x^{\lambda-1}.$$

It follows from (4.98) that for for $x \in [0, \alpha\nu]$, one has

$$x^{\lambda} - P(x) \leq x^{\lambda} + \nu \leq (\alpha\nu)^{\lambda} - \nu \leq 2\nu \leq 2\nu\, x^{\lambda-1},$$

and hence,

(4.104) $\qquad x^{\lambda} - P(x) < 2\nu\, x^{\lambda-1} \qquad$ for $x \in [0, 1]$.

From (4.99), the same method of proof yields for $x \in [0, 1 - \alpha\nu]$,

$$P(x) - x^{\lambda} \leq (x + \alpha\nu)^{\lambda} + \nu - x^{\lambda} = \alpha\nu\lambda(x + \theta\alpha\nu)^{\lambda-1} + \nu$$

$$\leq \alpha\nu\lambda x^{\lambda-1} + \nu \leq 2\nu\, x^{\lambda-1},$$

where $\theta \in (0, 1)$, and from (4.100) we have for for $x \in [1 - \alpha\nu, 1]$

$$P(x) - x^{\lambda} \leq 1 + \nu - x^{\lambda} \leq 1 + \nu - (1 - \alpha\nu)x^{\lambda-1} \leq \nu + \alpha\nu x^{\lambda-1} \leq 2\nu\, x^{\lambda-1}.$$

Hence,

(4.105) $\qquad P(x) - x^{\lambda} \leq 2\nu\, x^{\lambda-1} \qquad$ for $x \in [0, 1]$.

From (4.104) and (4.105), we see that for every $x \in [0, 1]$ the inequality $|x^\lambda - P(x)| < 2\nu\, x^{\lambda-1}$ holds. As this contradicts the main hypothesis of the lemma, the lemma is proved.

We come to our main result.

Theorem 4.13. The best Hausdorff approximations of the functions $\varphi(\lambda; x) = |x|^\lambda$ and $\psi(\lambda; x) = |x|^\lambda \operatorname{sgn} x$ for $0 < \alpha \le 1$ by algebraic polynomials satisfy the following inequalities

$$2^{-2-1/\lambda}/n < E(H_n, [-1, 1], \alpha; \varphi(\lambda)) < 2^{1/\lambda}/\alpha n\,, \tag{4.106}$$

$$2^{-5-1/\lambda(1-\lambda)}/n < E(H_n, [-1, 1], \alpha; \psi(\lambda)) < 2^{1/\lambda}/\alpha n\,. \tag{4.107}$$

Proof. By Lemma 4.16, we have the following relations for the function $g_n(\lambda; x)$:

$$\max_{0\le x\le 1}\ \min_{0\le \xi\le 1}\ \max\{\alpha^{-1}|x-\xi|, |x^\lambda - g_n(\lambda;\xi)|\} \tag{4.108}$$

$$= \max_{0\le x\le 1}\ \min_{0\le \xi\le 1}\ \max\{\alpha^{-1}|x-\xi|, |\xi^\lambda - g_n(\lambda;x)|\} \le 1/\alpha n\,.$$

We shall also need to consider

$$d = \max_{0\le x\le 1}\ \min_{0\le \xi\le 1}\ \max\{\alpha^{-1}|x-\xi|, |x^\lambda - h_n(\lambda;\xi)|\} \tag{4.109}$$

$$= \max_{0\le x\le 1}\ \min_{0\le \xi\le 1}\ \max\{\alpha^{-1}|x-\xi|, |\xi^\lambda - h_n(\lambda;x)|\}.$$

It follows from the monotonicity of the considered functions that

$$d = h_n(\lambda; n^{-1}) - (n^{-1} + \alpha d)^\lambda\,,$$

or $n^\lambda d + (1 + \alpha n d)^\lambda = 2$. From this equality, we have $(1 + \alpha n d)^\lambda \le 2$, yielding the following estimate of d:

$$d \le \frac{2^{1/\lambda}}{\alpha n}\,. \tag{4.110}$$

From (4.108), (4.109) and (4.110), it follows that for every polynomial $P \in H_n$ for which $g_n(\lambda; x) \le P(x) \le h_n(\lambda; x)$, we have the inequality

$$r([0, 1], \alpha; x^\lambda, P(x)) \le \frac{2^{1/\lambda}}{\alpha n}.$$

Keeping in mind that $g_n(\lambda; 0) = h_n(\lambda; 0)$, the right inequality in (4.106) follows from Lemma 4.16, and the right inequality in (4.107) follows from Lemma 4.17.

In order to establish the left inequalities in (4.106) and (4.107), we apply Lemma 4.19 and 4.20 respectively, using Lemma 4.22 in the process. We shall only verify the left inequalitiy in (4.106), leaving the similar computation for (4.107) to the reader.

Since $\varphi(\lambda; x) = |x|^\lambda$ is an even function, its algebraic polynomial of best approximation is also an even one, because of its uniqueness. According to Lemma 4.19, there does not exist an even polynomial $P \in H_n$ for which the inequality

$$|x^\lambda - P(x)| \le 2^{-1-1/\lambda} x^{\lambda-1} n^{-1}$$

holds for all $x \in (0, 1]$. Acording to Lemma 4.22, the inequality $r([-1, 1], \lambda; |x|^\lambda, P) > 2^{-2-1/\lambda} n^{-1}$ will hold for each even polynomial $P \in H_n$, i.e., the left inequality in (4.106) is valid. The theorem is now proved.

The following assertion is obtained in a similar way, by using Lemmas 4.18, 4.21, and 4.22.

Theorem 4.14. The inequalities

$$\frac{2^{-6-1/\lambda(1-\lambda)}}{n^2} < E(H_n, [0, 1], \alpha; x^\lambda) < \frac{2^{1/\lambda}}{\alpha n^2}$$

hold for $\alpha \in (0, 1]$.

Corollary 4.3. The best Hausdorff approximation of the function $\theta(\lambda; x) = (1 - x^2)^\lambda$, $\lambda \in (0, 1]$, by algebraic polynomials satisfies the inequality

(4.111) $$E(H_n, [-1, 1], \alpha; \theta(\lambda)) \le 2^{3+1/\lambda} \alpha^{-1} n^{-2}$$

for $\alpha \in (0, 1]$.

The proof follows directly from Theorem 4.14, replacing x by $1 - x^2$, and the degree of the approximating polynomial doubles. It is necessary to use Lemmas 2.8 and 2.9, as well.

From (4.111), setting $\alpha = \lambda = 1$, one can see that it is possible to inscribe the graph of an algebraic polynomial of degree no higher than n in the strip between two concentric semicircles above the x-axis with centers at the origin and radii $\rho_1 = 1 - 16n^{-2}$ and $\rho_2 = 1 + 16n^{-2}$, respectively. It would be of interest to find the smallest number ν for which the inscription property holds for semicircles with radii $\rho_1 = 1 - \nu n^{-2}$ and $\rho_2 = 1 + \nu n^{-2}$. At this time, we know that $1/9 < \nu < 5$ (Bl. Sendov [16]).

A more general (and unsolved) problem is to find the the asymptotic value of best approximation by algebraic polynomials for the three already considered functions $\varphi(\lambda; x) = |x|^\lambda$, $\psi(\lambda; x) = |x|^\lambda \operatorname{sgn} x$, and $\theta(\lambda; x) = (1 - x^2)^\lambda$ on the segment $[-1, 1]$.

4.1.2.6. *Approximation of Convex Functions* . As we already proved in §4.1.2.2, the best Hausdorff approximation of a bounded function by algebraic polynomials is of order $n^{-1}\ln n$, and this order cannot be improved in the class Lip λ, $0 < \lambda < 1$ (T. Boyanov [4]). It is natural to look for such classes of bounded functions for which this approximation is better, although the best uniform approximation is substantially worse. In this section, we shall show that such a class is the class of all convex functions (which is, by Defintion 1.5, the convex functions of order 2).

Let K_Δ^M, $\Delta = [a, b]$, denote set of all the convex functions on the closed interval Δ which are bounded by the constant M, i. e.,

$$K_\Delta^M = \{f : f \in A_\Delta,\ D^2(f; x) \geq 0 \text{ and } |f(x)| \leq M \text{ for } x \in \Delta\}.$$

We shall prove that each function f in K_Δ^M has best approximation by algebraic polynomials of degree no higher than n^{-1}. More precisely, the following theorem holds.

Theorem 4.15. (V. Popov [17]). There exists an absolute constant $c > 0$ such that for $f \in K_\Delta^M$, $\Delta = [a, b]$, we have the inequality

$$E(H_n, \Delta, \alpha; f) \leq c(b - a)\alpha^{-1}n^{-1} \ln (e + M). \tag{4.112}$$

Keeping in mind Lemmas 2.8 and 2.9, it is sufficient to prove (4.112) only for $\alpha = 1$ and $\Delta = [0, 1]$, i.e., we must only show

$$E(H_n, [0, 1], 1; f) \leq cn^{-1} \ln (e + M). \tag{4.113}$$

Also, since every function $f \in K_\Delta^M$ may be approximated by a continuous function to an arbitrary accuracy with respect to Hausdorff distance, we may confine ourselves to the consideration of continuous functions in K_Δ^M (note that functions in K_Δ^M may have discontinuities only at the endpoints of the segment Δ).

Following our earlier discussion, the **one-sided Hausdorff distance with parameter** α between continuous functions on the closed interval $\Delta = [a, b]$ is given by

$$h(\Delta, \alpha; f, g) = \max_{x \in \Delta} \min_{\xi \in \Delta} \max \{\alpha^{-1}|x - \xi|, |f(x) - g(\xi)|\}. \tag{4.114}$$

Paralleling (4.57) we of course have

$$r(\Delta, \alpha; f, g) = \max \{h(\Delta, \alpha; f, g), h(\Delta, \alpha; g, f)\}. \tag{4.115}$$

It is directly seen that if $c \in [a, b]$, then

$$h([a, b], \alpha; f, g) \le \max \{h([a, c], \alpha; f, g), h([c, b], \alpha; f, g)\}, \tag{4.116}$$

$$r([a, b], \alpha; f, g) \le \max \{r([a, c], \alpha; f, g), r([c, b], \alpha; f, g)\}. \tag{4.117}$$

The proof of Theorem 4.15 will be prefaced by several lemmas, following the original path of V. Popov [17]. In the sequel we write $h(\Delta; f, g)$ for $h(\Delta, 1; f, g)$.

Lemma 4.23. Let $f, g \in C_\Delta$, $\Delta = [a, b]$, where g is monotone on Δ. Then with $\nu(g, f; x) = \min_{\xi \in \Delta} \max \{|x - \xi|, |g(x) - f(\xi)|\}$, we have

$$h(\Delta; g, f) \le \max \{h(\Delta; f, g), p, q\}, \tag{4.118}$$

where

$$p = \nu(g, f; a), \quad q = \nu(g, f; b). \tag{4.119}$$

Proof. Without loss of generality, we may assume that the function g does not decrease monotonically. We write $A = \min \{f(\xi) : \xi \in \Delta\} = f(\xi_1)$, and $B = \max \{f(\xi) : \xi \in \Delta\} = f(\xi_2)$. From (4.119), it follows that $g(a) \ge A - p$, and $g(b) \le B + q$, and since g is monotonic, we also have $A - p \le g(x) \le B + q$ for every $x \in \Delta$. If $g(x) \in [A, B]$, then there exists $\xi(x) \in \Delta$ for which $f(\xi(x)) = g(x)$, so that in this case

(4.120) $\quad \nu(g, f; x) \leq b - a.$

If $g(x) \in [A - p, A]$, or $g(x) \in [B, B + q]$, then the inequalities

(4.121) $\quad \nu(g, f; x) \leq \max\{|x - \xi_1|, p\}, \qquad \nu(g, f; x) \leq \{|x - \xi_2|, q\},$

hold, repectively. From (4.120) and (4.121), we obtain

(4.122) $\quad h(\Delta; f, g) = \max_{x \in \Delta} \nu(g, f; x) \leq \max\{b - a, p, q\}.$

If $\delta = h(\Delta; f, g) \geq b - a$, then the conclusion of the lemma follows from (4.122). Next we consider the case $\delta < b - a$, where in addition $a + \delta \leq b - \delta$. If for some $x \in [a + \delta, b - \delta]$ we have $\nu(g, f; x) > \delta$, then either $f(\xi) > g(x)$ for $\xi \in [x - \delta, x + \delta]$, or $f(\xi) < g(x)$ for $\xi \in [x - \delta, x + \delta]$. Since g does not decrease monotonically, in the first case we have $\nu(g, f; x - \delta) > \delta$, and in the second case $\nu(g, f; x + \delta) > \delta$. This contradicts the equality $\max\{\nu(g, f; x) : x \in \Delta\} = h(\Delta; f, g) = \delta$. Consequently,

(4.123) $\quad \max\{\nu(g, f; x) : x \in [a + \delta, b - \delta]\} \leq \delta,$

(4.124) $\quad \nu(g, f; a + \delta) \leq \delta, \qquad \nu(g, f; b - \delta) \leq \delta.$

From (4.116) and (4.122), considering the one-sided Hausdorff distance $h(\Delta; f, g)$ on the closed intervals $[a, a + \delta]$, $[a + \delta, b - \delta]$, $[b - \delta, b]$, according to (4.123), we obtain (4.118).

It remains to consider the case $b - a > \delta > (b - a)/2$. This is handled by repeating the previous arguments for $x = a + \delta$, or $x = b - \delta$. The lemma is proved.

Lemma 4.24. Let $f_i, g_i \in C_\Delta$, $\Delta = [a, b]$, $i = 1, 2, 3, \ldots, m$, where each g_i is a monotone nondecreasing function on Δ, and $h(\Delta; f_i, g_i) \leq \delta_i$ for $i = 1, 2, 3, \ldots, m$. Let us write

$$\delta = \max\{\delta_i : i = 1, 2, 3, \ldots, m\},$$

$$\mu = \sum_{i=1}^m \mu_i \delta_i, \qquad f = \sum_{i=1}^m \mu_i f_i, \qquad g = \sum_{i=1}^m \mu_i g_i,$$

where $\mu_i \geq 0$ for $i = 1, 2, 3, \ldots, m$. Then $h(\Delta; f, g) \leq \max\{\mu, \delta\}$.

Proof. From the definition of $h(\Delta; f_i, g_i)$, we have

$$g_i(x - \delta'(x)) - \delta \leq I(\delta_i, g_i; x) - \delta_i \leq f_i(x) \leq S(\delta_i, g_i; x) + \delta_i \leq g_i(x + \delta''(x)) + \delta,$$

where $\delta'(x) = \min\{\delta, x - a\}$ and $\delta''(x) = \min\{\delta, b - x\}$. Hence,

$$I(\delta, g; x) - \mu = g(x - \delta'(x)) - \mu \leq f(x) \leq g(x + \delta''(x)) + \mu \leq S(\delta, g; x) + \mu.$$

But from the last inequality string, we see that

$$h(\Delta; f, g) \leq \delta + |\mu - \delta| = \max\{\mu, \delta\}, \tag{4.125}$$

and the lemma is proved.

Lemma 4.25. For every $M > 0$ and for every positive integer n for which

$$\lambda_n = \frac{(\ln(e + M))^2}{2n} < 1,$$

there exists an algebraic polynomial $s_n(M) = s_n$ of degree no higher than n, such that

$$|s_n(x)| \leq 1/n \qquad \text{for } x \in [0, (1 - \lambda_n)/(1 + \lambda_n)], \tag{4.126}$$

where s_n is monotonically increasing on $[(1 - \lambda_n)/(1 + \lambda_n), 1]$ and $s_n(1) \geq M$.

Proof. We consider the algebraic polynomial $s_n(x) = n^{-1} T_n((1 + \lambda_n)x/(1 - \lambda_n))$, where $T_n(x) = \cos(n \arccos x)$ is the usual Chebyshev polynomial. The inequality (4.126) clearly holds, and s_n is monotonically increasing for $x \geq (1 - \lambda_n)/(1 + \lambda_n)$. On the other hand, for those n for which $\lambda_n < 1$, we have

$$s_n(1) = n^{-1} T_n\left(\frac{1 + \lambda_n}{1 - \lambda_n}\right) \geq \frac{1}{2n}\left(\frac{1 + \sqrt{\lambda_n}}{1 - \sqrt{\lambda_n}}\right)^n \geq \frac{1}{2n}\exp(2n\sqrt{\lambda_n}) > e + M > M.$$

This completes the proof of the lemma.

Lemma 4.26. For $M > 0$ and each $\nu \in [0, 1)$, let $g(\nu; x)$ be the monotone nondecreasing function on $[-1, 1]$ defined by

$$g(\nu; x) = \max\{0, M(x - \nu)/(1 - \nu)\}.$$

There exists an absolute constant $c_1 > 0$ such that for every positive integer n, there exists an algebraic polynomial $p_n \in H_n$ for which $p_n(1) = g(\nu; 1)$ and

$$r([-1, 1]; p_n, g(\nu)) \le c_1 n^{-1} \ln(e + M). \tag{4.127}$$

Proof. If $\nu \ge 1 - n^{-1}$, then we take as our polynomial $p_n(x) = Ms_n((x + 1)/2)/s_n(1)$, where s_n is the polynomial described in Lemma 4.25. According to (4.126), we have $|p_n(x) - g(\nu; x)| = |p_n(x)| \le n^{-1}$ for $x \in [-1, 1/(1 + \lambda_n)]$, and since $p_n(1) = g(\nu; 1)$ and p_n increases monotonically on $[1/(1 + \lambda_n), 1]$, we have for all sufficiently large n

$$r([-1, 1]; p_n, g(\nu)) \le \max\{n^{-1}, 2\lambda_n/(1 + \lambda_n)\} \le n^{-1} \ln(e + M),$$

provided $\lambda_n < 1$. For $\lambda_n \ge 1$, we have $n^{-1} \ln(e + M) \ge 1/2$, and we obtain for $p(x) = M(x + 1)/2$

$$r([-1, 1]; p, g(\nu)) \le 2 \le 4n^{-1} \ln(e + M).$$

This establishes the assertion of our lemma for $\nu \ge 1 - n^{-1}$.

On the other hand, provided the lemma is proved for some $\nu_0 \in [0, 1)$, then the function $g(\nu)$ becomes, through the linear transformation $x = (1 - \rho)t + \rho$, $0 < \rho \le \nu$, a function $g(\nu')$ with $0 \le \nu' = \nu - \rho(1 - \nu)/(1 - \rho) < \nu$. Since the functions $g(\nu)$ are identically equal to zero on $[-1, 0]$, we have

$$r([-1, 1]; p_n, g(\nu)) \ge (1 - \rho) r([1 - 2\rho/(1 - \rho), 1]; p_n^*, g(\nu'))$$

$$= (1 - \rho) r([-1, 1]; p_n^*, g(\nu'))$$

where $p_n^*(t) = p_n((1 - \rho)t + \rho)$. Hence, if the lemma is proved for $\nu = 3/4$, then it holds for all $\nu \in [0, 3/4]$, with respect to a four-fold increase of the constant c_1. Thus, it remains to prove the lemma for $\nu \in [3/4, 1 - n^{-1}]$.

It follows from Theorem 4.9 that for every function $f \in C_{2\pi}$ and for every positive integer n, there exists an even trigonometric polynomial $q_n \in T_n$ for which

(4.128) $r(\alpha; f, q_n) \le c_2(\alpha n)^{-1} \ln (e + \alpha n \omega(f; n^{-1}))$.

Let us set $f(\nu; t) = g(\nu; \cos t)$. Since

$$\omega(f(\nu); n^{-1}) \le \omega(g(\nu); n^{-1}) \le M/n(1 - \nu),$$

then according to (4.128) there exists an even trigonometric polynomial $q_n \in T_n$ for which

(4.129) $r(1/(1 - \nu); f(\nu), q_n) \le c_2(1 - \nu)n^{-1}\ln (e + M(1 - \nu)^{-1}) \le 2c_2 n^{-1} \ln (e + M)$,

where we used the inequalitites

(4.130) $0 \le (1 - \nu) \ln (1 - \nu)^{-1} \le e^{-1}$

for each $\nu \in [0, 1]$. By the definition of Hausdorff distance, the inequality (4.129) is equivalent to the following conditions: for every t there exist points $\xi'(t)$ and $\xi''(t)$ such that

(4.131) $|t - \xi'(t)| \le c_2 n^{-1}\ln (e + M(1 - \nu)^{-2}) = \delta, \qquad |t - \xi''(t)| \le \delta;$

(4.132) $|f(\nu; t) - q_n(\xi'(t))| \le 2c_2 n^{-1}\ln (e + M);$

(4.133) $|f(\nu; \xi''(t)) - q_n(t)| \le 2c_2 n^{-1}\ln (e + M).$

The function $p_n(x) = q_n(\text{arc cos } x)$ is an algebraic polynomial of order no higher than n. We write $\varphi = \text{arc cos } \nu$ for $\nu \in [3/4, 1 - n^{-1}]$, so that $\varphi \in (0, \pi/4)$. Then $f(\nu; t) = g(\nu; \cos t) = 0$ for $t \in [\varphi, 2\pi - \varphi]$.

We now estimate the Hausdorff distance between p_n and $g(\nu)$ on the closed interval $[-1, 1]$. Let $x \in [-1, 1]$, and let $t = \text{arc cos } x$. If $t \in [\varphi + \delta, 2\pi - \varphi - \delta]$, then $\xi'(t)$ and $\xi''(t)$ lie in $[\varphi, 2\pi - \varphi]$ and $f(\nu; \xi'(t)) = f(\nu; \xi''(t)) = f(\nu; t) = 0$. As a result, for all such x, (4.132) and (4.133) give

(4.134) $|g(\nu; x) - p_n(x)| \le 2c_2 n^{-1}\ln (e + M).$

Consider the remaining x, corresponding to $t \in [0, \varphi + \delta] \cup [2\pi - \varphi - \delta, 2\pi]$, or, equivalently, $t \in [-\varphi - \delta, \varphi + \delta]$. According to (4.131), both $\xi'(t)$ and $\xi''(t)$ lie in

$[-\varphi - 2\delta, \varphi + 2\delta]$. Since $\varphi \in (0, \pi/4)$, if $\delta \le \pi/8$, we have $\varphi + 2\delta \le \pi/2$. Using (4.130) again, we obtain

$$(4.135) \qquad |\cos t - \cos \xi'(t)| \le |t - \xi'(t)| \sin (\varphi + 2\delta)$$

$$\le c_2 n^{-2} \ln^2 (e + Mn^2) + c_2 n^{-1} \sqrt{2(1 - \nu)} \ln (e + M(1 - \nu)^{-2})$$

$$\le c_3 n^{-1} \ln (e + M),$$

where c_3 is an absolute constant. We obtain similarly that

$$(4.136) \qquad |\cos t - \cos \xi''(t)| \le c_3 \ln (e + M).$$

From (4.134) - (4.136), we obtain (4.127) for $\delta \le \pi/8$. But $\delta \le \pi/8$ holds for sufficiently large n, and since we can assume for $n \ge 1$ that the left-hand side of (4.127) is less than 2, independent of the value of M, the proof of (4.127) is completed.

It remains to prove that the polynomial p_n in (4.127) may be chosen so that $p_n(1) = g(\nu; 1) = M$. If the last condition is not satisfied, then we can change p_n such that (4.127) will be preserved, and the required condition is satisfied.

Choose $\lambda \in [-1, 1]$ so that

$$\max \{| 1 - \lambda|, |M - p_n(\lambda)| \le c_1 n^{-1} \ln (e + M).$$

For sufficiently large n, we may assume that $\lambda > 1/2$ and $c_1 n^{-1} \ln (e + M) < 1/2$. We now set $p_n^*(x) = p_n(\lambda x) + M - p_n(\lambda)$. It is not hard to verify that $p_n^*(1) = M$ and p_n^* satisfies (4.127), with a doubled constant c_1. This completes the proof.

Lemma 4.27. Let f be a monotonically nondecreasing continuous, convex function on the segment $[0, 1]$. Suppose $f(1) = M$ and $f(x) = 0$ for $x \in [0, 2\delta_n]$, where $\delta_n = c_1 n^{-1} \ln (e + M)$ and c_1 is the constant in (4.127). Then

$$E(H_n, [0, 1], 1; f) \le c_1 n^{-1} \ln (e + M) .$$

Proof. For every $\varepsilon > 0$ one can find a convex combination of the functions $g(\nu_i)$ of Lemma 4.26

$$g(x) = \sum_{i=1}^{m} \mu_i g(\nu_i; x), \quad \sum_{i=1}^{m} \mu_i = 1, \quad \mu_i \ge 0, \ i = 1, 2, \ldots, m,$$

such that

(4.137) $\max\{|f(x) - g(x)| : x \in [0, 1]\} < \varepsilon.$

It follows from Lemma 4.26 that for every function $g(v_i)$, there exists an algebraic polynomial $p_{n,i} \in H_n$, $i = 1, 2, 3, \ldots, m$, for which $h([0, 1]); p_{n,i}, g(v_i)) \le \delta_n$, $|p_{n,i}(0)| \le \delta_n$, and $p_{n,i}(1) = M$. Using Lemma 4.24, we obtain for $p_n = \mu_1 p_{n,1} + \mu_2 p_{n,2} + \cdots + \mu_m p_{n,m} \in H_n$ the following relations:

(4.138) $h([0, 1); p_n, g) \le \delta_n,$ $\qquad |p_n(0)| \le \delta_n,$ $\qquad p_n(1) = M.$

Using Lemma 4.23, from (4.138), we see that $r([0, 1); p_n, g) \le \delta_n$, and according to (4.137), we have $r([0, 1); p_n, f) \le \delta_n + \varepsilon$. But since $\varepsilon > 0$ was chosen arbitrarily, the conclusion of the lemma follows from the compactness of the set of uniformly bounded polynomials in H_n with respect to the common domain [0, 1].

Proof of Theorem 4.15 . Let $f \in K_\Delta^M$, where $\Delta = [0, 1]$. As we already mentioned, we need only consider continuous functions, i.e., we may assume $f \in C_\Delta$. Also, without loss of generality, we may assume that $\min\{f(x) : x \in [0, 1]\} = f(v) = 0$ where $v \in [0, 1/2]$. Let us write $\delta_n = c_1 n^{-1} \ln(e + M)$, where c_1 is the constant of Lemmas 4.26 and 4.27. Since δ_n tends to zero with increasing n, we may assume $\delta_n \le 1/8$.

For the given function f, we shall define continuous functions f_n, $n \ge 4$, as follows: $f_n(x) = f(x)$ for $x \in [0, v]$, $f_n(x) = 0$ for $x \in [v, v + 2\delta_n]$, $f_n(x) = f(x - 2\delta_n)$ for $x \in [v + 2\delta_n, 1 - n^{-1}]$, and $f_n(x)$ linear on $[1 - n^{-1}, 1]$, with $f_n(1) = 1$. Obviously,

(4.139) $r([0, 1]; f, f_n) \le 2\delta_n + n^{-1}.$

Let us represent f_n as a sum of two functions : $f_n = g_n + h_n$ where $g_n(x) = 0$ for $x \in [0, v + \delta_n]$ and $h_n(x) = 0$ for $x \in [v + \delta_n, 1]$. The functions $g_n(x)$ and $\bar{g}_n(x) = h_n(1 - x)$ satisfy the hypotheses of Lemma 4.27, and hence there exist algebraic polynomials $q_n, \bar{q}_n \in H_n$ such that

(4.140) $r([0, 1]; g_n, q_n) \le \delta_n,$ $\qquad r([0, 1]; \bar{g}_n, \bar{q}_n) \le \delta_n.$

From the definitions of g_n and $\bar{g}_n$, and (4.140), it follows that

(4.141) $|q_n(x)| \le \delta_n$ for $x \in [0, \nu + \delta_n]$,

$$|\overline{q}_n(1 - x)| \le \delta_n \quad \text{for} \quad x \in [\nu + \delta_n, 1].$$

If we write $p_n(x) = q_n(x) + \overline{q}_n(1 - x)$, then from (4.115), the definitions of g_n and h_n, and (4.140), (4.141), we obtain

$$r([0, 1]; f_n, p_n) \le \max \{r([0, \nu + \delta_n]; f_n, p_n), r([\nu + \delta_n, 1]; f_n, p_n)\}$$

$$\le \max \{r([0, \nu + \delta_n]; g_n, q_n) + \delta_n, \; r([0, 1 - \nu - \delta_n]; \overline{g}_n, \overline{q}_n)\}$$

$$\le 2\delta_n .$$

From this last inequality, with the help of (4.139), we obtain

$$r([0, 1]; f, p_n) \le r([0, 1]; f, f_n) + r([0, 1]; f_n, p_n) \le 5\delta_n ,$$

Thus, (4.113) is established, and hence (4.112) is established as well. The theorem is proved.

It is natural to investigate the problem of finding the best constant in (4.113), and hence also in (4.112). V. Popov [17] made the conjecture that function $f(x) = |x|$ has the worst Hausdorff approximation among all the convex functions defined on $[-1, 1]$ with values in $[0, 1]$.

4.1.2.7. *An Analogue of Nikol'skii's Theorem* . S. Nikol'skii [1] was the first to note that for the best approximation of functions by algebraic polynomials with respect to the uniform distance on a finite closed interval, it is possible to choose the polynomial in such a way that the approximation at the end points of the interval is substantially improved. For example, if the function f satisfies a Lipschitz condition on the closed interval $[-1, 1]$, then there exists a constant $c > 0$ such that for every n, there exists an algebraic polynomial $p_n \in H_n$ for which

(4.142) $$|f(x) - p_n(x)| \le c(\sqrt{1 - x^2}\, n^{-1} + n^{-2}) .$$

We shall exhibit the same effect for the Hausdorff approximations. In order to introduce a local estimate of distance in the sense of Hausdorff, we shall first introduce the so-called Hausdorff difference. Let $f, g \in C_\Delta$, $\Delta = [a, b]$; the **Hausdorff difference**

with parameter α $|f(x) - g(x)|_\alpha$ between the functions f and g at the point x is the following nonnegative number:

$$|f(x) - g(x)|_\alpha = \max \{ \min_{\xi\in\Delta} \max \{\alpha^{-1}|x - \xi|, |f(x) - g(\xi)|, \min_{\xi\in\Delta} \max \{\alpha^{-1}|x - \xi|, |f(\xi) - g(x)|\}. \tag{4.143}$$

The Hausdorff distance is obtained from the Hausdorff difference as follows:

$$r(\Delta, \alpha; f, g) = \max_{x\in\Delta} |f(x) - g(x)|_\alpha . \tag{4.144}$$

We shall prove an analogue of (4.142) for the Hausdorff difference.

Theorem 4.16. (Bl. Sendov and V. Popov [7]). If $f \in C_\Delta$, $\Delta = [-1, 1]$, then for every positive integer n, there exists an algebraic polynomial $p_n \in H_n$ for which one has the inequality

$$|f(x) - p_n(x)|_\alpha \leq c\alpha^{-1} (\sqrt{1 - x^2}\, n^{-1}\ln(e + Mn) + n^{-2} \ln n) \tag{4.145}$$

for any $x \in [-1, 1]$, where c is an absolute constant.

Proof . According to the definition (4.143) and Lemma 2.8, it is sufficient to prove (4.145) for $\alpha = 1$. Consider the 2π-periodic even and continuous function $g(t) = f(\cos t)$. According to Theorem 4.9, there exists an absolute constant c_1 such that for every positive integer n and for every $\alpha > 0$, there exists an even trigonometric polynomial q_n of order at most n for which

$$r(\alpha; g, q_n) \leq c_1\alpha^{-1}n^{-1} \ln (e + \alpha n\omega(g; n^{-1})) . \tag{4.146}$$

By setting $\alpha = n$, $\omega(g; n^{-1}) \leq 2M$, where $M = \max_x |g(x)|$, from (4.144) and (4.146), we obtain for every t that

$$|g(t) - q_n(t)|_n \leq c_1 n^{-2} \ln (e + 2Mn^2)) = \delta_n . \tag{4.147}$$

If we set $t = \arccos x$, then $q_n(t) = q_n(\arccos x) = p_n(x)$ is an algebraic polynomial of degree no higher than n. According to (4.143), for every $x \in \Delta = [-1, 1]$ with $x = \cos t$, we get

(4.148) $$|f(x) - p_n(x)|_1 = \max \{\min_{\xi \in \Delta} \max \{|x - \xi|, |f(x) - p_n(\xi)|\}, \min_{\xi \in \Delta} \max \{|x - \xi|, |f(\xi) - p_n(x)|\}\}$$

$$= \max \{\min_{\tau} \max \{|\cos t - \cos \tau|, |g(t) - q_n(\tau)|\}, \min_{\tau} \max \{|\cos t - \cos \tau|, |g(\tau) - q_n(t)|\}\}.$$

But

(4.149) $$|\cos t - \cos \tau| = 2|\sin (t - \tau)/2 \cdot \sin (t + \tau)/2| \le |t - \tau| \cdot |\sin (t + (\tau - t)/2)|$$

$$\le |t - \tau| \sqrt{1 - x^2} + (t - \tau)^2 .$$

According to (4.147), for every t there exist points τ_1 and τ_2 such that

(4.150) $$n^{-1} |t - \tau_1| \le \delta_n, \qquad n^{-1} |t - \tau_2| \le \delta_n,$$

$$|g(t) - q_n(\tau_1)| \le \delta_n, \qquad |g(\tau_2) - q_n(t)| \le \delta_n .$$

From (4.149) and (4.159), it follows that

(4.151) $$\max \{|\cos t - \cos \tau_1|, |g(t) - q_n(\tau_1)|\} \le (n \sqrt{1 - x^2} + n\delta_n + 1) \delta_n ,$$

(4.152) $$\max \{|\cos t - \cos \tau_2|, |g(\tau_2) - q_n(t)|\} \le (n \sqrt{1 - x^2} + n\delta_n + 1) \delta_n .$$

But according to (4.147)

$$(n \sqrt{1 - x^2} + n\delta_n + 1) \delta_n \le c \sqrt{1 - x^2}\, n^{-1} \ln (e + Mn) + n^{-2} \ln n),$$

where c is an absolute constant. From the last inequality, (4.151), (4.152) and Lemma 2.8, we get (4.145). The theorem is proved.

The problem of finding the minimal constant for which (4.145) holds is still unsolved. It is known, however, that $c \le 6$ (Bl. Sendov and V. Popov [7, p. 103]).

4.1.2.8. *Comonotone Approximations* . In approximation theory, the problem of approximation of functions by algebraic polynomials, which preserves some global properties of the approximated functions, is considered, too. A problem of this kind is the approximation of a monotone function by monotone algebraic polynomials. The first to consider this subject were O. Shisha [1] and G. Lorentz [3], and a number of studies have been published (G. Lorentz and K. Zeller [1], J. Roulier [1], et al.). It turns out that the additional restriction of monotonicity on the approximating polynomial, when the function is monotone, does not affect the order of approximation in the case of uniform approximation. The proof of the same result for Hausdorff distance, from which, as corollaries, fall out some of the well-known results for the uniform distance, will be our aim here.

The proof makes use of a certain modification of the method of G. Lorentz and K. Zeller [1], by using the bell-shaped trigonometric functions. We say that a 2π-periodic function f is **bell-shaped**, if it is even and is nonincreasing on $[0, \pi]$. The graph of such a function is in the shape of a bell on the closed interval $[-\pi,\pi]$.

G. Lorentz and K. Zeller proved that there exists an absolute constant c_1, such that for every bell-shaped function $f \in C_{2\pi}$, there exists a bell-shaped trigonometric polynomial $q_n \in T_n$ for which

$$\max_x |f(x) - q_n(x)| \le c_1\, \omega(f; n^{-1}) . \tag{4.153}$$

From here it follows that for every monotone function f on $[-1, 1]$, there exists a monotone algebraic polynomial $p_n \in H_n$ for which

$$|f(x) - p_n(x)| \le c_2\, \omega(f; \Delta_n(x)) , \tag{4.154}$$

where $\Delta_n(x) = n^{-1}\sqrt{1 - x^2} + n^{-2}$.

We shall prove the following two statements.

Theorem 4.17. (V. Popov and Bl. Sendov [2]). For every bell-shaped function f, there exists a bell-shaped trigonometric polynomial $q_n \in T_n$ such that

$$r(\alpha; f, q_n) \le c_1\, \omega(f; n^{-1}) \frac{\ln (e + \alpha n\omega(f; n^{-1}))}{1 + \alpha n\omega(f; n^{-1})} , \tag{4.155}$$

where c_1 is an absolute constant.

It is not hard to see that (4.153) follows from (4.155), letting $\alpha \to +0$.

Theorem 4.18. For every monotone function f on [-1, 1], there exists a sequence of polynomials $\langle p_n \rangle_1^\infty$ monotone on [-1, 1] with $p_n \in H_n$, such that for every $x \in$ [-1, 1], we have the inequality

(4.156) $$|f(x) - p_n(x)|_\alpha \le c_2\, \omega(f; \Delta_m(x)) / (1 + \alpha\omega(f; \Delta_m(x))/\Delta_m(x)),$$

where c_2 is an absolute constant, $M = \sup\{|f(x)| : x \in [-1, 1]\}$, $m = m(\alpha, M, n) = \max\{p : p$ is an integer and $p \ln e^2(1 + \alpha M p^2) \le n\}$, and $\Delta_m(x) = m^{-1}\sqrt{1 - x^2} + m^{-2}$.

From (4.156), letting $\alpha \to +0$, one obtains (4.154). We shall now prove Theorem 4.17, by using the generalized Jackson operator

$$U_{m,r}(f; x) = c_{m,r} \int_{-\pi}^{\pi} f(x + t)\, \psi_{m,r}(t)\, dt \quad ,$$

where $\psi_{m,r}(t) = (\sin(mt/2)/\sin(t/2))^{2r}$, m and r are positive integers, and $c_{m,r}$ is defined by the equality

(4.157) $$c_{m,r} \int_{-\pi}^{\pi} \psi_{m,r}(t)\, dt = 1.$$

If f is a bell-shaped function, then in the general case the trigonometric polynomial $U_{m,r}(f)$ of order no higher than mr is not a bell-shaped one.

Lemma 4.28. If the function f is a bell-shaped step function with jumps at the points $k\pi/m$, $k = 0, \pm1, \pm2, \ldots$, so that f is equal to a constant on every interval of the form $(k\pi/m, (k+1)\pi/m)$, $k = 0, \pm1, \pm2, \ldots$, then the trigonometric polynomial

$$U_{m,r}(f; x) = c_{m,r} \int_{-\pi}^{\pi} f(x + t)\, \psi_{m,r}(t)\, dt$$

is a bell-shaped trigonometric polynomial of order no higher than mr.

Proof. Let v_k denote the jump of the step function f at $k\pi/m$, $k = 1, 2, \ldots, m - 1$. Since f does not increase on $[0, \pi]$, $v_k = f(k\pi/m - 0) - f(k\pi/m + 0) \ge 0$. According to the definition of $U_{m,r}(f)$, we have

$$U_{m,r}(f; x) = c_{m,r} \sum_{k=1}^{m-1} v_k \int_{-k\pi/m}^{k\pi/m} \psi_{m,r}(x - t)\, dt .$$

We claim that

$$\phi_k(x) = \int_{-k\pi/m}^{k\pi/m} \psi_{m,r}(x - t)\, dt \tag{4.158}$$

is a bell-shaped function. According to the definition of $\psi_{m,r}$, the function ϕ_k is both even and 2π-periodic. It remains to prove that ϕ_k does not increase on $[0, \pi]$. It follows from (4.158) that

$$\phi_k(x) = \int_{x-k\pi/m}^{x+k\pi/m} \psi_{m,r}(t)\, dt\,,$$

and consequently

$$\begin{aligned}\phi_k'(x) &= \psi_{m,r}(x + k\pi/m) - \psi_{m,r}(x - k\pi/m),\\ &= (\sin m\xi)^{2r}\,((\sin \xi)^{-2r} - (\sin (\xi - k\pi/m))^{2r}),\end{aligned} \tag{4.159}$$

where $\xi = (x + k\pi/m)/2 \in [0, \pi]$. Since $k\pi/m \in [0,\pi]$, we have $\sin \xi \geq |\sin (\xi - k\pi/m)|$, and it follows from (4.159) that $\phi_k'(x) \leq 0$ for $x \in [0,\pi]$, i.e., that ϕ_k is a bell-shaped function. But then $U_{m,r}(f)$ is also bell-shaped, because it is a linear combination of bell-shaped functions with nonnegative coefficients. The lemma is proved.

Proof of Theorem 4.17 . Let f be an arbitrary bell-shaped function. Write

$$t_k = 10k\pi/m,\ k = 0, 1, 2, \ldots, l + 1, \quad \text{where } l = \max\,\{\nu : 10(\nu + 2) \leq m\},$$

and consider this bell-shaped step function:

$$f_m(x) = \begin{cases} f(t_k) & \text{if } t_k \leq x < t_{k+1}, \quad k = 0, 1, 2, \ldots, l \\ f(\pi) & \text{if } t_{k+1} \leq x \leq \pi \end{cases}.$$

Obviously,

$$r(\alpha; f, f_m) \leq \min\{20\pi/\alpha m,\ \omega(f; 20\pi/m)\}, \tag{4.160}$$

and according to Lemma 4.28, $U_{m,r}(f_m)$ is a bell-shaped trigonometric polynomial of order no higher than mr.

From the definition of Hausdorff distance and the fact that both f_m and $U_{m,r}(f_m)$ are bell-shaped functions, it follows that in order to estimate $r(\alpha; f_m, U_{m,r}(f_m))$, it is sufficient to estimate $|f_m(x_i) - U_{m,r}(f_m; x_i)|$, where $x_i = 5(2i + 1)\pi/m$, for $i = 0, 1, 2, \ldots, l + 1$. As in Theorem 3.9, we may obtain

$$0 < c_{m,r} \le 2^{-2}\left(\frac{\pi}{2m}\right)^{2r-1} . \tag{4.161}$$

From (4.161) for $r \ge 2$ we have

$$|U_{m,r}(f_m; x_i) - f_m(x_i)| \le c_{m,r}\int_{-\pi}^{\pi} |f_m(x_i + t) - f_m(x_i)|\, \psi_{m,r}(t)\, dt$$

$$\le 2c_{m,r}\sum_{k=1}^{l+1} \omega(f; 2t_k)\int_{5k\pi/m}^{\pi} \psi_{m,r}(t)\, dt$$

$$\le \frac{\left(\frac{\pi}{2m}\right)^{2r-1}}{2r-1}\sum_{k=1}^{\infty} \omega(f; 2t_k)\, \pi^{2r}\left(\frac{m}{5k\pi}\right)^{2r-1}$$

$$\le \frac{\left(\frac{\pi}{10}\right)^{2r-1}\omega(f; 20\pi/m)}{2r-1}\sum_{k=1}^{\infty} k^{-2r+2} \le c_3 r^{-1} e^{-r}\, \omega(f; m^{-1})$$

for $i = 0, 1, 2, \ldots, l + 1$. Inserting

$$\beta = \ln e^2(1 + \alpha n\omega(f; n^{-1}))m, \qquad r = [\beta], \qquad m = 10([n/10r] + 1),$$

in the last inequalities, we conclude that there exists an absolute constant c_4 such that

$$|U_{m,r}(f_m; x_i) - f_m(x_i)| \le \delta_n = \frac{c_4\omega(f; n^{-1})}{1 + \alpha n\omega(f; n^{-1})}$$

for $i = 0, 1, 2, \ldots, l + 1$. It now follows from the definition of Hausdorff distance and the choice of the knots x_i that

$$r(\alpha; f_m, U_{m,r}(f_m)) \le \max\{\delta_n, \min\{20\pi/\alpha m, \delta_n + \omega(f; 20\pi/m)\}\} \tag{4.163}$$

$$\leq c_5\chi_n = c_5\omega(f;\, n^{-1})\frac{\ln e^2(1 + \alpha n\omega(f;\, n^{-1}))}{1 + \alpha n\omega(f;\, n^{-1})},$$

where c_5 is an absolute constant.

Keeping in mind the choice of m and r, we obtain from (4.160) that

$$r(\alpha;\, f, f_m) \leq c_6\chi_n, \tag{4.164}$$

where c_6 is an absolute constant. Finally, (4.155) follows from (4.163) and (4.164), and the theorem is proved.

Proof of Theorem 4.18 . Let g be an arbitrary nondecreasing function, defined on the segment $[-1, 1]$. Then the 2π-periodic function $f(t) = g(\cos x)$ will be a bell-shaped one. Using the notation in the proof of Theorem 4.17, we shall estimate $|f_m(x_i) - U_{m,r}(f_m;\, x_i)|$ for $i = 0, 1, 2, \ldots, l + 1$.

It follows from the definition of f_m that

$$|f_m(x_i + t) - f_m(x_i)| = |f(\xi' + t) - f(\xi'')|, \tag{4.165}$$

where $|\xi' - x_i| \leq 10\pi/m$ and $|\xi'' - x_i| \leq 10\pi/m$. For $t \geq 5\pi/m$ we have

$$|f(\xi' + t) - f(\xi'')| = |g(\cos(\xi' + t) - g(\cos \xi'')| \tag{4.166}$$

$$\leq \omega(g;\, |\cos(\xi' + t) - \cos \xi''|) \leq c_7\, \omega(g;\, |t\sin x_i| + t^2),$$

where c_7 is an absolute constant. Then from (4.161), (4.165), and (4.166), we obtain

$$|U_{m,r}(f_m;\, x_i) - f_m(x_i)| \leq c_{m,r}\int_{-\pi}^{\pi} |f_m(x_i + t) - f_m(x_i)|\, \psi_{m,r}(t)\, dt$$

$$\leq c_{m,r}\left(\int_{-\pi}^{-5\pi/m} + \int_{5\pi/m}^{\pi}\right) |f_m(x_i + t) - f_m(x_i)|\, \psi_{m,r}(t)\, dt$$

$$\leq 2c_{m,r}\, c_7 \int_{5\pi/m}^{\pi} \omega(g;\, |t\sin x_i| + t^2)\, \psi_{m,r}(t)\, dt$$

$$\leq 4c_{m,r}\, c_7\, \omega(g;\, 5\pi m^{-1}|\sin x_i| + (5\pi/m)^2)(5\pi m^{-1}|\sin x_i| + (5\pi/m)^2)$$
$$\times \int_{5\pi/m}^{\pi} (t\,|\sin x_i| + t^2)\, \psi_{m,r}(t)\, dt$$

$$\le 5\pi c_7(2r-3)^{-1} e^{-2r+1}\,\omega(g;\, 5\pi m^{-1}|\sin x_i| + (5\pi/m)^2),$$

or finally,

$$|U_{m,r}(f_m;\, x_i) - f_m(x_i)| \le c_8 e^{-2r}\,\omega(g;\, m^{-1}|\sin x_i| + m^{-2}), \tag{4.167}$$

where c_8 is an absolute constant. Setting

$$\beta = \ln e^2(1+\alpha M m^2), \qquad M = \sup\{|f(x)| : x \in [-1, 1]\}, \qquad r = [\beta],$$

we obtain from (4.167) that

$$|U_{m,r}(f_m;\, x_i) - f_m(x_i)| \le c_9\,\omega(g;\, \Delta_m(t_i))/(1+\alpha M m^2), \tag{4.168}$$

where $\Delta_m(t_i) = m^{-1}\sqrt{1-t_i^2} + m^{-2} = m^{-1}\sqrt{1-\cos^2 x_i} + m^{-2}$.

Let $p_n(t) = U_{m,r}(f_m;\, \arccos t)$. The algebraic polynomial p_n is a nondecreasing monotone polynomial on the segment $[-1, 1]$, since $U_{m,r}(f_m)$ is a bell-shaped trigonometric polynomial of order no higher than mr by Lemma 4.28. The degree of this algebraic polynomial is no higher than $m\ln e^2(1+\alpha M m^2)$.

Keeping in mind that for $t_k = \cos x_k$, we have

$$|\cos(10k\pi/m) - \cos(10(k+1)\pi/m)| \le 10\pi m^{-1}\sqrt{1-t_k^2} \le 10\pi\Delta_m(t_k),$$

and using the construction of f_m, we obtain that

$$|g_m(t_i) - g(t_i)| \le c_{10}\min\{\alpha^{-1}\Delta_m(t_i),\, \omega(f;\, \Delta_m(t_i))\}, \tag{4.169}$$

where c_{10} is an absolute constant. From (4.168) and (4.169), we obtain

$$|g(t_i) - p_n(t_i)|_\alpha \le |g(t_i) - g_m(t_i)|_\alpha + |g_m(t_i) - p_n(t_i)|_\alpha$$

$$\le |g(t_i) - g_m(t_i)|_\alpha + |U_{m,r}(f_m;\, x_i) - f_m(x_i)|$$

$$\le c_{11}(\omega(g;\, \Delta_m(t_i))/(1+\alpha M m^2) + \min\{\alpha^{-1}\Delta_m(t_i),\, \omega(g;\, \Delta_m(t_i))\},$$

or

(4.170) $$|g(t_i) - p_n(t_i)|_\alpha \le c_{12}\, \omega(g; \Delta_m(t_i))/(1 + \alpha\omega(g; \Delta_m(t_i))/\Delta_m(t_i)),$$

where c_{11} and c_{12} are absolute constants.

Let $x \in [t_i, t_{i+1}]$. From (4.170) and the monotonicity of g and p_n, we have

$$|g(x) - p_n(x)|_\alpha \le \min\{\alpha^{-1}|t_i - t_{i+1}|, \omega(g; |t_i - t_{i+1}|)\}$$

$$+ \max\{|g(t_i) - p_n(t_i)|_\alpha, |g(t_{i+1}) - p_n(t_{i+1})|_\alpha\}$$

$$\le c_2\, \omega(g; \Delta_m(x))/(1 + \alpha\omega(g; \Delta_m(x))/\Delta_m(x)),$$

where c_2 is an absolute constant. Since the polynomial p_n is monotonic and its degree n satisfies $n \le m\ln e^2(1 + \alpha M m^2)$, the theorem is proved.

The problem of finding the best constants c_1 and c_2 in Theorem 4.18 and in Theorem 4.18 remains unsolved. Even estimates for these constants are unknown.

The problem of comonotone approximations is generalized in a natural way for partially monotone functions (D. Newman, E. Passov, and L. Raymon [1]). If the function f is defined on a certain segment Δ, and this segment is divided into a finite number of segments Δ_i, where $\Delta = \bigcup_{i=1}^m \Delta_i$, such that the function is monotonic on each of the segments Δ_i, then the problem arises of estimating the best approximation of f by algebraic polynomials with the same type of monotonicity exhibited by f on every segment Δ_i, $i = 1,2,3,\ldots, m$. G. Iliev [4] proved that for $m = 2$, this order is not worse than the order of approximation without the condition of comonotonicity. This holds both for uniform distance and for Hausdorff distance.

§ 4.2. Best approximation by rational functions

The set of rational functions $\varphi = p/q$ where $p, q \in H_n$, q not identically zero, will be denoted by R_n. A function $\varphi \in R_n$, which is a quotient of two algebraic polynomials of degree at most n, will be called a rational function of order at most n. The problem of approximation by rational functions is quite old, going back to the works of G. Zolotarev and P. Chebyshev. Newman's paper [1] on approximation of $|x|$ by rational functions with respect to uniform distance has aroused great interest in rational approximation. Many papers have been published considering classes of functions for which the best rational approximations are substantially better than the corresponding polynomial approximations. A substantial contribution to the establishment of final results in this field has been made by V. Popov [18,19] and his student P. Petrusev [1,2,3].

In connection with our considerations, it should be noted that Hausdorff approximations with rational functions are actually used in certain proof techniques when studying uniform approximations with rational functions, as for example, in Goncar's paper [1].

4.2.1. UNIVERSAL ESTIMATES FOR BOUNDED FUNCTIONS

Let $f \in A_\Delta^M$ be an arbitrary bounded function, defined on $\Delta = [a, b]$. Consistent with our notation, we denote the best Hausdorff approximation of f by rational functions by $E(R_n, \Delta, \alpha; f)$. Since $H_n \subset R_n$, in view of the results of §4.1.2.2., we have

$$E(R_n, \Delta, \alpha; f) \le E(H_n, \Delta, \alpha; f) \le O(n^{-1} \ln n).$$

Can we find the best estimate of $E(R_n, \Delta, \alpha; f)$? P. Petrusev has obtained a result that is, in a certain sense, final in this direction, which we will present in the sequel.

Definition 4.6. (V. Popov [16]). The **modulus of variation** $\chi(f; n)$ of a function $f \in A_\Delta^M$ is the number defined by

$$\chi(f; n) = \sup\left\{\sum_{i=1}^{n} |f(x_i) - f(x_{i-1})| \; : \; a \le x_0 \le x_1 \le \cdots \le x_n \le b\right\}. \tag{4.171}$$

Z. Canturia [1] used a similar characteristic, when studying the convergence of Fourier series.

Theorem 4.19. (P. Petrusev [3]). For every function f bounded on $\Delta = [a, b]$ and for every $\alpha \in (0, 1]$, the inequality

$$E(R_n, \Delta, \alpha; f) \le c(b - a)\, \alpha^{-1} n^{-1} \ln(e + \chi(f; n)) \tag{4.172}$$

holds, where c is an absolute constant.

Keeping in mind Lemma 2.8 and Lemma 2.9, it suffices to prove that

$$E(R_n, [0, 1], 1; f) = E(R_n; f) \le c n^{-1} \ln(e + \chi(f; n)), \tag{4.173}$$

where c is an absolute constant. The proof of (4.173) is based on some quite complicated constructions, following Popov's method [18]. At first, we shall obtain a refinement of Popov's theorem on uniform rational approximations.

Recall that the **total variation** of a function f on the segment [a, b] is determined by the equality

(4.174) $$V_a^b(f) = \lim_{n\to\infty} \chi(f; n),$$

and for every positive integer n, we have the inequality

(4.175) $$\chi(f; n) \le V_a^b(f).$$

Definition 4.7. Denote by $V_r^M = V_r^M[a, b]$ the set of functions f defined on [a, b] with $V_a^b D^r(f) \le M$, i.e., functions which have S-derivatives of rth order with total variation no higher than M.

Theorem 4.20. (P. Petrusev [3]). There exists a constant $D > 1$ such that for $r \ge 1$ and for every positive integer $n \ge 9r$, the inequality

$$\sup\{R_n([a, b]; f) : f \in V_r^M\} \le D^r (b - a)^r M n^{-r-1}$$

holds, where $R_n([a, b]; f)$ is the best uniform approximation of the function f by rational functions, i.e.,

$$R_n([a, b]; f) = \inf\{\sup\{|f(x) - \varphi(x)| : x \in [a, b]\} : \varphi \in R_n\}.$$

Keeping in mind Theorem 2.8, obviously

$$R_n([a, b]; f) = \lim_{\alpha\to+0} E(R_n, [a, b], \alpha; f).$$

To prove Theorem 4.20, we require a few preliminary results. To this end, we introduce some notation. In what follows, N is assumed nonnegative but not necessarily integral.

$$R_N = R_{[N]}, \qquad R_N([a, b], f) = R_{[N]}([a, b], f),$$

$$V_{r,0}^M[a, b] = V_{r,0}^M = \{f : f \in V_r^M,\ f^{(s)}(a) = 0 \text{ for } s = 0, 1, 2, \ldots, r\},$$

$$\phi_r(N) = \sup\{R_N([0, 1]; f) : f \in V_{r,0}^1[0, 1]\},$$

$$\phi_r(N,A) = \sup\{\inf\{\|f - q\|_{[0, 1]} : q \in R_N,\ \|q\|_{(-\infty, \infty)} \le A\} : f \in V_{r,0}^1\},$$

where $\|f\|_{[a,b]} = \sup \{|f(x)| : x \in [a, b]\}$.

Lemma 4.29. If n and r are arbitrary positive integers, $M > 0$ is prescribed, and $[a, b]$ is a finite closed interval, then

$$\sup \{R_{n+r}([a, b]; f) : f \in V_r^M[a, b]\} \le M(b - a)^r \phi_r(n).$$

Proof. If $f \in V_r^M[a, b]$, then the function

$$g(x) = M^{-1}(b - a)^{-r} \left(f(a + (b - a)x) - \sum_{\nu=0}^{r} \frac{1}{\nu!}(b - a)^\nu x^\nu f^\nu(a)\right) \tag{4.176}$$

belongs to $V_{r,0}^1[0, 1]$ and consequently, there exists a rational function $q \in R_n$ for which

$$\phi_r(n) \ge \|g - q\|_{[0,1]}$$

$$= M^{-1}(b - a)^{-r} \max_{a \le x \le b} \left|f(x) - \sum_{\nu=0}^{r} \frac{1}{\nu!}(x - a)^\nu f^\nu(a) - M(b - a)^r q\left(\frac{x - a}{b - a}\right)\right|$$

$$\ge M^{-1}(b - a)^{-r} R_{n+r}([a, b]; f).$$

Since the function f has been chosen arbitrarily from $V_r^M[a, b]$, the conclusion of the lemma follows from the last inequality.

Lemma 4.30. If r is a positive integer, $N \ge 0$, $M \ge 0$, $A \ge 0$, and $\varepsilon > 0$, and $[a,b]$ is a finite closed interval, then

$$\sup \{\inf \|f - q\|_{[a,b]} : q \in R_{N+2r}, \|q\|_{(-\infty,\infty)} \le AM(b - a)^r + 1/2\sqrt{\varepsilon}\} :$$
$$f \in V_r^M[a, b],\ \|f^{(s)}\|_{[a,b]} \le 1, \text{ for } s = 0, 1, 2, \ldots, r\}$$

$$\le M(b - a)^r \phi_r(N,A) + \varepsilon \left(\sum_{\nu=0}^{r} \frac{1}{\nu!}(b - a)^\nu\right).$$

Proof. If $f \in V_r^M[a, b]$ and $\|f^{(s)}\|_{[a, b]} \leq 1$, for $s = 0, 1, 2, \ldots, r$, then the function $g(x)$ as defined in (4.176) belongs to $V_{r,0}^1[0, 1]$ and, consequently, there exists a rational function $q \in R_N$ for which $\|q\|_{(-\infty, \infty)} \leq A$ and

$$\phi_r(N,A) \geq \|g - q\|_{[0, 1]}$$

$$= M^{-1}(b - a)^{-r} \max_{a \leq x \leq b} |f(x) - \sum_{\nu=0}^{r} \frac{1}{\nu!}(x - a)^{\nu} f^{\nu}(a) - M(b - a)^r q\left(\frac{x - a}{b - a}\right)|$$

Let us set

$$p(x) = \sum_{\nu=0}^{r} \frac{1}{\nu!}(x - a)^{\nu} f^{\nu}(a), \qquad u(x) = M(b - a)^r q\left(\frac{x - a}{b - a}\right),$$

$$v(x) = \frac{p(x)}{1 + \varepsilon p^2(x)} .$$

Then $v \in R_{2r}$, $\|v\|_{(-\infty,\infty)} \leq 1/2\sqrt{\varepsilon}$, $\|p\|_{[a, b]} \leq \sum_{\nu=0}^{r} \frac{1}{\nu!}(b - a)^{\nu}$, and $\|p - v\|_{[a, b]} \leq \varepsilon\|p\|^3_{[a, b]}$. As a result,

$$\|f - u - v\|_{[a, b]} \leq \|f - p - u\|_{[a, b]} + \|p - v\|_{[a, b]}$$

$$\leq M(b - a)^r \|g - q\|_{[0, 1]} + \|v - p\|_{[a, b]}$$

$$\leq M(b - a)^r \phi_r(N,A) + \varepsilon \left(\sum_{\nu=0}^{r} \frac{1}{\nu!}(b - a)^{\nu}\right),$$

and

$$\|u + v\|_{(-\infty, \infty)} \leq \|u\|_{(-\infty, \infty)} + \|v\|_{(-\infty, \infty)} \leq AM(b - a)^r + 1/2\sqrt{\varepsilon} .$$

From these inequalities, and the fact that $u + v \in R_{N+2r}$, the lemma follows.

Lemma 4.31. For every $\varphi \in C_{[0, 1]}$ such that $V_0^1 \varphi \leq 1$ and for each natural number r, there exists $x_1 \in [1/4, 3/4]$ such that

$$(x_i - x_{i-1})^r \, V_{x_{i-1}}^{x_i} \varphi \le 2^{-r-1}, \qquad i = 1, 2,$$

where $x_0 = 0$ and $x_2 = 1$.

Proof. Without loss of generality, we may assume that $V_0^1 \varphi = 1$. Under this condition, the numbers α and β are uniquely determined from the inequalities

$$(1 - \alpha) \, V_\alpha^1 \varphi = 2^{-r-1}, \qquad \beta^r \, V_0^\beta \varphi = 2^{-r-1}.$$

If we can show that the inequalities

$$\beta \ge 1/4, \qquad \alpha \le 3/4, \qquad \alpha \le \beta$$

all hold, then every number x_1 in $[1/4, 3/4] \cap [\alpha, \beta]$ satisfies the lemma condition. We shall first show that $\beta \ge 1/4$.

$$\beta = 2^{-1-1/r} (V_0^\beta \varphi)^{-1/r} \ge 2^{-1-1/r} \ge 1/4.$$

In a similar way, one can prove that $\alpha \le 3/4$. With $a = V_0^\beta \varphi$, we finally have

$$\alpha - \beta = 1 - 2^{-1-1/r} a^{-1/r} - 2^{-1-1/r} (1 - a)^{-1/r}$$

$$= 1 - 2^{-1-1/r} \left[\frac{1}{a^{1/r}} + \frac{1}{(1 - a)^{1/r}} \right] \ge 0.$$

This completes the proof of the lemma.

Lemma 4.32. (A. Goncar [2]). Let $\sigma(x) = 0$ for $x < 0$ and $\sigma(x) = 1$ for $x > 0$. For every even n and every $\varepsilon \in (0, 1/2)$ there exists a rational function $\sigma_n(\varepsilon; x) \in R_n$ for which

$$(4.177) \qquad |\sigma(x) - \sigma_n(\varepsilon; x)| \le 6e^{n/6 \ln \varepsilon} \qquad \text{for } x \in [-1, -\varepsilon] \cup [\varepsilon, 1],$$

$$(4.178) \qquad 0 \le \sigma_n(\varepsilon; x) \le 1 \qquad \text{for } x \in (-\infty, \infty).$$

Proof. Applying Newman's method [1], we shall consider the polynomial $p(x) = \prod_{i=1}^{n}(x - \varepsilon^{i/n})$. It is directly verified that for all $x \in [0, 1]$, we have

$$\prod_{i=1}^{n} \frac{|x - \varepsilon^{i/n}|}{|x + \varepsilon^{i/n}|} \leq \prod_{i=1}^{n} \frac{1 - \varepsilon^{i/n}}{1 + \varepsilon^{i/n}} .$$

It follows that

$$\left|\frac{p(x)}{p(-x)}\right| \leq \exp \sum_{i=1}^{n} \ln \left(\frac{1 - 2\varepsilon^{i/n}}{1 + \varepsilon^{i/n}}\right) < \exp \left(-\sum_{i=1}^{n} e^{i/n} \right)$$

$$= \exp\left(\frac{-\varepsilon^{1/n}(1 - \varepsilon)}{1 - \varepsilon^{1/n}}\right),$$

or

$$\left|\frac{p(x)}{p(-x)}\right| \leq \exp\left(\frac{n(1 - \varepsilon)\varepsilon^{1/n}}{\ln \varepsilon}\right), \qquad x \in [\varepsilon, 1]. \tag{4.179}$$

Let us consider the rational function $s(x) = p(x)/p(-x)$. For $n \geq - \ln \varepsilon$ we have $(1 - \varepsilon)e^{1/n} > 1/6$, and from (4.179) we obtain $|s(x)| \leq e^{n/6\ln \varepsilon}$ for $x \in [\varepsilon, 1]$. For $n < - \ln \varepsilon$, we have $n/6\ln \varepsilon > 1/6$ and

$$|s(x)| \leq 1 \leq \exp (n/6\ln \varepsilon - 1/6), \qquad x \in [\varepsilon, 1]. \tag{4.180}$$

Hence, (4.80) holds for all positive integers n.

According to the definition of s, we have $s(-x) = s^{-1}(x)$, so that (4.180) yields

$$|s(x)| \geq \exp (-n/6\ln \varepsilon - 1/6), \qquad x \in [-1, -\varepsilon]. \tag{4.181}$$

We now write $\bar{\sigma}_n(x) = 1/(1 + s(x))$. This is a rational functional satisfying $|\bar{\sigma}_n(x)| \leq 1/(|s(x)| - 1)$ for $x \in [-1, -\varepsilon]$, and $|1 - \bar{\sigma}_n(x)| \leq 1/(|s(-x)| - 1)$ for $x \in [\varepsilon, 1]$. Keeping in mind (4.181), from the last inequalities we obtain $|\sigma(x) - \bar{\sigma}_n(x)| \leq 3\exp (n/6\ln \varepsilon)$ for $x \in [-1, -\varepsilon] \cup [\varepsilon, 1]$, and $0 < \bar{\sigma}_n(x) < 1$ for $x \in (-\varepsilon, \varepsilon)$. Since n is even, we have

$s(n) \geq 0$ for $|x| \geq 1$, and consequently, $0 \leq \bar{\sigma}_n(x) \leq 1$ for $|x| \geq 1$. Then the rational function

$$\sigma_n(x) = (1 + 6e^{n/6\ln \varepsilon})(\bar{\sigma}_n(x) + 3e^{n/6\ln \varepsilon})$$

satisfies both (4.177) and (4.178). The lemma is proved.

Corollary 4.3. For every positive integer r and for $N \geq 2$, there exists a rational function $\sigma_n \in R_n$ with $n \leq 200r^2 \ln^2 N$ for which

$$|\sigma(x) - \sigma_n(x)| \leq N^{-3r-3} \qquad \text{for } x \in [-1, -N^{-r-2}] \cup [N^{-r-2}, 1],$$

and $0 \leq \sigma_n(x) \leq 1$ for $x \in (-\infty, \infty)$.

Proof. It is sufficient to take $\varepsilon = N^{-r-2} < 1/2$, and $n = 2[10(r+2)^2 \ln^2 N < 200r^2 \ln^2 N$. Then

$$6e^{n/6\ln \varepsilon} \leq 6\exp\left(\frac{-10(r+2)^2 \ln^2 N}{3(r+2)\ln N}\right)$$

$$6 \leq \exp(-3(r+2)\ln N) \leq N^{-3r-3}.$$

Lemma 4.33. There exist absolute constants $\beta, \gamma > 0$ such that if $r \geq 1$, $N > \beta r^3$, $k = (N - 200r^2 \ln^2 N)/2 - 2r$ and

(4.182) $$\phi_r(k, k^{(r+2)/2}) \leq \eta(k) k^{-r-1},$$

where $\eta(k) \geq 1$, then the inequality

(4.183) $$\phi_r(N, N^{(r+2)/2}) \leq \eta(k) N^{-r-1} (1 + \gamma r^2 N^{-1} \ln^2 N)^{r+1}$$

holds.

Proof. Let r be an arbitrary positive integer, select constants $\beta, \gamma > 2$ be such that whenever $N \geq \beta r^3 > 2$, we have the following inequalities

(4.184) $k = (N - 200r^2 \ln^2 N)/2 - 2r \geq 1,$

(4.185) $(2k)^{-r-1} + 30N^{-r-2} \leq N^{-r-1}(1 + \gamma r^2 N^{-1} \ln^2 N)^{r+1}.$

Let $f \in V^1_{r,0}[0, 1]$. Without loss of generality, we may assume that $f^{(r)}$ is continuous on the interval $[0, 1]$, since the set of functions f for which $f^{(r)} \in C_{[0, 1]}$ and $f^{(s)} = 0$ for $s = 0, 1, 2, \ldots, r$ is dense in $V^1_{r,0}[0, 1]$. By Lemma 4.31, there exists $x_1 \in [1/4, 3/4]$ such that

(4.186) $(x_i - x_{i-1})^r V_{x_{i-1}}^{x_i} \varphi \leq 2^{-r-1},$

for $i = 1, 2,$ where $x_0 = 0$ and $x_2 = 1$.

We now write $\Delta_1 = [0, x_1 + N^{-r-2}]$ and $\Delta_2 = [x_1 - N^{-r-2}, 1]$. Since $N \geq 2$, both $\Delta_1 \subset [0, 1]$ and $\Delta_2 \subset [0, 1]$. Let φ_1, φ_2 be linear transformations that map Δ_1, Δ_2 into $[0, x_1]$ and $[x_1, 1]$, respectively. Then we have

(4.187) $|\varphi_i(x) - x| \leq N^{-r-2}$ for $x \in \Delta_i$, $i = 1, 2.$

From Lemma 4.30, (4.183) and (4.186), there must exist two rational functions $q_1, q_2 \in R_{k+2r}$ such that for $i = 1, 2$

(4.188) $\|f - q_i\|_{[x_{i-1}, x_i]} \leq \eta(k)(2k)^{-r-1} + 27N^{-r-2},$

$\|q_i\|_{(-\infty, \infty)} \leq N^{(r+2)/2},$

Keeping in mind that $\left(\sum_{\nu=0}^{r} 1/\nu!\right)^3 \leq e^3 < 27$ and that according to (4.184), $k < N/2$, it is sufficient to set $\varepsilon = N^{-r-2}$, $A = k^{(r+2)/2}$, and $M = 2^{-r-1}$. It is not difficult to see that for $f \in V^1_{r,0}[0, 1]$, we have $|f(x) - f(\xi)| \leq |x - \xi|$ for $x, \xi \in [0, 1]$. From (4.187) and (4.188), we now obtain

(4.189) $\max_{x \in \Delta_i} |f(x) - q_i(\varphi_i(x))|$

$$\leq \max_{x \in \Delta_i} |f(x) - f(\varphi_i(x))| + \max_{x \in \Delta_i} |f(\varphi_i(x)) - q_i(\varphi_i(x))|$$

$$\le \max_{x\in\Delta_i} |x - \varphi_i(x)| + \|f - q_i\|_{[x_{i-1},x_i]}$$

$$\le \eta(k)(2k)^{-r-1} + 28N^{-r-2},$$

where $\|q_i\|_{(-\infty,\infty)} \le N^{(r+2)/2}$, for $i = 1, 2$.

Now we shall proceed to the "sewing" of the two rational functions q_1 and q_2, using the function σ_n of Lemma 4.32, where $n \le 200r^2 \ln^2 N$. To this end, let us consider the rational function

$$u(x) = q_1(\varphi_1(x)) + \sigma_n(x - x_1)(q_2(\varphi_2(x)) - q_1(\varphi_1(x)))$$

of order no higher than $2(k + 2r) - 200r^2 \ln^2 N = N$, according to (4.184). We now estimate $|f(x) - u(x)|$ for $x \in [0, 1]$ by applying (4.189) and Corollary 4.3. We consider three cases for x:

(a) If $x \in [0, x_1 - N^{-r-2}]$, then

$$|f(x) - u(x)| \le |f(x) - q_1(\varphi_1(x))| + \sigma_n(x - x_1)(|q_1(\varphi_1(x))| + |q_2(\varphi_2(x))|)$$

$$\le \eta(k)(2k)^{-r-1} + 28N^{-r-2} + 2N^{(r+2)/2}\,N^{-3r-3}$$

$$\le \eta(k)(2k)^{-r-1} + 30N^{-r-2}.$$

(b) If $x \in [x_1 - N^{-r-2}, x_1 + N^{-r-2}]$, then

$$|f(x) - u(x)| \le (1 - \sigma_n(x - x_1))|f(x) - q_1(\varphi_1(x))| + \sigma_n(x - x_1)|f(x) - q_2(\varphi_2(x))|$$

$$\le \eta(k)(2k)^{-r-1} + 28N^{-r-2}.$$

(c) If $x \in [x_1 + N^{-r-2}, 1]$, then

$$|f(x) - u(x)| \le |f(x) - q_2(\varphi_2(x))| + (1 - \sigma_n(x - x_1))(|q_1(\varphi_1(x))| + |q_2(\varphi_2(x))|)$$

$$\le \eta(k)(2k)^{-r-1} + 28N^{-r-2} + 2N^{(r+2)/2}\,N^{-3r-3}$$

$$\le \eta(k)(2k)^{-r-1} + 30N^{-r-2}.$$

Consequently, according to (4.185) and since $\eta(k) \ge 1$, we have

$$\|f - u\|_{[0,1]} \le \eta(k)(2k)^{-r-1} + 30N^{-r-2} \le \eta(k)N^{-r-1}(1 + \gamma r^2 N^{-1}\ln^2 N)^{r+1}.$$

On the other hand, for all x,

$$|u(x)| \le (1 - \sigma_n(x - x_1))\,|q_1(\varphi_1(x))| + \sigma_n(x - x_1))|q_2(\varphi_2(x))| \le N^{(r+2)/2}.$$

Hence, $\|u\|_{(-\infty,\infty)} \le N^{(r+2)/2}$. This completes the proof of the lemma.

Proof of Theorem 4.20. According to Lemma 4.29 and the obvious inequality $\phi_r(N) \le \phi_r(N, N^{(r+2)/2})$ valid for $N > 0$ and r any positive integer, it follows that to prove Theorem 4.20, it suffices to show that for every positive integer r and $N \ge 8r$, the following inequality is valid for some absolute constant D:

$$\phi_r(N, N^{(r+2)/2}) \le D^r N^{-r-1}.$$

Indeed, we shall have

$$\sup\{h_n([a,b]; f) : f \in V_r^M[a,b]\} \le M(b-a)^r \phi_r(n-r)$$

$$\le M(b-a)^r \phi_r(n-r, (n-r)^{(r+2)/2}) \le M(b-a)^r D^r (n-r)^{-r-1}$$

$$\le (2D)^r (b-a)^r M n^{-r-1},$$

for $n \ge 9r$ and $r \ge 1$.

Let us set $N_0 = \lambda^r$, where $\lambda \ge e$ is an absolute constant, chosen so that the inequalities

(4.190) $\qquad N_0 > \beta r^3, \qquad N_0 > \gamma r^4,$

(4.191) $\qquad (N - 200r^2\ln^2 N)/2 - 2r > 8r \qquad$ for $N \ge N_0$

hold, where β, γ are the constants in Lemma 4.33.

First, we shall consider the case $8r \le N \le N_0$. Let $f \in V_{r,0}^1[0,1]$. Then both $\|f^{(r)}\|_{[0,1]} \le 1$ and $\|f\|_{[0,1]} \le 1$. By Jackson's theorem, there exists an algebraic polynomial p of degree no higher than $[N/2]$ such that $\|p\|_{[0,1]} \le 2$ and

$$\|f - p\|_{[0,1]} \le D_1^r \|f^{(r)}\|_{[0,1]} ([N/2] - r)^r \le (4D_1)^r N^{-r}$$

$$\le (4D_1\lambda)^r N^{-r-1} = D_2^r N^{-r-1},$$

where D_1 and D_2 are absolute constants.

We introduce the rational function $q = p/(1 + N^{-r-1}p^2)$ of order at most N for which $\|q\|_{(-\infty,\infty)} \le N^{(r+1)/2} < N^{(r+2)/2}$, and

$$\|f - q\|_{[0,1]} \le \|f - p\|_{[0,1]} + N^{-r-1} \|f\|_{[0,1]} \|p\|^2_{[0,1]}$$

$$\le D_2^r N^{-r-1} + 4N^{-r-1} \le D_3^r N^{-r-1},$$

where D_3 is an absolute constant. Hence,

$$\phi_r(N, N^{(r+2)/2}) \le D_3^r N^{-r-1}, \qquad \text{for} \quad 8r \le N \le N_0 = \lambda^r. \tag{4.192}$$

Let $N \ge N_0$. We consider the functions defined recursively by

$$y_0(x) = x, \;\; y_1(x) = (x - 200r^2 \ln^2 x)/2 - 2r,$$

$$y_m(x) = y_1(y_{m-1}(x)) \qquad \text{for } m = 1, 2, 3, \dots .$$

Clearly, $y_1(x) < x/2$ and as a result, there exists a positive integer s such that $8r < y_{s+1}(N) < N_0 \le y_s(N)$. Then from (4.192) we obtain

$$\phi_r(y_{s+1}(N), (y_{s+1}(N))^{(r+2)/2}) \le D_3^r (y_{s+1}(N))^{-r-1}. \tag{4.193}$$

Now denote $y_m(N)$ by ξ_m. Applying Lemma 4.33 consecutively s times, beginning with inequality (4.193), we get

$$\phi_r(\xi_{s+1}, \xi_{s+1}^{(r+2)/2}) \le D_3^r \xi_{s+1}^{-r-1},$$

$$\phi_r(\xi_s, \xi_s^{(r+2)/2}) \le D_3^r \xi_s^{-r-1} (1 + \gamma r^2 \xi_s^{-1} \ln^2 \xi_s)^{r+1},$$

$$\cdot \quad \cdot \quad \cdot \quad \cdot \quad \cdot \quad \cdot \quad \cdot$$

$$\phi_r(N, N^{(r+2)/2}) \le D_3^r N^{-r-1} \prod_{m=0}^{s} (1 + \gamma r^2 \xi_m^{-1} \ln^2 \xi_m)^{r+1}. \tag{4.194}$$

Keeping in mind that the function $x^{-1}\ln^2 x$ monotonically decreases on $[N_0, \infty)$ and that $N_0 > \gamma r^4$, $\lambda > \varepsilon$, and $y_1(x) < x/2$ for $x > 0$, we obtain

$$\prod_{m=0}^{s} (1 + \gamma r^2 \xi_m^{-1} \ln^2 \xi_m) \le \prod_{m=0}^{s} (1 + \gamma r^2 2^{-s+m} \xi_s^{-1} \ln^2 (2^{s-m}\xi_s))$$

$$\le \prod_{i=0}^{\infty} (1 + \gamma r^2 2^{-i} N_0^{-i} \ln^2 (2^i N_0)) \le \prod_{i=0}^{\infty} (1 + \gamma r^2 2^{-i} (\gamma 2^4)^{-1} \ln^2 (2^i \gamma r))$$

$$\le \prod_{i=0}^{\infty} (1 + 2^{-i} r^{-2} (i + r \ln \lambda)^2) \le \prod_{i=0}^{\infty} (1 + 2^{-i} (i + 1)^2 \ln^2 \lambda)$$

$$= D_4 < \infty.$$

Consequently, (4.194) can be written in the form

$$\phi_r(N, N^{(r+2)/2}) \le D_5^r N^{-r-1} ,$$

where $D_5 = D_3 D_4^2$ is an absolute constant. Theorem 4.20 is proved.

We are almost ready to prove Theorem 4.19. First, we need to establish

Lemma 4.34. For every $f \in A_\Delta$ with $\Delta = [0, 1]$, and for every positive integer m, there exists a step function s with jumps at the points $x_i = i/2m$, $i = 1, 2, 3, \ldots, 2m - 1$ such that

$$V_0^1 s \le \chi(f; 2m) \qquad \text{and} \qquad r([0, 1], 1; f, s) \le m^{-1} .$$

Proof. For each i let m_i and M_i be the following constants:

$$m_i = \inf \{f(x) : x \in [x_{i-1}, x_{i+1}]\}, \qquad M_i = \sup \{f(x) : x \in [x_{i-1}, x_{i+1}]\}.$$

There exist ξ_i' and $\xi_i'' \in [x_{i-1}, x_{i+1}]$ such that $I(f; \xi_i') = m_i$ and $S(f; \xi_i'') = M_i$. If $\xi_{2i+1}' \le \xi_{2i+1}''$, we define

$$s(x) = \begin{cases} m_{2i+1} & \text{if } x_{2i} \le x \le x_{2i+1} \\ M_{2i+1} & \text{if } x_{2i+1} \le x \le x_{2i+2} \end{cases},$$

and if $\xi'_{2i+1} \ge \xi''_{2i+1}$, we set

$$s(x) = \begin{cases} M_{2i+1} & \text{if } x_{2i} \le x \le x_{2i+1} \\ m_{2i+1} & \text{if } x_{2i+1} \le x \le x_{2i+2} \end{cases}.$$

It is straight forward to check that $V_0^1 s \le \chi(f;\ 2m)$ and $r([0,\ 1],\ 1;\ f,\ s) \le m^{-1}$, completing the proof.

Proof of Theorem 4.19 . Let n be an arbitrary positive integer and let $f \in A_\Delta$ with $\Delta = [0,\ 1]$. Without loss of generality, we may assume that $\sup\{|f(x) - f(\xi)| : x, \xi \in \Delta\} = 2\|f\|_\Delta$. First we consider the case when $\|f\|_\Delta \ge e^{n/B}$ where $B = 32eD$, and D is the constant in Theorem 4.20. In view of the obvious inequality $2\|f\|_\Delta \le \chi(f;\ n)$, we get

$$\text{(4.195)} \qquad E(R_n, [0,\ 1], 1;\ f) = E(R_n;\ f) \le E(R_1;\ f) \le 1$$

$$= Bn^{-1} \ln e^{n/B} \le 32eDn^{-1} \ln (e + \chi(f;\ n)).$$

We can take $n > B^2$, since for $n \le B^2$ the inequalities

$$\text{(4.196)} \qquad E(R_n, f) = E(R_1;\ f) \le 1 \le B^2 n^{-1} \le (32eD)^2 n^{-1} \ln (e + \chi(f;\ n))$$

are obviously satisfied.

For $\chi(f;\ n) \le e$, according to Lemma 4.34, there exists a step function s with jumps at the points $x_i = i/2n$, $i = 1, 2, 3, \ldots, 2n-1$, for which $V_0^1 s \le \chi(f;\ 2n)$ and $r([0,\ 1],\ 1;\ f,\ s) \le n^{-1}$. Let us set $s(x) = s(1)$ for $x > 1$, and consider the Steklov function

$$s_h(x) = \frac{1}{h}\int_0^h s(x + t)\, dt$$

with $h = 1/2n$. Clearly,

$$V_0^1 s'_h = \frac{1}{h} V_0^1 (s(x + h) - s(x)) \le \frac{4}{h}\chi(f;\ n) \le 8en,$$

and $r(s, s_h) \le h = 1/2n$. Since $n > B^2 > 9$, we have by Theorem 4.20 that

$$R_n([0, 1]; s_h) \le Dn^{-2} V_0^1 s'_h \le 8eDn^{-1},$$

and consequently

$$E(R_n; f) \le r([0, 1]; f, s) + r([0, 1]; s, s_h) + R_n([0, 1]; s_h) \tag{4.197}$$

$$\le 1/n + 1/2n + 8eD/n \le \frac{9eD \ln (e + \chi(f; n))}{n}.$$

It remains to consider the case when $e \le \chi(f; n)$ and $\|f\|_\Delta \le e^{n/B}$, $B = 32eD$, and $n > B^2$. Set $\nu = [\ln \chi(f; n)]$, $h = 2eDn^{-1}$, and $m = [1/2\nu h]$. Here we have $2n\|f\|_\Delta \ge \chi(f; n)$, and since $x^{-1/2} \ln x < 1$ for $x > e^2$, we obtain

$$2\nu h = 2 [\ln \chi(f; n)] 2eDn^{-1} \le 4eDn^{-1} \ln (2n\|f\|_\Delta) \tag{4.198}$$

$$\le 4eDn^{-1} \ln (2ne^{n/32eD}) \le 1/2.$$

Hence,

$$m \ge 2, \qquad \frac{1}{2\nu h} \ge m \ge \frac{1}{2\nu h} - 1 \ge \frac{1}{4\nu h}, \qquad \nu h \le \frac{1}{2m}. \tag{4.199}$$

By Lemma 4.34 there exists a step function s with jumps at the points $x_i = i/2m$, $i = 1, 2, 3, \ldots, 2m - 1$ such that

$$V_0^1 s \le \chi(f; 2m) \qquad \text{and} \qquad r([0, 1]; f, s) \le m^{-1}. \tag{4.200}$$

Let $s(x) = s(1)$ for $x > 1$ and consider the following Steklov function:

$$s_h(x) = h^{-\nu} \int_0^h \int_0^h \cdots \int_0^h s(x + t_1 + t_2 + \cdots + t_\nu)\, dt_1\, dt_2 \cdots dt_\nu .$$

By (4.199) $\nu h \le 1/2m$; so,

$$r(s, s_h) \le \nu h. \tag{4.201}$$

On the other hand,

$$s_h^{(\nu)} = h^{-\nu} \sum_{i=0}^{\nu} (-1)^i \binom{\nu}{i} s(x + ih)$$

and consequently $V_0^1 s_h^{(\nu)} \le 2^\nu h^{-\nu} \chi(f; 2m)$. From (4.198) we have $\nu \le 1/4h = n/8eD < n/9$, so that $n \ge 9\nu$. By Theorem 4.20, we have for $\nu = [\ln \chi(f; n)]$:

(4.202) $\qquad R_n(s_h) \le D^\nu 2^\nu h^{-\nu} n^{-\nu-1} \chi(f; 2m).$

From (4.200) - (4.202), we have

$$E(R_n; f) \le r(f, s) + r(s, s_h) + R_n(s_h) \le m^{-1} + \nu h + D^\nu 2^\nu h^{-\nu} n^{-\nu-1} \chi(f; 2m).$$

From this inequality, keeping in mind (4.199), the definitions of ν, h, and m, and that $\chi(f; 2n)$ is a nondecreasing function of n, we obtain

(4.203) $\qquad E(R_n; f) \le 4\nu h + \nu h + (2D)^\nu h^{-\nu} n^{-\nu-1} \chi(f; 2m)$

$$\le 10eDn^{-1} \ln \chi(f; 2m) + (2D)^\nu (hn)^{-\nu} n^{-1} \chi(f; 2m)$$

$$\le 10eDn^{-1} \ln \chi(f; n) + n^{-1} e^{-\nu} \chi(f; n)$$

$$\le 10eDn^{-1} \ln \chi(f; n) + en^{-1} \le 11eDn^{-1} \ln (e + \chi(f; n)).$$

Finally, from (4.195) - (4.197) and (4.203), it follows that for $c = 32eD^2$, we have

$$E(R_n, [0, 1], 1; f) \le cn^{-1} \ln (e + \chi(f; n))$$

for all positive integers n. Thus (4.173) is established and the proof of Theorem 4.19 is completed.

Corollary 4.4. (V. Popov [18]). If f is a function of bounded variation on [a, b], then

$$E(R_n, [a, b], \alpha; f) = O(n^{-1}).$$

More precisely, there exists an absolute constant c for which

$$E(R_n, [a, b], \alpha; f) \le c\alpha^{-1}(b - a)n^{-1} \ln (e + V_a^b f)$$

It should be noted that V. Popov [10] was the first to obtain for the best Hausdorff approximation of functions with bounded variation by rational functions an estimate which is better than the estimate $O(n^{-1}\ln n)$, replacing $\ln n$ by $\ln (\ln n)$.

4.2.2. UNIMPROVABILITY OF THE UNIVERSAL ESTIMATE

We shall show that the estimate in Theorem 4.19 is an unimprovable one in the class of bounded functions with unbounded variation. More precisely, the following result is valid.

Theorem 4.21. (P. Petrushev [3]). For every $\chi(n) = \chi(g; n)$, where g is a bounded function on $[0, 1]$ with unbounded variation, i.e., $\chi(n) \to \infty$ as $n \to \infty$, there exists a bounded function f on $[0, 1]$, a positive constant c, and a sequence of positive integers $n_1 < n_2 < n_3 \le \cdots$ such that $\chi(f; n_i) \le \chi(n_i)$, $i = 1, 2, 3, \ldots$, and

$$E(R_{n_i}, [0, 1], 1; f) = E(R_{n_i}; f) > cn_i^{-1} \chi(n_i), \qquad i = 1, 2, 3, \ldots$$

The proof of Theorem 4.21 is based on the use of points of maximum deviation. It will be preceeded by some auxiliary lemmas.

Lemma 4.35. For every positive integer k there exists a rational function s_k of order k such that

$$|\operatorname{sgn} x_i - s_k(x_i)| > e^{-4k}, \quad i = -k, -k + 1, \ldots, -1, 0, 1, \ldots, k,$$

where $e^{-1} = x_0 < x_1 < x_2 < \cdots < x_k = 1$ and $x_{-i} = -x_i$. Moreover, the function $\operatorname{sgn} x - s_k(x)$ changes its sign on each of the closed intervals $[-x_i, -x_{i-1}]$, $[x_{i-1}, x_i]$, $i = 1, 2, \ldots, k$, and s_k increases monotonically on the closed interval $[-e^{-1}, e^{-1}]$.

Proof. Once again, we use Newman's method [1], considering the rational function $s_k(x) = (x + \xi_1)(x + \xi_2) \cdots (x + \xi_k)$, where $\xi_i = e^{-(2i-1)/2k}$. Since s_k is an odd function, it suffices to establish its properties only for nonegative x. We now set $x_i = e^{-(k-i)/k}$, $i = 0, 1, 2, \ldots, k$. Clearly, this choice of x_i satisfies the condition $e^{-1} = x_0 < x_1 < x_2 < \cdots < x_k = 1$. Also, for $x > 0$,

$$|\operatorname{sgn} x - s_k(x)| = \left|\frac{2p_k(-x)}{p_k(x) + p_k(-x)}\right| = \frac{2}{1 + p_k(x)/|p_k(-x)|}$$

$$\geq \frac{|p_k(-x)|}{p_k(x)} = \prod_{i=1}^{k} \frac{|x - \xi_i|}{x + \xi_i} .$$

Using an identity of Oiler,

$$(*) \qquad |\operatorname{sgn} x_j - s_k(x_j)| \geq \prod_{i=1}^{k} \frac{|e^{-(k-j)/k} - e^{-(2i-1)/2k}|}{e^{-(k-j)/k} + e^{-(2i-1)/2k}}$$

$$\geq \prod_{i=1}^{k} \frac{(1 - e^{-(2i-1)/2k})^2}{(1 + e^{-(2i-1)/2k})^2} \geq (1 - e^{-1/2k})^3 .$$

From $(*)$ and the well-known inequality $1 - x \geq \exp(1/\ln x)$ for $x \in (0, 1)$, we obtain

$$|\operatorname{sgn} x_j - s_k(x_j)| \geq e^{-4k} .$$

The proof is completed, using $p_k(-x) = 0$ for $x = \xi_i$, $i = 1, 2, \ldots, k$, and the fact that s_k is monotonically increasing on $[0, e^{-1}]$.

Lemma 4.36. For every positive integer m, there exists an algebraic polynomial p_l of degree $l = 100m$ and points a_ν, $\nu = 0, 1, 2, \ldots, l - 2m$ where $0 = a_0 < a_1 < a_2 < \cdots < a_{l-2m} = 1$ with the following properties:

(a) $p_l(a_\nu) = (-1)^\nu$, $\nu = 0, 1, 2, \ldots, l - 2m$;

(b) p_l is monotonic on every closed interval $[a_{\nu-1}, a_\nu]$, $\nu = 1, 2, 3, \ldots, l - 2m$;

(c) $\|p_l\|_{[0, 1]} = 1$ and $\|p'_l\|_{[0, 1]} \leq e^{13m}$;

(d) the real roots ξ_ν of p_l on $[0, 1]$ are arranged as follows:

$$0 = a_0 < \xi_1 < a_1 < \xi_2 < \cdots < \xi_{l-2m} < a_{l-2m} = 1.$$

Proof. Consider the Chebyshev polynomial $T_l(x) = \cos(l \arccos x)$. The derivative polynomial T_l' has roots $\eta_\nu = \cos(\nu\pi/l)$, $\nu = 1, 2, 3, \ldots, l-1$. Our polynomial p_l will be

$$p_l(x) = (-1)^m T_l(2x\eta_m - \eta_m).$$

We also set $a_\nu = \eta_{l-m-\nu}/2\eta_m + 1/2$, for $\nu = 0, 1, 2, \ldots, l-2m$. Since $l = 100m$, we have $p_l(a_\nu) = (-1)^m T_l(\eta_{l-m-\nu}) = (-1)^{l-\nu} = (-1)^\nu$. Since $a_0 = \eta_{l-m}/2\eta_m + 1/2 = 0$ and $a_{l-2m} = \eta_m/2\eta_m + 1/2 = 1$, condition (a) holds. Conditions (b) and (d) follow directly from properties of the Chebyshev polynomial. To establish (c), we must consider

$$p_l'(x) = (-1)^m 2\eta_m T_l'(2x\eta_m - \eta_m).$$

Using the Bernstein inequality we obtain

$$\|p_l'\|_{[0,1]} \le 2\eta_m \|T_l'\|_{[-\eta_m, \eta_m]} \le 2l\eta_m (1 - \eta_m^2)^{-1/2} = 200m \cot(\pi/100).$$

From the inequality $200 \cot(\pi/100) < 365,000 < e^{13}$, the lemma is proved.

In what follows, along with modulus of variation $\chi(f; n)$ of a given bounded function, which we have been using all along, we shall need a different modulus of variation (Z. Canturia [1]), which we now introduce. Let $f \in A_\Delta$, where $\Delta = [a, b]$. Then

$$\nu(f; n) = \sup\left\{\sum_{i=1}^n |f(x_{2i-2}) - f(x_{2i-1})| : a \le x_0 \le x_1 \le \cdots \le x_{2n-1} \le b\right\}.$$

The two characteristics satisfy the easily verfied relation

$$\chi(f; n) \le \nu(f; n) \le 2\chi(f; n), \quad n = 1, 2, 3, \ldots . \tag{4.204}$$

It is seen that $\nu(f; n)$ shares these properties with $\chi(f; n)$: $0 \le \nu(f; n) \le \nu(f; n+1) \le V_a^b f$, $\nu(f; kn) \le k\nu(f; n)$, and $\nu(f; n) \le 2n\|f\|_{[a,b]}$. The basic difference between $\nu(f; n)$ and $\chi(f; n)$ is that $\nu(f; n)$ is a concave function (Z. Canturia [1]), whereas $\chi(f; n)$ need not be.

Proof of Theorem 4.21 . Let us write $\chi(n)$ for $\chi(g; n)$, $n = 1, 2, 3, \ldots$, where g is an arbitrary bounded function on $[0, 1]$ with unbounded variation, i.e., $\chi(n) \to 0$ as $n \to \infty$. The function f, whose existence is asserted in Theorem 4.21, will be constructed as a sum of a series:

$$f(x) = \sum_{i=1}^{\infty} f_{n_i}(x) , \tag{4.205}$$

where $f_n = n^{-1} \chi^{1/2}(n) \operatorname{sgn}(p_l(x))$, $l = 100m$, $m = [n/k]$, $k = [2^{-4} \ln \chi(n)]$, the polynomial p_l is as described in Lemma 4.36, and the indices $n_1 < n_2 < n_3 < \cdots$ are chosen to satisfy the following conditions:

(1) $\ln \chi(n_1) \geq e^{20}$;

(2) $\dfrac{n_i}{[2^{-4} \ln n_i]} \geq 1, \quad i = 1, 2, 3, \ldots;$

(3) $E(R_{n_i}; \sum_{j=1}^{i-1} f_{n_j}) < \frac{1}{n_i}, \quad i = 2, 3, 4, \ldots;$

(4) $200 \sum_{j=1}^{i-1} \left[\dfrac{n_j}{[2^{-4} \ln \chi(n_j)]}\right] \leq \left[\dfrac{n_i}{[2^{-4} \ln \chi(n_i)]}\right]$ for $i = 2, 3, \ldots;$

(5) $\sum_{j=i+1}^{\infty} \dfrac{\chi^{1/2}(n_j)}{n_j} < \dfrac{1}{n_i}, \quad i = 1, 2, 3, \ldots;$

(6) $194 \left[\dfrac{n_i}{[2^{-4} \ln \chi(n_i)]}\right] [2^{-4} \ln \chi(n_i)] \geq 102 n_i + 1, \quad i = 1, 2, 3, \ldots;$

(7) $400 \sum_{j=1}^{i} \left[\dfrac{n_j}{[2^{-4} \ln \chi(n_j)]}\right] \dfrac{\chi^{1/2}(n_j)}{n_j} + 4 n_i \sum_{j=i-1}^{\infty} \dfrac{\chi^{1/2}(n_j)}{n_j} \leq \chi(n_i),$
$i = 1, 2, 3, \ldots;$

(8) $4 \sum_{j=1}^{\infty} \dfrac{\chi^{1/2}(n_j)}{n_j} \leq \chi(1);$

(9) $98 \sum_{j=1}^{i} [n_j [2^{-4} \ln \chi(n_j)]] \leq n_i, \quad i = 1, 2, 3, \ldots;$

$$(10)\quad \chi(n_i) + 4(n_{i+1} - n_i) \sum_{j=i+1}^{\infty} \frac{\chi^{1/2}(n_j)}{n_j} \le \chi(n_{i+1}), \quad i = 0, 1, 2, \ldots, \; n_0 = 1.$$

The existence of the sequence of indices $\langle n_j \rangle_1^\infty$ satisfying conditions (1) - (10) is guaranteed because $\chi(n) \to 0$ as $n \to \infty$, and $\chi(n) = \chi(g; n) \le 2n\|g\|_{[0, 1]}$.

First, we shall prove $\chi(f; n) \le \chi(n)$ for $n = 1, 2, 3, \ldots$. Let us write $m_j = [n_j[2^{-4} \ln \chi(n_j)]]$. It follows from (4.205) that

$$\chi(f; n_i) \le \sum_{j=1}^{i} V_0^1 f_{n_j} + 2n_i \| \sum_{j=i+1}^{\infty} f_{n_j} \|_{[0, 1]}$$

$$\le 196 \sum_{j=1}^{i} m_j \frac{\chi^{1/2}(n_j)}{n_j} + 2n_i \sum_{j=i+1}^{\infty} \frac{\chi^{1/2}(n_j)}{n_j},$$

and according to (7) we obtain

$$(4.206)\qquad \chi(f; n_i) \le \frac{\chi(n_i)}{2}, \quad i = 1, 2, 3, \ldots .$$

With $n_0 = 1$ and $\Delta = [0, 1]$, (8) yields

$$(4.207)\qquad \chi(f; n_0) \le 2\|f\|_\Delta \le 2 \sum_{j=1}^{\infty} \|f_{n_j}\|_\Delta \le 2 \sum_{j=1}^{\infty} \frac{\chi^{1/2}(n_j)}{n_j}$$

$$\le \frac{\chi(1)}{2} = \frac{\chi(n_0)}{2}.$$

For $i \ge 0$, let n be an arbitrary positive integer with $n_i \le n \le n_{i+1}$. For some $\lambda \in [0, 1]$ we may write $n = n_i + \lambda(n_{i+1} - n_i)$. Since by (9) we have $98 \sum_{j=1}^{i} m_j \le n_i$, for $i = 1, 2, 3, \ldots,$ from (4.205) - (4.207) we get

$$(4.208)\qquad \chi(f; n) \le \chi(f; n_i) + 2\lambda(n_{i+1} - n_i) \sum_{j=i-1}^{\infty} \|f_{n_j}\|_\Delta$$

$$\le \frac{\chi(n_i)}{2} + 2\lambda(n_{i+1} - n_i) \sum_{j=i+1}^{\infty} \frac{\chi^{1/2}(n_j)}{n_j}.$$

On the other hand, (10) yields

(4.209) $$\frac{\chi(n_i)}{2} + 2\lambda(n_{i+1} - n_i)\sum_{j=i+1}^{\infty} \frac{\chi^{1/2}(n_j)}{n_j} \le \frac{\chi(n_{i+1})}{2}.$$

From (4.204), (4.208), and (4.209), and the concavity of $\nu(g; n)$, we obtain

$$2\chi(g; n) \le (1 - \lambda)\chi(n_i) + \lambda\chi(n_{i+1}) \le (1 - \lambda)\nu(g; n_i) + \lambda\nu(g; n_{i+1})$$

$$\le \nu(g; n_i + \lambda(n_{i+1} - n_i)) = \nu(g; n) \le 2\chi(n).$$

Hence, $\chi(f; n) \le \chi(n)$, for $n = 1, 2, 3, \ldots$.

It remains to prove that there exists an absolute constant $c > 0$ for which

(4.210) $$E(R_{n_i}, [0, 1], 1; f) \ge cn_i^{-1} \ln \chi(n_i), \qquad i = 1, 2, 3, \ldots .$$

The inequality (4.210) will be obtained using the principle for points of maximal deviation.

In connection with the function f_n, where n is one of the indices n_i, let us consider the rational function $u_\rho(x) = n^{-1}\chi^{1/2}(n)\, s_k(p_l(x))$ of order $\rho \le 100n$, where s_k is the rational function of order $k = [2^{-4}\ln \chi(n)]$ in Lemma 4.35, and again, p_l is the polynomial of degree $l = 100m$, $m = [n/k]$, in Lemma 3.36.

Let $[a_{\nu-1}, a_\nu]$, $\nu = 1, 2, 3, \ldots, l - 2m$ be one of the intervals as specified in Lemma 4.36. There are two possibilities. Suppose $p_l(a_{\nu-1}) = -1$, $p_l(a_\nu) = 1$, and p_l is monotonically increasing on the closed interval $[a_{\nu-1}, a_\nu]$. Denote by y_ν and $z_\nu \in (a_{\nu-1}, a_\nu)$ the points at which $p_l(y_\nu) = -e^{-1}$ and $p_l(z_\nu) = e^{-1}$. Then $a_{\nu-1} < y_\nu < z_\nu < a_\nu$. For the possibility that p_l decreases on $[a_{\nu-1}, a_\nu]$, we set $p_l(y_\nu) = e^{-1}$ and $p_l(z_\nu) = -e^{-1}$. By Lemma 4.36, $\|p'_l\|_{[0, 1]} \le e^{13m}$. Applying (1), we obtain

(4.211) $$\min\{y_\nu - a_{\nu-1}, \xi_\nu - y_\nu, z_\nu - \xi_\nu, a_\nu - z_\nu\} \ge e^{-14}m^{-1}$$

$$\ge e^{-14}n^{-1}[2^{-4}\ln \chi(n)] \ge e^{-18}n^{-1}\ln \chi(n).$$

From Lemma 4.35, there exists points $a_{\nu-1} = x_{\nu,0} < x_{\nu,1} < \cdots < x_{\nu,k} = y_\nu$ and $z_\nu = x_{\nu,k+1} < x_{\nu,k+2} < \cdots < x_{\nu,2k+1} = a_\nu$ such that

(4.212) $$|f_n(x_{\nu,j}) - u_\rho(x_{\nu,j})| > n^{-1}\chi^{1/2}(n)\,\chi^{-1/4}(n) = n^{-1}\chi^{1/4}(n)$$

for $j = 0, 1, 2, \ldots, 2k + 1$, and function $f_n(x) - u_\rho(x)$ is alternating in sign at the points $x_{\nu,j}$, $j = 0, 1, 2, \ldots, 2k + 1$. In addition, u_ρ is monotonic on $[y_\nu, z_\nu]$.

It follows from condition (3) that for every index n_i of the sequence, there exists a rational function v_i of order not exceeding n_i for which

$$(4.213) \qquad r([0, 1], 1; F_i, v_i) < n_i^{-1},$$

where $F_i = \sum_{j=1}^{i-1} f_{n_j}$. The function F_i is step-like, and it has at most $98 \sum_{j=1}^{i-1} m_j$ discontinuities. From (4.211) and (1), it follows that $a_\nu - a_{\nu-1} \geq e^{-18} n_i^{-1} \ln \chi(n_i) \geq 2/n_i$ for $\nu = 1, 2, 3, \ldots, l - 2m_i$. By (4) there exist at most $196 \sum_{j=1}^{i-1} m_j < m_i$ segments $[a_{\nu-1}, a_\nu]$, $1 \leq \nu \leq 98m_i$, which have nonempty intersection with the neighborhood of radius $1/n_i$ of the discontinuities of F_i. Let J_i index these segments. The number of elements in J_i does not exceed m_i.

From (4.213) and the definition of Hausdorff distance, it follows that for $x \in \bigcup \{[a_{\nu-1}, a_\nu] : 1 \leq \nu \leq 98m_i,\ \nu \notin J_i\}$, we have

$$(4.214) \qquad |F_i(x) - v_i(x)| < \frac{1}{n_i},$$

and from (5) we have

$$(4.215) \qquad \left\| \sum_{j=i-1}^{\infty} f_{n_j} \right\|_{[0,1]} \leq \sum_{j=i-1}^{\infty} \frac{\chi^{1/2}(n_j)}{n_j} < n_i^{-1}.$$

We claim that every rational function q_i of order no higher than n_i satisfies the inequality

$$(4.126) \qquad r([0, 1], 1; f, q_i) > e^{-20} n_i^{-1} \ln \chi(n_i).$$

Indeed, let us assume this fails, namely that for a certain rational function q^* of order no higher than n_i, we have

$$(4.217) \qquad r(f, q^*) \leq e^{-20} n_i^{-1} \ln \chi(n_i).$$

Then, by (4.211) and (5), for every

$$(4.218) \qquad x \in \bigcup \{[a_{\nu-1}, y_\nu] \cup [z_\nu, a_\nu] : 1 \leq \nu \leq 98m_i,\ \nu \notin J_i\},$$

we have

(4.219) $$\left|\sum_{j=1}^{i} f_{n_j}(x) - q^*(x)\right| \le r(f, q^*) + \sum_{j=i-1}^{\infty} n_j^{-1} \chi^{1/2}(n_j)$$

$$\le e^{-20} n_i^{-1} \ln \chi(n_i) + n_i^{-1} \le 2e^{-20} n_i^{-1} \ln \chi(n_i) .$$

From (4.214) and (4.219), we have for all x defined by (4.218),

(4.220) $$|f_{n_i}(x) - q^*(x) - v_i(x)| \le \left|\sum_{j=1}^{i} f_{n_j}(x) - q^*(x)\right| - |F_i(x) - v_i(x)|$$

$$\le 3e^{-20} n_i^{-1} \ln \chi(n_i).$$

Let $w(x) = u_\rho(x) - q^*(x) - v_i(x)$, a rational function of order no higher than $\rho + 2n_i = 102n_i$. From (4.212) and (4.220), it follows that for every $x \in \{x_{\nu,j} : 1 \le \nu \le 98m_i, \nu \notin J_i, j = 0, 1, 2, \ldots, 2k_i - 1\}$, we have

$$|w(x)| \ge |f_{n_i}(x) - u_\rho(x)| - |f_{n_i}(x) - q^*(x) - v_i(x)| > 0.$$

Since $f_{n_i}(x) - u_\rho(x)$ is alternating in sign, w has at least $2k_i$ zeros in each segment $[a_{\nu-1}, a_\nu]$, $1 \le \nu \le 98m_i$, $\nu \notin J_i$. Consequently, by (6), w has at least $2k_i(98m_i - m_i) = 194[n_i/k_i]\, k_i \ge 102n_i + 1$ zeros. But this is impossible, since w has order no higher than $102n_i$. This contradiction proves (4.216), completing the proof of Theorem 4.21.

Remark. It is not difficult to construct a continuous function f for which the statement of Theorem 4.21 holds. For this purpose we need only to slightly modify the step functions f_{n_i} involved in the definition of f.

Corollary 4.5. There exist a bounded function f on [0, 1], a constant $c > 0$, and a sequence of positive integers $n_1 < n_2 < \cdots$ such that

$$E(R_{n_i}, [0, 1], 1; f) > cn_i^{-1} \ln n_i^{-1} , \qquad i = 1, 2, 3, \ldots .$$

It suffices to take $\chi(n) = n$, and it is not hard to construct a function g for which $\chi(g; n) = n$.

4.2.3. APPROXIMATION OF ANALYTIC FUNCTIONS WITH SINGULARITIES ON THE BOUNDARY OF A CLOSED INTERVAL

Here, we shall consider the problem of rational approximation, studied by A. Goncar [2], by using Hausdorff distance instead of uniform distance. The main result of A. Goncar on this topic is the following.

Theorem 4.22. (A. Goncar [2]). Let f be a continuous function on the closed interval $[0, 1]$, and suppose f coincides with a bounded analytic function defined on the disc $\{z : |z - 1| < 1\}$ for all $x \in (0, 1]$. Then there exists a positive constant c such that

$$R_n([0, 1]; f) = O(\delta_n)$$

where $\delta_n = \inf\limits_{t>1} te^{-cn/t} + \omega(f; e^{-t})$.

In the course of his proof, Goncar established the following statement.

Lemma 4.37. (A. Goncar [2]). Let f be continuous on $[0, 1]$, admitting an analytic (not necessarily bounded) continuation to the disc $\{z : |z - 1| < 1\}$. Then for every $t > 1$,

$$R_n([e^{-t}, t]; f) \leq c_1 M(e^{-t})\, te^{-cn/t},$$

where $M(h) = \sup \{|f(z)| : z \in \Gamma_h\}$ where Γ_h is a circle symmetric with respect to the real axis, and which intersects the axis at $h/2$ $(h > 0)$ and $3/2$, and where c and c_1 are positive constants.

In the theory of functions of a complex variable, the boundary properties of functions analytic in a disc and belonging to the classes A and H_δ are well-studied (I. Privalov [1, p.78]). Let us recall that a function f, analytic in the unit disc, belongs to the class A, provided

$$A(f) = \lim_{\delta \to 1-0} \frac{1}{2\pi}\int_0^{2\pi} \ln^+ |f(\rho e^{i\theta})|\, d\theta < \infty,$$

where $\ln^+ a = \max\{0, \ln a\}$, whereas f belongs to H_δ $(\delta > 0)$ provided

$$H_\delta(f) = \lim_{\delta \to 1-0} \frac{1}{2\pi}\int_0^{2\pi} |f(\rho e^{i\theta})|^\delta\, d\theta < \infty.$$

It is well-known (I. Privalov [1, p.84]) that each $f \in A$ satisfies the inequality

(4.221) $$\max\{|f(\rho e^{i\theta})| : \theta \in [0, 2\pi]\} \le e^{2A(f)/(1-\rho)}, \quad \text{for} \quad 0 < \rho < 1,$$

whereas each $f \in H_\delta$ satisfies

(4.222) $$\max\{|f(\rho e^{i\theta})| : \theta \in [0, 2\pi]\} \le \left(\frac{2H_\delta(f)}{1 - \rho}\right)^{1/\rho}.$$

Our aim here will be to prove the following statement.

Theorem 4.23. Let f be continuous and real valued on $(0, 1]$, admitting an analytic continuation to the disc $D = \{z : |z - 1| < 1\}$. Assume further that f is either bounded below or above on $(0, 1]$. If $f(z - 1)$ belongs to the class A, then

$$E(R_n, (0, 1]; 1; f) = O(n^{-1}\ln n),$$

and if $f(z - 1)$ belongs to the class H_δ, then for some $c > 0$ independent of δ,

$$E(R_n, (0, 1]; 1; f) = O(e^{-c\sqrt{n}}).$$

It follows from Theorem 4.23 that

(4.223) $$E(R_n, (0, 1]; 1; x^s) = O(e^{-c\sqrt{n}})$$

for arbitrary real s, not just positive s. It follows from Goncar's result [2] that the order in (4.223) cannot be improved, but the exact value of c in (4.223) remains to be found.

The proof of Theorem 4.23 requires a preliminary lemma.

Lemma 4.38. Let f be a continuous real valued function on the interval $(0, 1]$ that is bounded from below, i.e.,

(4.224) $$\liminf_{x \to +0} f(x) > -\infty.$$

Let us assume that f can be extended to an analytic function on the disc $D = \{z : |z - 1| < 1\}$. For $h \in (0, 1)$, write $M(f; h) = \sup\{|f(z)| : z \in C_h\}$, where C_h is the circle $|z - 1| = 1 - h$. Then the inequality

$$E(R_n, (0, 1]; 1; f) \le e^{-t} + c_2 M(f; e^{-t})\, te^{-cn/t}$$

holds for every $t > 1$, where c and c_2 are positive constants.

Proof. Initially, we will use some of Goncar's arguments in the proof of Theorem 4.22. Let $h \in (0, e^{-1})$ be held fixed, and set $\xi = \xi(x) = h + (1 - h)x$ for $x \in (0, 1]$, so that $\xi(x) \in (h, 1]$. The function $f_1(x) = f(\xi(x))$, $x \in (0, 1]$, will be considered as function of the independent variable $\xi \in (h, 1]$. By virtue of Lemma 4.37, there exists a rational function v_n of order no higher than n such that

(4.225) $$|f(\xi) - v_n(\xi)| \le c_1 M(e^{-t})\, te^{-cn/t}, \quad \xi \in [e^{-t}, 1], \quad e^{-t} = h,$$

where $M(h) = M(f; h)$.

For this fixed h and the rational function v_n, we shall construct a rational function $u(x)$ of at most fourth order as follows:

(a) if $\limsup_{x \to +0} f(x) = \infty$, then

$$u(x) = \frac{h^2}{2x} - Ah^3\,(2A(2x - h)^2 + 1),$$

where $A \ge 0$ is determined as follows: if

$$\inf\{f(x) : x \in (0, h]\} \ge \inf\{v_n(\xi(x)) : x \in (0, h]\}, \text{ then } A = 0;$$

if

$$\inf\{f(x) : x \in (0, h]\} < \inf\{v_n(\xi(x)) : x \in (0, h]\},$$

then A is determined from the equality

$$\inf\{v_n(\xi(x)) - u(x) : x \in (0, h]\} = \inf\{f(x) : x \in (0, h]\}.$$

Such a choice of A is possible, because $u(h/2) = h - Ah^{-3}$.

(b) If $\limsup_{x \to +0} f(x) < \infty$, then

$$u(x) = \frac{Ah^3}{2Ax^2 + 1} - \frac{Bh^3}{2B(2x - h)^2 + 1},$$

where $A \geq 0$ and $B \geq 0$ are determined as follows:

(b1) if $\sup \{f(x) : x \in (0, h]\} \leq \sup \{v_n(\xi(x)) : x \in (0, h]\}$, then $A = 0$;

(b2) if $\sup \{f(x) : x \in (0, h]\} > \sup \{v_n(\xi(x)) : x \in (0, h]\}$,

then A is determined from the equality

$$\sup \{v_n(\xi(x)) - u(x) : x \in (0, h]\} = \sup \{f(x) : x \in (0, h]\}.$$

(b3) if $\inf \{f(x) : x \in (0, h]\} \geq \inf \{v_n(\xi(x)) : x \in (0, h]\}$, then $B = 0$;

(b4) if $\inf \{f(x) : x \in (0, h]\} < \inf \{v_n(\xi(x)) : x \in (0, h]\}$,

then B is determined from the equality

$$\inf \{v_n(\xi(x)) - u(x) : x \in (0, h]\} = \inf \{f(x) : x \in (0, h]\}.$$

The constants A and B are determined independently of each other in the cases (b1,b3), (b1,b4), (b2,b4), and the existence of a solution of the posed problem is obvious. In the case (b2,b4) we obtain a system of two equations. It has a solution, too, since it is possible to choose the minimum and maximum of u arbitrarily in the closed interval $[0, h]$, in view of the inequalities

$$u(0) = Ah^3 - \frac{Bh^3}{4Bh^2 + 1} > Ah^3 - h,$$

$$u\left(\frac{h}{2}\right) = \frac{Ah^3}{Ah^2 + 1} - Bh^3 < -Bh^3 + h.$$

This completes the construction of $u(x)$. It is directly verified that

(4.226) $$|u(x)| \leq h = e^{-t} \qquad \text{for } x \in [h, 1].$$

We intend to show that

(4.227) $$r((0, 1], 1; f, W_n) \leq e^{-t} + c_1 M(e^{-t})\, te^{-cn/t}, \qquad e^{-t} = h,$$

where

(4.228) $\qquad W_n(x) = v_n(\xi(x)) + u(x).$

In view of (4.225), (4.226), and (4.228), to establish (4.227), it suffices to show that

(1) for every $x \in (0, e^{-t}]$ there exists $\eta \in (0, e^{-t}]$ such that $f(x) = W_n(\eta)$, and

(2) for every $x \in (0, e^{-t}]$ there exists $y \in (0, x - e^{-t}]$ such that $|W_n(x) - f(y)| \le e^{-t} + c_1 M(e^{-t})\, te^{-cn/t}$.

Statement (1) follows directly from the choice of the function u. We now prove (2). For every $x \in (0, e^{-t}]$, either there exists $y \in (0, e^{-t}]$ such that $|W_n(x) - f(y)| \le e^{-t}$ (from which (2) follows), or no such y exists. Thus either

(4.229) $\qquad W_n(x) > \sup\{f(y) : y \in (0, h]\},$

or

(4.230) $\qquad W_n(x) < \sup\{f(y) : y \in (0, h]\}.$

We consider the case (4.229); the case (4.230) is analagous and is left to the reader. If (4.229) holds, then according to the construction of the function u, we have $u(x) \le 0$ for $x \in [0, 1]$, so that

$$\sup\{f(y) : y \in (0, h]\} < W_n(x) = v_n(\xi(x)) + u(x) \le v_n(\xi(x)).$$

But according to (4.211), $|f(\xi(x)) - v_n(\xi(x))| \le c_1 M(e^{-t})\, te^{-cn/t}$, which means that $v_n(\xi(x)) \le f(\xi(x)) + c_1 M(e^{-t})\, te^{-cn/t}$. Hence,

$$f(h) < W_n(x) \le f(\xi(x)) + c_1 M(e^{-t})\, te^{-cn/t}.$$

Condition (2) follows from this last inequality. Thus, (4.227) is established, and the proof of the lemma is complete.

Proof of Theorem 4.23. From Lemma 4.38 and (4.221), it follows that if $f \in A$, then for every $t > 1$,

(4.231) $\qquad E(R_n, (0, 1]; 1; f) \le e^{-t} + c_1 t \exp(c_2 e^t - cn/t),$

where c, c_1 and c_2 are positive constants. If in (4.217) we set $t = \ln(n/\lambda \ln n)$, where $\lambda = \max\{1, 2c_2/c_1\}$, then for $n > 3$ we obtain

$$E(R_n, (0,1]; 1; f) \le \lambda n^{-1} \ln n + c_1 \ln(n/\lambda \ln n) \exp\left(\frac{c_2 n}{\lambda \ln n} - \frac{cn}{\ln(n/\lambda \ln n)}\right)$$

$$\le \lambda n^{-1} \ln n + c_1 \exp\left(\frac{-cn}{2\ln n}\right) \ln n = O(n^{-1} \ln n).$$

If $f \in H_\delta$, then from Lemma 4.38 and (4.222) it follows that for every $t > 1$, we have

(4.232) $$E(R_n, (0,1]; 1; f) \le e^{-t} + c_2 t\,(c_1 t)^{1/\delta} e^{-cn/t},$$

where c, c_1, and c_2 are positive constants. If in (4.232) we set $t = \sqrt{cn}$, then

$$E(R_n, (0,1]; 1; f) \le e^{-\sqrt{cn}} + c_2 \exp(\ln \sqrt{cn} + \delta^{-1} \ln(c_1 \sqrt{cn}) - \sqrt{cn})$$

$$= O(e^{-\sqrt{cn}}).$$

This completes the proof of Theorem 4.23.

Remark. The condition (4.224) of boundedness from below (or above) is essential. For any $\delta > 0$, there exists a function $\varphi(z+1)$ in H_δ which takes real values on $(0, 1]$ such that both $\limsup_{x \to +0} \varphi(x) = \infty$ and $\liminf_{x \to +0} \varphi(x) = -\infty$. For example, such a function is

$$\varphi(z) = z^{-1/\delta} \prod_{k=1}^{\infty} \frac{2^k z - 1}{(2^k - 1)z + 1}.$$

§ 4.3. Best approximation by spline functions

Let $S_{k,n} = S_{k,n}[a, b]$ denote the set of all spline functions of order (k,n), defined on the closed interval $\Delta = [a, b]$. The function s belongs to $S_{k,n}$ if $s \in C_\Delta^{k-1}$ i.e., if it has continuous derivatives up to order $k - 1$ on Δ, and there exists $n + 1$ knots $a = x_0 < x_1 < x_2 < \cdots < x_n = b$ such that the function s coincides with an algebraic polynomial of degree no higher than k on every closed interval $[x_{i-1}, x_i]$, $i = 1, 2, \ldots, n$. For $k = 0$, the set $S_{0,n}$ consists of step functions, and for $k = 1$, the set $S_{1,n}$ consists of piecewise linear continuous functions (polygonal curves). The elements of $S_{0,n}$ may be regarded as segment functions, whose value at each knot x_i is a segment of length equal to the jump of the function at this knot.

Denote

$$(x_i - x)_+^k = \begin{cases} (x_i - x)^k & \text{if } x < x_i \\ 0 & \text{if } x \geq x_i \end{cases}$$

The following basic result is well-known, and is proved by induction on k, using the fact that if $s \in S_{k,n}$, then $s' \in S_{k-1,n}$

Lemma 4.39. Every function $s \in S_{k,n}$ can be represented in the form

$$s(x) = p(x) + \sum_{i=0}^{n} a_i \, (x_i - x)_+^k$$

where $a_0, a_1, \ldots, a_n$ are constants and p is an algebraic polynomial of degree no higher than $k - 1$.

The following lemma, which was proved long ago by L. Cakalov [1], and later by I. Schoenberg [1], will be needed in the sequel.

Lemma 4.40. The spline functions

$$\varphi_{i,k}(x) = \sum_{j=i}^{i+k+1} \frac{(k+1)(x_j - x)_+^k}{w_{i,k+1}'(x_j)}, \quad i = 0, 1, 2, \ldots, n-k-1,$$

where $w_{i,k+1}(x) = (x - x_i)(x - x_{i+1}) \cdots (x - x_{i+k+1})$, satisfy the following conditions:

$$\varphi_{i,k}(x) = 0 \quad \text{for} \quad x \in [a, x_i] \cup [x_{i+k+1}, b],$$

$$\varphi_{i,k}(x) > 0 \quad \text{for} \quad x \in (x_i, x_{i+k+1}),$$

$$\int_a^b \varphi_{i,k}(x)\, dx = 1.$$

We use Lemma 4.40 to prove the next statement about spline functions in $S_{k,n}$.

Lemma 4.41. If $\varphi \in S_{k,n}$ and

(4.233) $\quad |\varphi(x)| \le \delta \qquad$ for $x \in [x_{i-1}, x_i] \cup [x_{i+k+1}, x_{i+k+2}], \quad i + k + 2 < n,$

then there exists a constant c, independent of δ, such that either $\varphi(x) \le c\delta$ for $x \in [x_i, x_{i+k+1}]$, or $\varphi(x) \ge -c\delta$ for $x \in [x_i, x_{i+k+1}]$.

Proof. Let us assume that such a constant does not exist. Then for every positive number M, there exists a spline function $\psi_M \in S_{k,n}$ for which (4.233) holds, and in addition

$$\max \{\psi_M(x) : x \in [x_i, x_{i+k+1}]\} \ge M, \quad \min \{\psi_M(x) : x \in [x_i, x_{i+k+1}]\} \le -M.$$

Then there exists $\psi_0 \in S_{k,n}$ for which $\psi_0(x_\nu) = \psi_0'(x_\nu) = \cdots \psi_0^{(k-1)}(x_\nu)$, for $\nu = i$, $i + k + 1$, and ψ_0 takes both positive and negative values on the closed interval $[x_i, x_{i+k+1}]$.

Consider the spline function $s \in S_{k+1,n}$ defined by

$$s(x) = 0 \text{ for } x \in [a, x_i] \cup [x_{i+k+1}, b], \text{ and}$$

$$s(x) = \int_{x_i}^{x} (\psi_0(t) + \lambda\varphi_{i,k}(t))\, dt \quad \text{for} \quad x \in [x_i, x_{i+k+1}],$$

where $\varphi_{i,k} \in S_{k,n}$ is the function defined in Lemma 4.40, and the constant λ is defined by the condition $s(x_{i+k+1}) = 0$. In view of the properties of the function ψ_0, it is not hard to see that s really belongs to $S_{k+1,n}$, and that it is not identically zero. According to the definition of spline functions, we have $s(x_\nu) = s'(x_\nu) = s''(x_\nu) = \cdots = s^{(k)}(x_\nu) = 0$ for $\nu = 0, 1, 2, \ldots, i, i + k + 1, i + k + 2, \ldots, n$. Using Lemma 4.39, it follows from these equalities that s is identically equal to zero. The obtained contradiction concludes the proof of the lemma.

We shall mention one quite general and strong result of C. Fitzgerald and L. Schumaker [1], which in a special case, pertains to splines.

Lemma 4.42. (C. Fitzgerald and L. Schumaker [1, p. 126]). Suppose $k \ge 2$, and the sequence of $n + k$ numbers $l_0, l_1, l_2, \ldots, l_{n+k-1}$ satisfy the conditions

(4.234) $\quad (-1)^i(l_i - l_{i-1}) < 0 \qquad i = 1, 2, 3, \ldots, n + k - 1.$

Then there exists a unique spline function $s \in S_{k,n}$ and a unique set of points $a = t_0 < t_1 < \cdots < t_{n+k-1} = b$ such that

(4.235) $$s(t_i) = l_i, \qquad i = 0, 1, 2, 3, \ldots, n + k - 1,$$

(4.236) $$s'(t_i) = 0, \qquad i = 1, 2, 3, \ldots, n + k - 2.$$

Moreover, the points $\{t_i : 0 \le i \le n + k - 1\}$ satisfy the following conditions:

(4.237) $$t_i \in (x_{i-k+1}, x_i), \qquad i = 1, 2, 3, \ldots, n + k - 2,$$

where $x_i = x_0 = a$ for $i < 0$ and $x_i = x_n = b$ for $i > n$.

4.3.1 SPLINE FUNCTIONS WITH EQUIDISTANT KNOTS

We shall consider here only those spline functions on the interval $\Delta = [a, b]$ with equidistant knots, i.e., $x_i = a + (b - a)i/n$, $i = 0, 1, 2, \ldots, n$. We denote the set of all spline functions of order (k,n) with equidistant knots by $\bar{S}_{k,n}$.

Lemma 4.43. Let $\bar{F}_\Delta$, $\Delta = [a, b]$ be the set of all bounded functions $f \in F_\Delta$. Then

$$E(\bar{S}_{k,n}, \Delta, \alpha, \bar{F}_\Delta) = \sup\,\{E(\bar{S}_{k,n}, \Delta, \alpha; f) : f \in \bar{F}_\Delta\} \ge \frac{(k+3)(b-a)}{2\alpha n}$$

for $n \ge k + 5$.

Proof. It suffices to prove that for every $\varepsilon > 0$, there exists $f_\varepsilon \in \bar{F}_\Delta$ such that $\{E(\bar{S}_{k,n}, \Delta, \alpha; f_\varepsilon) \ge (k + 3)(b - a)/2\alpha n$. This is easily verified, taking for f_ε the segment function

$$f_\varepsilon(x) = \begin{cases} 0 & \text{if } a \le x \le b \text{ and } x \ne \xi \\ [-M, M] & \text{if } x = \xi \end{cases},$$

where $\xi = x_\nu$ for odd k and $\xi = x_\nu + (b - a)/2n$ for even k, where $\nu = n/2$. Let us now write $\delta = (k + 3)(b - a)/2\alpha n$, and assume to the contrary that $E(\bar{S}_{k,n}, \Delta, \alpha; f_\varepsilon) < \delta$. This means that some $s \in S_{k,n}$ satsifies the condition

$$\max \{|s(x)| : x \in [\xi - \alpha\delta, \xi + \alpha\delta]\} \le \delta,$$

and s takes sufficiently large values with opposite signs on the segment $[\xi - \alpha\delta, \xi + \alpha\delta]$. Then by Lemma 4.41, we have

$$\min \{i : x_i \ge \xi + \alpha\delta\} - \max \{i : x_i \le \xi - \alpha\delta\} \ge k + 2, \tag{4.238}$$

provided $\xi - \alpha\delta \ge x_1$ and $\xi + \alpha\delta \le x_{n-1}$. The last two inequalities hold, since $n \ge k + 5$. From (4.237), it follows that $\alpha\delta \ge (k + 1)(b - a)/n$, or $(k + 3)/2 \ge k + 1$. Thus, $k \le 1$. Thus, the conclusion of the lemma holds for all positive integers $k \ge 2$. For $k = 1$ and $n \ge 2$, the conclusion is obtained by again considering the same function f_ε for $\xi = x_1$ and for sufficiently large M. This completes the proof.

Theorem 4.24. For every function $f \in \bar{\mathbf{F}}_\Delta$, $\Delta = [a, b]$, and for arbitrary positive integers k and n, we have the inequality

$$E(\bar{S}_{k,n}, \Delta, \alpha; f) \le \frac{(k + 3)(b - a)}{2\alpha n}.$$

Proof. First we shall prove the theorem for $k = 1$. Let f be an arbitrary element in $\bar{\mathbf{F}}_\Delta$. Set $m_i = I(\delta, f; x_i)$ and $M_i = S(\delta, f; x_i)$ for $i = 0, 1, 2, \ldots, n$, where $\delta = (b - a)/n$.

Let $s \in \bar{S}_{1,n}$ be the piecewise linear function whose values at the knots are given by

$$s(x_0) = M_0, \quad s(x_1) = m_1, \quad s(x_2) = M_2, \quad s(x_3) = m_3, \quad \ldots .$$

The graph of a typical f and the corresponding s appear in Figure 4.5. It is directly verified that $r(\Delta, \alpha; f, s) \le 2(b - a)/\alpha n$, i.e., the theorem is proved for $k = 1$.

Now let $k \ge 2$ and let $f \in \bar{\mathbf{F}}_\Delta$ be arbitrary. For $j = 1, 2, 3, \ldots$, we shall introduce numbers $m_{2j}(f)$ and $m_{2j+1}(f)$, that depend on the location of j, as follows:

(a) for $j < \frac{k-3}{4}$, $m_{2j}(f) = \min \{y : y \in f(a)\}$, $m_{2j+1}(f) = \max \{y : y \in f(a)\}$;

(b) for $\frac{k-3}{4} \le j \le \frac{k-3}{4} + \frac{n}{2}$,

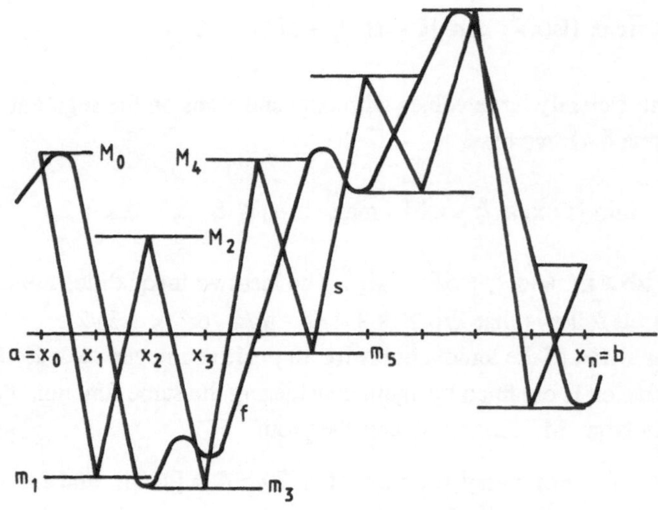

FIGURE 4.5

$$m_{2j}(f) = \min\{y : y \in f(x),\ x \in [x_{2j} - \lambda, x_{2j+2} - \lambda]\},$$

$$m_{2j+1}(f) = \max\{y : y \in f(x),\ x \in [x_{2j} - \lambda, x_{2j+2} - \lambda]\},$$

where $\lambda = (k + 1)(b - a)/2n$;

(c) for $j > \dfrac{k-3}{4} + \dfrac{n}{2}$, $m_{2j}(f) = \min\{y : y \in f(b)\}$, $m_{2j+1}(f) = \max\{y : y \in f(b)\}$.

Fix $\varepsilon \in (0, (b - a)/\alpha n)$ and define numbers l_i by

(4.239) $$l_i = m_i(f) - (-1)^i\varepsilon, \qquad i = 0, 1, 2, \ldots, n + k - 1.$$

Obviously, the sequence $l_0, l_1, \ldots, l_{n+k-1}$ satisfies the hypotheses of Lemma 4.42. Let $s_f \in \overline{S}_{k,n}$ be the spline function whose existence and uniqueness is ensured by Lemma 4.42 with respect to the numbers determined by (4.239). The proof of theorem will be completed, if we manage to demonstrate that

(4.240) $$r(\Delta, \alpha; f, s_f) \leq \frac{(k+3)(b-a)}{2\alpha n}.$$

Let $\{t_i : 0 \leq i \leq n+k-1\}$ be the points that satisfy the conditions (4.235) - (4.237). It is not hard to see that s_f is monotonic on $[t_i, t_{i+1}]$, $i = 0, 1, 2, \ldots, n+k-2$. In order to prove (4.240), we first take an arbitrary point (x,y), $x \in \Delta$, $y \in f(x)$, and show that there exists a point $(\xi, s_f(\xi))$ with $\xi \in \Delta$ such that

(4.241) $$\max\{\alpha^{-1}|x-\xi|, |y - s_f(\xi)|\} \leq \frac{(k+3)(b-a)}{2\alpha n}.$$

Then we shall take an arbitrary $x \in \Delta$ and show that there exists a point (ξ,η) with $\xi \in \Delta$ and $\eta \in f(\xi)$ for which

(4.242) $$\max\{\alpha^{-1}|x-\xi|, |s_f(x) - \eta|\} \leq \frac{(k+3)(b-a)}{2\alpha n}.$$

Obviously, (4.240) follows from Lemma 2.4, (4.241) and (4.242).

Let (x,y), $x \in \Delta$, $y \in f(x)$, be arbitrary. Select an index j such that $x \in [x_{2j} - \lambda, x_{2j+2} - \lambda]\}$. Then $y \in [l_{2j}, l_{2j+1}] = [s_f(t_{2j}), s_f(t_{2j+1})]$, where $t_{2j} \in [x_{2j-k+1}, x_{2j}]$ and $t_{2j} + 1 \in [x_{2j-k+2}, x_{2j+1}]$. Hence, there exists $\xi \in [x_{2j-k+1}, x_{2j+1}]$ for which $s_f(\xi) = y$. Then

$$\max\{\alpha^{-1}|x-\xi|, |y - s_f(\xi)|\} = \alpha^{-1}|x-\xi|$$

$$\leq \alpha^{-1} \max\{\lambda + \frac{b-a}{n}, \frac{(k+1)(b-a)}{n} - \lambda\} = \frac{(k+3)(b-a)}{2\alpha n},$$

i.e., (4.241) holds.

For (4.242), let $x \in \Delta$ be arbitrary, and choose i with $x \in [t_i, t_{i+1}]$. Consider the case i even, with $i = 2j$ (the case i odd is handled in the same way). Then $s_f(x) \in [l_{2j}, l_{2j+1}] = [m_{2j}(f) - \varepsilon, m_{2j+1}(f) + \varepsilon]$. It follows from the definition of $m_i(f)$ that for every $\eta \in [m_{2j}(f), m_{2j+1}(f)]$ there exists $\xi \in [x_{2j} - \lambda, x_{2j+2} - \lambda]$ such that $\eta \in f(\xi)$. Let η satisfy the condition

(4.243) $$|\eta - s_f(x)| \leq \varepsilon < \frac{b-a}{\alpha n}.$$

On the other hand, (4.223) yields $x \in [t_{2j}, t_{2j+1}] \subset [x_{2j-k+1}, x_{2j+1}]$, and consequently

$$(4.244) \qquad |x - \xi| \le \frac{(k+3)(b-a)}{2\alpha n}.$$

Evidently, (4.242) follows from (4.243) and (4.244), completing the proof.

Remark. The case $k = 1$ of Theorem 4.24 was proved by E. Dolzenko and E. Sevastianov [1,2].

The exact value of the best Hausdorff approximation for the class $\overline{F}_\Delta$ by spline functions for many n and k is obtained as a direct consequence of Theorem 4.24 and Lemma 4.43.

Theorem 4.25. Let $\Delta = [a, b]$. For every positive integer k and for every positive integer $n \ge k + 5$ we have the equality

$$(4.245) \qquad E(\overline{S}_{k,n}, \Delta, \alpha; \overline{F}_\Delta) = \frac{(k+3)(b-a)}{2\alpha n}.$$

Actually, the equality (4.245) holds for all positive values of n when $k = 1$, as shown by E. Dolzenko and E. Sevastianov [1,2]. The cases $2 \le n \le k + 4$, $k = 2, 3, 4, \ldots$ remain open.

Let us next consider the problem of the approximation of elements of $\overline{F}_\Delta$ by step functions, i.e., by spline functions of the class $S_{0,n}$. In this case we choose the knots in this way: $x_0 = a$, $x_1 = a + h/2$, $x_2 = x_1 + h$, ..., $x_{n-1} = x_{n-2} + h$, and finally $x_n = b = x_{n-1} + h/2$, where $h = (b - a)/(n - 1)$. The set of functions in $S_{k,n}$ for such a choice of the knots will be denoted by $\hat{S}_{k,n}$.

Theorem 4.26. For every positive integer $n \ge 3$, the equality

$$E(\hat{S}_{0,n}, \Delta, \alpha; \overline{F}_\Delta) = \frac{3(b-a)}{2\alpha(n-1)}$$

holds, where $\Delta = [a, b]$.

Proof. Let f be an arbitrary element of $\overline{F}_\Delta$. Write $h = (b - a)/(n - 1)$, and let m_i, M_i be the following numbers, defined for $i = 1, 2, 3, \ldots, n - 1$:

$$m_i = \min \{y \in f(x) : \; x \in [x_i - h/2, x_i + h/2]\},$$

$$M_i = \max \{y \in f(x) : \; x \in [x_i - h/2, x_i + h/2]\}.$$

The step function $s \in \hat{S}_{0,n}$ will be defined as follows (see Figure 4.6):

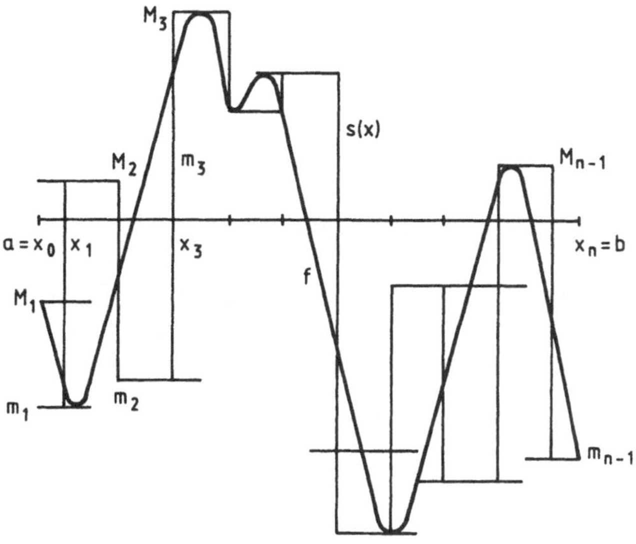

FIGURE 4.6

$s(x) = m_1$ for $x \in [x_0, x_1]$,

$s(x) = \max \{M_1, M_2\}$ for $x \in [x_1, x_2]$,

$s(x) = \min \{m_2, m_3\}$ for $x \in [x_2, x_3]$,

$s(x) = \max \{M_3, M_4\}$ for $x \in [x_3, x_4]$, ...

It is directly verified that

$$r(\alpha, \Delta; f, s) \leq \frac{3(b - a)}{2\alpha(n - 1)} . \tag{4.246}$$

On the other hand, if $n \geq 3$, then for the segment function

$$f_0(x) = \begin{cases} [-M, M] & \text{if } x = a \\ 0 & \text{if } a < x \leq b \end{cases},$$

and for every $s \in \hat{S}_{0,n}$, the inequality

$$r(\Delta, \alpha; f_0, s) \geq \frac{3(b-a)}{2\alpha(n-1)} \tag{4.247}$$

holds, provided $M \geq 3(b-a)/2(n-1)$. The conclusion of the theorem follows from (4.246) and (4.247).

4.3.2. SPLINE FUNCTIONS WITH FREE KNOTS

If we extend our approximation apparatus and, instead of spline functions with equidistant knots, we consider spline functions whose knots are arbitrarily spaced, an improvement of the corresponding approximation should be expected. From Theorem 4.24, the following corollary is obtained, noting $\bar{S}_{k,n} \subset S_{k,n}$.

Corollary 4.6. For every function $f \in \bar{F}_\Delta$, $\Delta = [a, b]$, and for arbitrary positive integers k and n, we have

$$E(S_{k,n}, \Delta, \alpha; f) \leq \frac{(k+3)(b-a)}{2\alpha n}.$$

On the other hand, having in mind that every function $s \in S_{k,n}$ can change its sign in the segment $[a, b]$ no more than $n + k - 1$ times, since $s^{(k)}$ is a step function with at most $n - 1$ jumps, then for every $\varepsilon > 0$, we can find an $M > 0$ such that for every $s \in S_{k,n}$, the inequality

$$r(\Delta, \alpha; g_M, s) \geq (b-a)/2\alpha(n+k-1) - \varepsilon \tag{4.248}$$

holds, where $g_M(x) = [-M, M]$ for all $x \in [a, b]$. From (4.248), we obtain the following result.

Theorem 4.27. The inequality

$$E(S_{k,n}, \Delta, \alpha; \bar{F}_\Delta) \geq \frac{b - a}{2\alpha(n + k - 1)}$$

holds for arbitrary positive integers n and k, and in the case $n \geq k - 1$, we have

$$E(S_{k,n}, \Delta, \alpha; \bar{F}_\Delta) \geq \frac{b - a}{4\alpha n}.$$

It follows from Theorem 4.27 that the extended apparatus of approximation $S_{k,n}$ cannot yield anything more than $\bar{S}_{k,n}$ does with respect to the order on the entire class $\bar{F}_\Delta$. In other words the estimate $E(S_{k,n}, \Delta, \alpha; f) = O(n^{-1})$ is unimprovable in the class $\bar{F}_\Delta$. As for subclasses of $\bar{F}_\Delta$, there are certain cases when $S_{n,k}$ improves the order in comparison with $\bar{S}_{k,n}$. One such example is the subclass of functions of bounded variation. This is a special case of the approximation of rectifiable curves by inscribed polygonal paths in the plane, which we will consider in Chapter 7. As a special case, we obtain the following statement.

Theorem 4.28. If the function $f \in F_\Delta$ has bounded variation then $E(S_{1,n}, \Delta, \alpha; f) = o(n^{-1})$.

The proof of the aforementioned general statement, from which Theorem 4.28 follows, is in turn based on a theorem of N. Korneicuk and A. Polovina [1], which states that each Lipschitz function can be approximated uniformly by a polygonal curve with n components with error $o(n^{-1})$. Certainly, it would be of interest to obtain a direct proof of Theorem 4.28. Then the theorem of Korneicuk and Polovina would be be a corollary of Theorem 4.28, in view of Theorem 2.3.

§ 4.4. Best approximation by piecewise monotone functions

Until now, we have considered approximations with respect to the Hausdorff distance by using as means of approximation algebraic polynomials, trigonometric polynomials, rational functions, and spline functions. All of these functions, used to approximate other functions, have the common characteristic property of being piecewise monotonic. A number of estimates reflect the order of this piecewise monotonicity. Dolzenko and

Sevastianov [1,2] formulated and studied the problem of Hausdorff approximation by piecewise monotone functions. We shall present here a number of their results.

Following these authors, we introduce the set $\mathbf{M}_n = \mathbf{M}_{n,\Delta}$ of piecewise monotone segment functions of order no higher than n, defined on $\Delta = [a, b]$, where $-\infty \leq a < b \leq \infty$. The constants will be called piecewise monotone functions of zero order. As a point of departure, by $M_n = M_{n,\Delta}$, we mean the set of all functions $f \in C_\Delta$ such that there exists a partition of the closed interval Δ by points $a = x_0 < x_1 < x_2 < \cdots < x_n = b$ into n subintervals, such that f is monotonic on each subinterval $[x_{j-1}, x_j]$.

Definition 4.8. The set $\mathbf{M}_n = \mathbf{M}_{n,\Delta} \subset \mathbf{F}_\Delta$ of piecewise monotone functions of order no higher than n consists of all the functions $f \in \overline{\mathbf{F}}_\Delta$ which can be approximated with arbitrary accuracy by functions in $M_n = M_{n,\Delta}$ with respect to Hausdorff distance, i.e., $\mathbf{M}_n$ is the closure of M_n with respect to Hausdorff distance.

The compactness of each set of uniformly bounded piecewise monotone functions of order no higher than n follows directly from the defintion of piecewise monotone function. Of course, this implies the existence of an element of best approximation by piecewise monotone functions of order no higher than n for every function $f \in \overline{\mathbf{F}}_\Delta$ with respect to Hausdorff distance, i.e., the following assertion holds.

Lemma 4.44. For every $f \in \overline{\mathbf{F}}_\Delta$, there exists $\varphi^* \in \mathbf{M}_n$ for which $E(\mathbf{M}_n, \Delta, \alpha; f) = r(\Delta, \alpha; f, \varphi^*)$.

Let us now consider some of the properties of piecewise monotone functions. It follows directly from the definition that if $f \in \mathbf{M}_{n,\Delta}$, then there exists points $a = x_0 < x_1 < x_2 < \cdots < x_m = b$ with $m \leq n$ such that f is monotonic on every open interval (x_{i-1}, x_i), $i = 1, 2, \ldots, m$. Monotonicity need not hold throughout the closed interval, as can be seen from the following example: let $\varphi \in M_{3,[-1,1]}$ be defined by

$$\varphi(x) = \begin{cases} [-1, 1] & \text{if } x = 0 \\ 0 & \text{otherwise} \end{cases}$$

The function φ is monotonic on $(-1, 0)$ and $(0, 1)$, but it is not monotonic on $[-1, 0]$ or on $[0, 1]$.

To characterize the functions in $\mathbf{M}_n$ we shall introduce the concept of monotone multiplicty of the point $x \in \Delta$ for $f \in \mathbf{F}_\Delta$.

Definition 4.9. Let $f \in \mathbf{F}_\Delta$, $\Delta = [a, b]$. The point $\xi \in (a, b)$ has monotone multiplicity $\kappa(f; \xi)$ with regard to f which is (a) equal to 1, if there exists $\delta > 0$ such that f is monotonic either on the segment $[\xi - \delta, \xi] \cap \Delta$ or on the segment $[\xi, \xi + \delta] \cap \Delta$, or (b) equal to 2, if for every $\delta > 0$ the function is monotonic neither on $[\xi - \delta, \xi] \cap \Delta$ nor on $[\xi, \xi + \delta] \cap \Delta$. The point a (resp. b) has monotone multiplicity 0 if there exists $\delta > 0$ such that the function f is monotonic on the segment $[a, a + \delta] \cap \Delta$ (resp. $[b - \delta, b] \cap \Delta$). Otherwise, the point a (resp. b) has monotone multiplicity 1.

The correctness of the following statements are easily verified.

Lemma 4.45. If $f \in \mathbf{F}_\Delta$, $\Delta = [a, b]$, let the points $x_i = a + i(b - a)/n$, $i = 0, 1, 2, \ldots, n$ partition $[a, b]$ into n equal intervals, and let $m_i = \min \{y : y \in f(x),\ x \in [x_{i-1}, x_i]\}$, $M_i = \max \{y : y \in f(x),\ x \in [x_{i-1}, x_i]\}$, $i = 1, 2, 3, \ldots n$. Define a function φ as follows:

$$\varphi(x) = m_1 + (M_1 - m_1)\frac{x - a}{x_1 - a} \quad \text{for } x \in [a, x_1),$$

$$\varphi(x_1) = M_1 \vee M_2 = \begin{cases} [M_1, M_2] & \text{if } M_1 \le M_2 \\ [M_2, M_1] & \text{if } M_2 \le M_1 \end{cases},$$

$$\varphi(x) = M_2 + (m_2 - M_2)\frac{x - x_1}{x_2 - x_1} \quad \text{for } x \in (x_1, x_2),$$

$$\varphi(x_2) = m_2 \vee m_3,$$

$$\varphi(x) = m_3 + (M_3 - m_3)\frac{x - x_2}{x_3 - x_2} \quad \text{for } x \in [a, x_1),$$

and so on, then $\varphi \in \mathbf{M}_{n,\Delta}$.

Theorem 4.29. A necessary and sufficient condition for the function $f \in \overline{\mathbf{F}}_\Delta$ to be piecewise monotonic of order no higher than n is the existence of knots $a = x_0 < x_1 < x_2 < \cdots < x_m = b$ such that f is monotonic on each of the intervals (x_{i-1}, x_i), $i = 1, 2, 3, \ldots, m$, and the sum of the monotone multiplicity of these knots at most $n - 1$, i.e., $1 + \sum_{i=0}^{m} \kappa(f; x_i) \le n$.

Now we shall consider the estimate for best approximation of functions in $\bar{\mathbf{F}}_\Delta$ by piecewise monotone functions.

Theorem 4.30. (E. Dolzenko and E. Sevastianov [1,2]). The inequality

$$E(\mathbf{M}_n, \Delta, \alpha; f) \leq \frac{b - a}{\alpha n}$$

holds for every function $f \in \bar{\mathbf{F}}_\Delta$, $\Delta = [a, b]$.

Proof. Let the points $x_i = a + i(b - a)/n$, $i = 0, 1, 2, \ldots, n$ partition $[a, b]$ into n equal intervals, and let $m_i = \min \{y : y \in f(x),\ x \in [x_{i-1}, x_i]$, $M_i = \max \{y : y \in f(x),\ x \in [x_{i-1}, x_i]\}$, $i = 1, 2, 3, \ldots, n$. Let the function φ be as described in Lemma 4.45. We claim that $r(\Delta, \alpha; f, \varphi) \leq (b - a)/\alpha n$. Indeed, for every point (x,y) where $x \in \Delta$ and $y \in f(x)$, there exists an index $i \in \{1, 2, 3, \ldots, n\}$ such that $x \in [x_{i-1}, x_i]$. But since φ attains all values between m_i and M_i on the closed interval $[x_{i-1}, x_i]$, there exists $\xi \in [x_{i-1}, x_i]$ for which $\varphi(\xi) = y \in f(x)$. Therefore,

$$h(\Delta, \alpha; f, \varphi) \leq \alpha^{-1}(x_i - x_{i-1}) = \frac{b - a}{\alpha n}. \tag{4.249}$$

On the other hand, for $x \in \Delta$ and $y \in \varphi(x)$ arbitrary, we consider first the case that x is not one of the knots. Then there is an index i with $x \in (x_{i-1}, x_i)$ and $y = \varphi(x) \in [m_i, M_i]$. According to Lemma 1.6, there exists $\xi \in [x_{i-1}, x_i]$ such that $\eta = \varphi(x) \in f(\xi)$.. If x coincides with one of the knots x_i, $i = 1, 2, \ldots, \cdot - 1$, then such a ξ can be found in either $[x_{i-1}, x_i]$ or $[x_i, x_{i+1}]$. Hence ,we see in either case that $h(\Delta, \alpha; f, \varphi) \leq (b - a)/\alpha n$. From this and (4.249), it follows that $r(\Delta, \alpha; f, \varphi) \leq (b - a)/\alpha n$, completing the proof.

Let us show that for $n = 1, 2, 3, \ldots$, there exists a function $f_n \in \bar{\mathbf{F}}_\Delta$, $\Delta = [a, b]$, such that

$$E(\mathbf{M}_n, \Delta, \alpha; f_n) = \frac{b - a}{\alpha n}. \tag{4.250}$$

To this end let $x_i = a + i(b - a)/n$, $i = 0, 1, 2, \ldots, n$ and let $h > 2(b - a)$, and let $m = [(n + 1)/2]$. We shall define $f_n \in \bar{\mathbf{F}}_\Delta$ in the following way:

$$f_n(x_{2k-1}) = [-h, h] \qquad \text{for} \quad k = 1, 2, 3, \ldots, m,$$

$$f_n(x) = 0 \qquad \text{for} \quad x \notin \{x_1, x_3, x_5, \ldots, x_{2m-1}\}.$$

Let φ be an arbitrary piecewise monotone function, for which $r(\Delta, \alpha; f_n, \varphi) \leq (b - a)/\alpha n = \delta$ holds. Then at every knot x_{2k}, $k = 0, 1, 2, \ldots, m$ and for every $\eta \in \varphi(x_{2k})$ the inequality $|\eta| < (b - a)/n < h/2$ will hold. On the other hand, for every knot x_{2k-1}, $k = 1, 2, \ldots, m$, there exist points $\xi'_{2k-1}, \xi''_{2k-1} \in [x_{2k-1} - \delta, x_{2k-1} + \delta]$ and $\eta'_{2k-1} \in \varphi(\xi'_{2k-1})$, $\eta''_{2k-1} \in \varphi(\xi''_{2k-1})$, such that $\eta'_{2k-1} > h - \alpha\delta > h/2$ and $\eta''_{2k-1} < -h + \alpha\delta < -h/2$. Consequently, the function φ changes its direction of monotonicity at least twice on the interval (x_{2k}, x_{2k+1}), that is, in this interval there is a point of order of monotonicity 2 with respect to φ. If n is odd, then the function φ changes the direction of its monotonicity at least once on $(x_{n-1}, b]$. Then for every partition $a = \xi_0 < \xi_1 < \cdots < \xi_p = b$ for which φ remains monotonic in every interval (ξ_{i-1}, ξ_i), $i = 1, 2, 3, \ldots, p$, we have

$$\sum_{i=0}^{p} \kappa(\varphi; \xi_i) \geq n + 1.$$

Therefore φ does not belong to $\mathbf{M}_n$. Hence, (4.250) is established.

From (4.250) and Theorem 4.30 follows

Theorem 4.31. (E. Dolzenko and E. Sevastianov [1,2]). The inequality

$$E(\mathbf{M}_n, \Delta, \alpha; \bar{\mathbf{F}}_\Delta) = \frac{b - a}{\alpha n}$$

holds for every positive integer n, where $\Delta = [a, b]$.

The determination of the exact value of best Hausdorff approximation by piecewsie monotone functions for the class $\bar{\mathbf{F}}_\Delta$ raises the question of finding subclasses of $\bar{\mathbf{F}}_\Delta$ for which better estimates hold. E. Dolzenko and E. Sevastianov [2] obtained quite precise results of a negative nature in this direction. They proved that the modulus of continuity of a function f cannot in essence be used to obtain an improved rate of Hausdorff approximation in this context.

Recall that a function $\omega(\delta)$, $\delta \geq 0$, is called a **function of modulus of continuity type** if it is continuous, nondecreasing, subadditive, and $\lim_{\delta\to 0} \omega(\delta)/\delta = 0$.

Theorem 4.32. (E. Dolzenko and E. Sevastianov [2]). Let $\omega(\delta)$ be a function of modulus of continuity type, and let $\Delta = [a, b]$. Then there exists a function $f \in C_\Delta$ with modulus of continuity $\omega(f; \delta) \leq \omega(\delta)$ such that for a certain sequence of positive integers $m(1) < m(2) < m(3) < \cdots$, the inequality

$$E(\mathbf{M}_{m(k)}, \Delta, \alpha; f) = \frac{(1 - \varepsilon_k)(b - a)}{\alpha m(k)}$$

holds, where $\lim_{k \to \infty} \varepsilon_k = 0$.

Theorem 4.33. (E. Dolzenko and E. Sevastianov [2]). For every function $\omega(\delta)$ of modulus of continuity type satisfying $\int_0^1 \frac{1}{\omega(\tau)} dt < \infty$, and for any $\Delta = [a, b]$, there exist $f \in C_\Delta$ with modulus of continuity $\omega(f; \delta) \leq \omega(\delta)$ such that for each positive integer n, we have the inequality

$$E(\mathbf{M}_n, \Delta, \alpha; f) \geq \frac{c(b - a)}{\alpha n},$$

where $c > 0$ is independent of n, Δ, and α.

All the approximation means that have been considered up to this point- algebraic polynomials, trigonometric polynomials, rational functions, and spline functions- are piecewise monotone functions. Hence, the two theorems mentioned above are highly relevant. For the proof of these theorems, we shall use the following lemma.

Lemma 4.46. (E. Dolzenko and E. Sevastianov [2]). Let $E \subset \Delta$ be the union of a finite number of closed intervals, and let $\Delta_1, \Delta_2, \ldots, \Delta_\nu$ be subintervals of E at least $2\alpha d$ distant from the complement of E and from each other. Set $\delta = \max\{|\Delta_i| : i = 1, 2, \ldots, \nu\}$, and let $\varphi \in F_\Delta$ be a function such that

$$\min\{y : y \in \varphi(x),\ x \in \Delta_i\} \leq m, \quad \max\{y : y \in \varphi(x),\ x \in \Delta_i\} \geq M,$$

for $i = 1, 2, 3, \ldots, \nu$, where $M - m > 2d$. Then the inequality

$$E(\mathbf{M}_{2\nu-1}, \Delta, \alpha; \varphi + \psi) \geq d$$

holds for every $\psi \in \mathbf{F}_\Delta$ for which $V_\xi^\eta \psi \leq M - m - 2d$ where $\xi, \eta \in E$ and $0 \leq \eta - \xi \leq 2\alpha d + \delta$.

Proof . Without loss of generality, we can assume that φ and ψ are continuous functions. Let us assume that the conclusion of the lemma fails, namely, that $E(\mathbf{M}_{2\nu-1}, \Delta, \alpha; \varphi + \psi) < d$. Let $g \in \mathbf{M}_{2\nu-1}$ and $r(\Delta, \alpha; \varphi + \psi, g) = \rho < d$, where one can assume that g is also continuous. Denote by x_i^- and x_i^+ the points of the interval Δ situated to the left and tho the right of the interval Δ_i respectively, and at a distance αd away from it. Since the total variation of ψ on $[x_i^-, x_i^+]$ is less than or equal to $M - m - 2d$ and the total variation of φ on $[x_i^-, x_i^+]$ is at least $M - m$, the function g has to change its direction of monotonicity (from increasing to decreasing) at least once on Δ_i. But then g will change its direction of monotonicity at least 2ν times on Δ, so that g cannot belong to $\mathbf{M}_{2\nu-1}$. Thus, $E(\mathbf{M}_{2\nu-1}, \Delta, \alpha; \varphi + \psi) < d$ cannot occur, and the lemma is proved.

Proof of Theorem 4.32 . To simplify notation, we shall set the parameter α of Hausdorff distance equal to 1, which, in view of Lemma 2.8, is no restriction. Also, without loss of generality, we can also assume that $\omega(f; 0) = 0$. The construction of the function f whose existence was asserted in the statement of the theorem, is based on the use of the function

$$\lambda(\omega, h; x) = \max \{0, h - \omega(|x|)\}, \tag{4.251}$$

where ω is the function of modulus of continuity type in the hypotheses of the theorem. One can verify that the function of (4.251) is equal to zero outside the inverval $[-\delta(h), \delta(h)]$, where $\omega(\delta(h)) = h$, and hence

$$\lim_{h \to +0} \frac{\delta(h)}{h} = 0, \tag{4.252}$$

since by the hypotheses of the theorem we have $\omega(\delta)/\delta \to 0$ as $\delta \to 0$. In addition, from the definition of λ, it follows that

$$\max_{x} \lambda(\omega, h; x) = \lambda(\omega, h; 0) = h \quad \text{and} \quad \omega(\lambda; \delta) \le \omega(\delta). \tag{4.253}$$

Let $x_k(n) = a + (b - a)(2k - 1)/2n$ and let $h(n) = (b - a)/n$. Define for $\nu \ge 1$ fixed the function $\varphi(n)$ by

$$\varphi(n; x) = \varphi(n, \omega, \nu; x) = \sum_{k=1}^{n} \lambda(2^{-k}\omega, 2\nu h(n), x - x_k(n)). \tag{4.254}$$

The graph of the function in (4.254) consists of n teeth, and according to (4.252), if n is sufficiently large, the teeth of the function are above the nonintersecting equal intervals $\Delta_k(n) = (x_k(n) - \delta(2\nu h(n)), x_k(n) + \delta(2\nu h(n))$, outside of which the function is equal to zero, and on each of which $\varphi(n)$ attains the maximum value of $\nu h(n)$. In addition, according to (4.252), the ratio of the length of an interval $\Delta_k(n)$ to the distance betweeen the vertices of adjacent teeth can be made arbitrarily small.

The desired function f will be constructed as a sum of a series, namely,

$$f(x) = \sum_{i=1}^{\infty} \varphi(n_i, 2^{-i}\omega, \nu_i; x) , \tag{4.255}$$

where the sequences $\langle\nu_i\rangle_1^\infty$ and $\langle n_i\rangle_1^\infty$ are defined recursively by a procedure that we now describe.

Let $\nu_1 = n_1 = 1$, and denote by A_1 the empty set. Now assume $\nu_1, \nu_2, \ldots, \nu_{i-1}$ and $n_1, n_2, \ldots, n_{i-1}$, $i \ge 2$, have been constructed. Denote by A_i the finite set of points $x_k(n_j)$, $j = 1, 2, 3, \ldots, i - 1$, $k = 1, 2, \ldots, n_j$. The set $A_i \subset (a, b)$ contains the vertices of all the teeth of the functions $\varphi(n_1), \varphi(n_2), \ldots, \varphi(n_{i-1})$. Let us cover A_i with a finite number of nonintersecting open intervals without endpoints in common of total length $(1 - 2^{-1/i})(b - a)$. Denote by E_i the set of those $x \in \Delta$ which are not included in any of these intervals. Then E_i consists of a finite number of closed intervals with total length $2^{-1/i}(b - a)$, and the partial sum function

$$S_i(x) = \sum_{j=1}^{i-1} \varphi(n_j, 2^{-j}\omega, \nu_j; x)$$

satisfies a Lipschitz condition with constant $L_i > 1$ on E_i. We now set $\nu_i = 4L_i$ and choose n_i so large that these conditions all hold:

(a) $\nu_i h(n_i) < \nu_{i-1} h(n_{i-1})/9$;

(b) $\dfrac{\delta(2^{i+1}\nu_i h(n_i))}{h(n_i)} \le 2^{-i}$;

(c) the number of points $x_k(n_i)$, $k = 1, 2, 3, \ldots, n_i$, in E_i and at a distance at least $h(n_i)$ from $\Delta\setminus E_i$ is greater than $2^{-1/i}\,\mathrm{mes}\,(E_i)/h(n_i) + 2 = 2^{-2/i}n_i + 2$.

The sequences $\langle\nu_i\rangle_1^\infty$ and $\langle n_i\rangle_1^\infty$ are now defined. Since $0 \le \varphi(n_i, 2^{-i}\omega, \nu_i; x)) \le \nu_i h(n_i)$, both the uniform convergence of the series of (4.255) and the continuity of its limit f follow from (a). For the modulus of continuity of the function f, we have

$$\omega(f; \delta) \le \sum_{i=1}^{\infty} \omega(\varphi(n_i, 2^{-i}\omega, \nu_i); \delta) \le \sum_{i=1}^{\infty} 2^{-i}\,\omega(\delta) = \omega(\delta).$$

Let $2d = h(n_i)(1 - \gamma(n_i))$, where

$$\gamma(n_i) = \frac{2\delta(2^{i+1}\nu_i h(n_i))}{h(n_i)} .$$

By (b), $\gamma(n_i) \to 0$ as $i \to \infty$. Obviously, the distance between the adjacent intervals $\Delta_k(n_i) = (x_k(n_i) - \delta(2^{i+1}\nu_i h(n_i)), x_k(n_i) + \delta(2^{i+1}\nu_i h(n_i)))$ is equal to 2d. Let us set $E_i^* = E_i\setminus U'\,\Delta_k$ where U' denotes the union over k of those intervals $\Delta_k(n_i)$ that contain points whose distance from $\Delta\setminus E_i$ is less than $2d < h(n_i)$. According to (c), the set E_i^* contains more than $2^{-2/i}n_i + 1$ intervals $\Delta_k(n_i) = \Delta_k$. The set of these indices k will be denoted by J_i. As a result, the number of elements of J_i, denoted by $N(J_i)$, satisfies the inequality

(4.256) $\qquad N(J_i) \ge 2^{-2/i}n_i + 1.$

The total variation of the function S_i on an arbitrary closed interval of length $h(n_i)$, lying in the set E_i^*, is at most

$$L_i h(n_i) = \frac{\nu_i h(n_i)}{4} = \frac{1}{8}\max_x \varphi(n_i, 2^{-i}\omega, \nu_i; x).$$

On the other hand, by (a), the function

$$R_i(x) = \sum_{j=i+1}^{\infty} \varphi(n_j, 2^{-j}\omega, \nu_j; x)$$

satisfies the inequality

$$0 \le R_i(x) \le 2\nu_i h(n_i)\,(3^{-2} + 3^{-4} + 3^{-6} + \cdots) = \frac{1}{8} \max_x \varphi(n_i, 2^{-i}\omega, \nu_i; x).$$

Therefore, the total variation of the function $\psi_i = S_i + R_i$ on an arbitrary closed interval of length

$$h(n_i) = h(n_i)(1 - \gamma(n_i)) + h(n_i)\gamma(n_i) = 2d + 2\delta(2^i \nu_i h(n_i))$$

$$= 2d + \max\{|\Delta_k(n_i)| : k \in J_i\} = 2d + \delta$$

lying in the set E_i^* is at most $\nu_i h(n_i)/2$. Further, for every $k \in J_i$, if we define M and m by

$$M = \max\{\varphi(n_i, 2^{-i}\omega, \nu_i; x) : x \in \Delta_k\} = 2\nu_i h(n_i),$$

$$m = \min\{\varphi(n_i, 2^{-i}\omega, \nu_i; x) : x \in \Delta_k\} = 0,$$

then $M - m = 2\nu_i h(n_i) \ge \nu_i h(n_i)/2 + h(n_i)(1 - \gamma(n_i))$, so that $\nu_i h(n_i)/2 \le M - m - 2d$. Now set $m(i) = 2N(J_i) - 1 \ge 2^{1-2/i} n_i + 1$. By Lemma 4.46 and (4.256), we have

$$E(M_{m(i)}, \Delta, 1; f) \ge d = h(n_i)(1 - \gamma(n_i))/2$$

$$\ge \frac{(b - a)(1 - \gamma(n_i))m(i)}{2n_i m(i)} \ge \frac{2^{-2/i}(b - a)(1 - \gamma(n_i))}{m(i)}$$

$$= \frac{(1 - \varepsilon_i)^i (b - a)}{m(i)},$$

where $\varepsilon_i \to 0$ as $i \to \infty$. The theorem is proved.

Proof of Theorem 4.33. Integration by parts of $\int dt/\omega(t)$ permits us to write the condition on $\omega(\delta)$ in the statement of the theorem in the form

$$\int_0^1 \frac{t\omega'(t)}{\omega^2(t)}\,dt < \infty.$$

Substituting $t = \delta(h)$, where $\omega(\delta(h)) = h$, we obtain the equivalent condition

$$\int_0^1 \frac{\delta(x)}{x^2}dx < \infty,$$

which in turn is equivalent to the convergence of the series $\sum_{k=1}^{\infty} 5^k\delta(2h_k)$, where $h_k = (b - a)/5^k$. Therefore, there exists a positive integer $q \geq 4$ such that

$$\sum_{k=q}^{\infty} 5^k \cdot 2\delta(2h_k) < \frac{b - a}{10}. \tag{4.257}$$

We shall recursively construct sets E_k, B_k, and G_k, for $k = q, q+1, \ldots$. For E_q we take the empty set, and for B_q we take the set of points $x_{i,q} = a + (2i - 1)h_q/2$, where $i = 1, 2, 3, \ldots, 5^q$. For G_q, we take the union of all intervals with centers in B_q and with length $2\delta(2(b - a)/5^q)$. Obviously, $\text{mes}\,(G_q) = 5^q \cdot 2\delta(2h_q)$.

Suppose now that $E_q, B_q, G_q, \ldots, E_{k-1}, B_{k-1}, G_{k-1}$ have already been constructed. Then we set $E_k = \Delta \setminus \bigcup \{G_i : i = 1, 2, \ldots, k - 1\}$. The set B_k will be composed of all the points $x_{i,k} = a + (2i - 1)h_k/2$, where $i = 1, 2, 3, \ldots, 5^k$, lying in the set E_k and at a distance $\delta(2h_k)$ from its complement. The set of indices i for which $x_{i,k} \in B_k$ will be denoted by J_k. For G_k we shall take the union of all the intervals of length $2\delta(2h_k)$ with centers at the points of the set B_k. Thus, the sets E_k, B_k, and G_k are defined for all $k \geq q$.

Let us single out the following properties of these sets:

(1) the sets G_k are pairwise disjoint;

(2) $\text{mes}\,(G_k) \leq 5^k \cdot 2\delta(2h_k)$, so that by (4.257), $\sum_{k=q}^{\infty} \text{mes}\,(G_k) < \frac{b - a}{10}$;

(3) for $k \geq q + 1$, the set E_k consists of at most 5^{k+1} (possibly degenerate) closed intervals.

Property (3) follows from the fact that the points of the sets $B_q, B_{q+1}, \ldots, B_{k-1}$ are contained among the points $x_{i,k-1}$, and

$$\text{mes}\,(E_k) = b - a - \sum_{k=q}^{\infty} \text{mes}\,(G_i) > \frac{9(b - a)}{10}.$$

We now define φ_k by

$$\varphi_k(x) = \sum_{i \in J_k} \lambda(\omega, 2h_k; x - x_{i,k}),$$

where $\lambda(\omega, h; x)$ is defined by (4.251). Then $\varphi_k(x) = 0$ exterior to G_k, $\varphi_k(x) > 0$ on G_k, and for each $i \in J_k$, we have $\max_x \varphi_k(x) = \varphi_k(x_{i,k})$.

In view of these properties, the function

$$f(x) = \sum_{k=q}^{\infty} \varphi_k(x)$$

is continuous on Δ with modulus of continuity $\omega(f; \delta) \le \omega(\delta)$, and $\omega(f; \delta) = \omega(\delta)$ for $0 < \delta < \delta(2h_q)$. Let us estimate $E(M_n, \Delta, 1; f)$ from below.

It is easy to see that if one of the closed intervals has length l, and on the axis Ox there is a network of points with step h, then more than $l/h - 2$ points of the network belong to this closed interval, whatever its location, and $(l - 2h)/h - 2 = l/h - 4$ of these points will be at a distance $\ge h$ from the complement of this closed interval. Let $\nu(k)$ be the number of points in $B_k \subset E_k$ that are at a distance h_k from the complement of E_k. From the above observations and property (3), we obtain the inequality

$$\nu(k) \ge \frac{\text{mes}\,(E_k)}{h_k} - 4(5^{k-1} + 1) > \frac{9}{10} 5^k - 4\cdot 5^{k-1} - 4 \ge 2\cdot 5^{k-2} + 1,$$

since $k \ge q \ge 4$. Since for $x \in E_k$ we have $\varphi_q(x) = \varphi_{q+1}(x) = \cdots = \varphi_{k-1}(x) = 0$, then for such x we also have $f(x) = \varphi_k(x) + \psi_k(x)$, where $0 \le \psi_k(x) \le \max\{\varphi_i(x) : x \in \Delta, i \ge k+1\} = \max\{\varphi_{k+1}(x) : x \in \Delta\} = \max\{\varphi_k(x) : x \in \Delta\}/5$.

The intervals $\Delta_{i,k}$, $i = 1, 2, \ldots, \nu(k)$ composing the set $G_k \subset E_k$, are at a distance no less than $2d = h_k - 2\delta(2h_k) = h_k(1 - \gamma_k)$ from the complement of E_k and from each other, where $\gamma_k = 2\delta(2h_k)/h_k < 1/10$ (see (4.257)). The function $\varphi_k(x)$ attains the value $M = 2h_k$ on each of the intervals $\Delta_{i,k}$ and for $x \in E_k \backslash G_k$, we have $\varphi_k(x) = 0$. The oscillation of $\psi_k(x)$ on the closed interval Δ does not exceed $M/5 < M - 2d$. All of the conditions of Lemma 4.46 are satisfied by the functions φ_k and ψ_k for $m = 0$. Since $2\nu(k) - 1 > 4\cdot 5^{k-2}$, we have

$$E(M_{4\cdot 5^{k-2}}, \Delta, 1; f) \ge E(M_{2\nu(k)-1}, \Delta, 1; f) > h_k.$$

If $n \ge 4\cdot 5^{q-3}$, then we can find $k = k(n) \ge q$ such that $4\cdot 5^{k-3} \le n \le 4\cdot 5^{k-2}$, and

$$E(M_n, \Delta, 1; f) \ge h_k \ge \frac{4\cdot 5^{-3}(b - a)}{n}.$$

Replacing, if necessary, the constant $4 \cdot 5^{-3}$ by a smaller one, we achieve the fulfillment of the inequality $E(\mathbf{M}_n, \Delta, 1; f) \geq c(b - a)/n$. Hence, the theorem is proved for $\alpha = 1$. For arbitrary $\alpha > 0$, the theorem follows from the case $\alpha = 1$ and Lemma 2.8.

For functions of bounded variation, Theorem 4.28 yields

Corollary 4.7. If $f \in \mathbf{F}_\Delta$ is of bounded variation, then $E(\mathbf{M}_n, \Delta, \alpha; f) = o(n^{-1})$.

Proof . This is immediate by virtue of the following identity (E. Dolzenko and E. Sevastianov [2]):

$$\sum_{n=0}^{\infty} M_n(\Delta, f) = \frac{1}{2} V_a^b f,$$

where $\Delta = [a, b]$, $M_n(\Delta, f) = \inf \{\|f - \varphi\|_\Delta : \varphi \in \mathbf{M}_n\}$, since $E(\mathbf{M}_n, \Delta, \alpha; f) \leq M_n(\Delta, f)$ for $\alpha > 0$.

Remarks . The first proof of the universal estimate for best Hausdorff approximation by algebraic polynomials was based on the Hausdorff approximation of the jump (Bl. Sendov [1,7]; see also V. Veselinov [4]). The best Hausdorff approximation with trigonometric polynomials can be determined by a direct proof, since the trigonometric case does not follow immediately from the algebraic one (V. Veselinov [1,2,3]). The equivalence of the algebraic and trigonometric cases was proved by V. Popov and V. Veselinov [1]. The best weighted approximations were studied by V. Veselinov [20,23].

The universal estimate for the functions of several variables can be obtained by analogy with the one-dimensional case (V. Popov [1,15] and Bl. Sendov and V. Popov [5]), but the exact constants in this case have not been found.

The best approximation of complex functions with respect to Hausdorff distance has been considered by M. Kasciev [1], and the approximation of abstract functions in Hausdorff metric was studied by V. Popov and S. Troyanski [1].

B. Boyanov [1], V. Veselinov [25], and S. Yakov [1] have studied the properties of algebraic polynomials of best Hausdorff approximation.

As it was demonstrated in §4.1.2.3, the classical Jackson's theorem on best uniform approximation by algebraic polynomials can be obtained as a corollary of the estimates for best Hausdorff approximation. In a certain sense, the converse statement was proved by B. Boyanov and V. Veselinov [3]. The exact aymptotics in (4.77) in approximation by algebraic and trigonometric polynomials was established by T. P. Boyanov [6].

The best Hausdorff approximation of concrete functions by algebraic polynomials was considered in the papers of B. Boyanov and M. Zvetanov [1] and Bl. Sendov [15,16,22, 27].

Distribution of the points of maximal deviation for the polynomials of best Hasudorff approximation is considered in S. Tashev [3].

The unimprovability of the universal estimate of best Hausdorff approximation, even in the class Lip α, was established by T. Boyanov [4] (see also T. Boyanov [2] and T. P. Boyanov and L. B. Geshev [1]).

Best Hausdorff approximation by algebraic polynomials with positive coefficients have been studied by V. Veselinov [24], and G. Iliev [3]. P. Petrusev and V. Hristov [2] considered best Hausdorff approximation by Müntz polnomials.

A detailed study of Hausdorff approximation by rational functions can be found in the monograph of P. Petrusev and V. Popov [1]. Best Hausdorff approximation by rational functions can be found in the papers of A. Andreev and V. Popov [2], P. Petrusev [4,5], Bl. Sendov [17,18], T. P. Boyadzieva and P. G. Boyadziev [1], and E. S. Moskona [1]. Exact aymptotics of the best Hausdorff approximation of the function sgn (x) by rational functions was considered by N. V. Vladov [1].

Best Hausdorff approximation by step functions and by spline functions were considered by Bl. Sendov [8,19,28,29], V. Popov and Bl. Sendov [1], and Bl. Sendov and V. Popov [3]. The approximation of functions of two variables by polyhedral functions was studied by T. Martynjuk and V. Storcai [1]. V. Popov [4,5] has studied the approximation of convex functions with polygons with respect to Hausdorff distance.

One-sided Hausdorff approximation was considered by S. Markov [4].

Other questions, relative to this chapter, have been addressed in the following papers: E. Dolzenko and E. Sevastianov [3], V. Popov [11,13], D. Dimitrov and V. Popov [1], S. Tasev [1], and E. Dolzenko and P. Ulyanov [1].

Chapter 5

Converse Theorems

Converse theorems in approximation theory have their origins in the classical investigations of Bernstein. These theorems are of two types. Some are concerned with the existence of a function with preassigned approximations. In others, the rate of approximation is prescribed, from which one derives certain properties of the approximated function, e. g., membership to a given class of functions.

In this chapter, we shall show that in contrast to uniform distance, the converse theorems of the first type are negative in nature with respect to Hausdorff distance. As for the converse theorems of the second type, in the case of Hausdorff distance, the constants in the front of the order of approximation as well as the order itself are significant.

§ 5.1. Existence of a function with preassigned best approximations

The following theorem (I. Natanson [1, p. 174]) is well-known.

Theorem 5.1. Let $\langle A_n \rangle_0^\infty$ be a nonincreasing sequence of nonnegative reals with $\lim_{n \to \infty} A_n = 0$. Then on every segment $[a, b]$, there exists a continuous function f having for each n the constant A_n as its best uniform approximation by algebraic polynomials of degree n. A similar statement is valid for the approximation of 2π-periodic functions by trigonometric polynomials.

It turns out that there does not exist an analogue of Theorem 5.1 in the case of Hausdorff approximation. Keeping in mind the universal estimate obtained in §4.1.2.2, it is natural to pose the following question. Let $\langle A_n \rangle_0^\infty$ be a nonincreasing sequence of nonnegative reals with $0 \le A_n \le Cn^{-1}\ln n$, where C is some fixed constant. For $\alpha > 0$, does there exists and interval $\Delta = [a, b]$ and a function $f \in \mathbf{F}_\Delta$ for which the best Hausdorff approximations by algebraic polynomials coincide with the A_n, i.e., $A_n = E(H_n, \Delta, \alpha; f)$ for $n = 0, 1, 2, \dots$? Similarly does there exist a 2π-periodic function $f \in \mathbf{F}_{2\pi}$ such that $E(T_n, \Delta, \alpha; f) = A_n$ for each n?

We show that the answers to these questions are negative. More precisely, we establish the following statement.

Theorem 5.2. (V. Popov [8]). There exists a nonincreasing sequence $\langle A_n \rangle_0^\infty$ of nonnegative reals with $0 \le A_n \le n^{-1}\ln(e+n)$ such that at least one of the conditions

$$E(H_n; f) = E(H_n, \Delta, 1; f) = A_n, \qquad n = 0, 1, 2, \ldots$$

fails for each $f \in \mathbf{F}_\Delta$ and each $\Delta = [a, b]$. Similarly, there does not exist a 2π-periodic function $f \in \mathbf{F}_{2\pi}$ for this same sequence for which

$$E(T_n; f) = E(T_n; (-\infty, \infty), 1; f) = A_n, \; n = 0, 1, 2, \ldots .$$

Proof. Let $\langle A_n \rangle_0^\infty$ be defined in the following way:

(5.1) $$A_0 = 1, \quad A_n = k!e^{-k}, \quad \text{for} \quad 2^{(k-1)!} \le n < 2^{k!}, \quad n = 1, 2, 3, \ldots .$$

Evidently, $\langle A_n \rangle_0^\infty$ satisfies the hypotheses of the theorem. Suppose there exist $\Delta = [a, b]$ and $f \in \mathbf{F}_\Delta$ for which

(5.2) $$E(H_n; f) = A_n, \qquad n = 0, 1, 2, \ldots .$$

In §4.1.2.4, we showed that for some constant c and each n

(5.3) $$E(H_n; f) \le cn^{-1}\ln(e + n\omega(f; n^{-1})).$$

Denote by $p_n \in H_n$ one of the polynomials of best Hausdorff approximation of degree n for f. According to (5.2), for every $x \in [a, b]$, there exists $\xi(x) \in [a, b]$ and $y \in f(x)$ for which both $|x - \xi(x)| \le A_n$ and $|y - p_n(\xi(x))| \le A_n$.

Since $A_0 = 1$, we have $\max\{p_n(x) : x \in [a, b]\} \le 2$. Then, if $x', x'' \in [a, b]$ with $|x' - x''| \le A_n$, and $y' \in f(x')$ and $y'' \in f(x'')$, we obtain from A. Markov's inequality for polynomials that

$$|y' - y''| \le |y' - p_n(\xi(x'))| + |p_n(\xi(x')) - p_n(\xi(x''))| + |p_n(\xi(x'')) - y''|$$

$$\le 2A_n + \frac{3A_n \cdot 4n^2}{b - a} \le \nu n^2 A_n,$$

where $\nu = 3 + 12/(b - a)$. From this inequality and $n^{-1} \le A_n$, we obtain an estimate for the modulus of continuity of f:

(5.4) $\qquad \omega(f; n^{-1}) \leq vn^2A_n$.

Set $m = 2^{k!} - 1$ and $n = 2^{(k-1)!}$. From (5.3) and (5.4), we have for large k

$$A_m = k!2^{-k!} = E(H_m; f) \leq cm^{-1}\ln(e + m\omega(f; m^{-1}))$$

$$\leq 2c2^{-k!}\ln(e + v2^{2(k-1)!}) < k!2^{-k!},$$

an obvious contradiction. Thus, the part of the theorem pertaining to algebraic polynomials is proved. The proof for trigonometric polynomial is similar and is left to the reader.

We intend to establish a similar negative result for best Hausdorff approximation by piecewise monotone functions.

Theorem 5.3. (E. Dolzenko and E. Sevastianov [2]). There does not exist $f \in \bar{\mathbf{F}}_\Delta$, $\Delta = [a, b]$, for which $E(\mathbf{M}_1, \Delta, 1; f) = b - a$, $E(\mathbf{M}_2, \Delta, 1; f) = (b - a)/2$, and $E(\mathbf{M}_3, \Delta, 1; f) = (b - a)/3$.

In §4.4, we proved that for every function $f \in \bar{\mathbf{F}}_\Delta$, the bound $E(\mathbf{M}_n, \Delta, 1; f) \leq (b - a)/n$ holds, and that this bound is attained for every positive integer n. Theorem 5.3 as stated says in particular that there cannot exist a single $f \in \bar{\mathbf{F}}_\Delta$ for which $E(\mathbf{M}_n, \Delta, 1; f) = A_n$, where $0 \leq A_n \leq (b - a)/n$, $n = 1, 2, 3, \ldots$ for all sequences $\langle A_n \rangle_1^\infty$.

Following the presentation of Dolzenko and Sevastianov [2], we will establish a few preliminary results. For $f \in \bar{\mathbf{F}}_\Delta$, $P(f)$ will denote the minimal closed rectangle with sides parallel to the coordinate axes in which f can be inscribed. We shall say that $f \in \bar{\mathbf{F}}_\Delta$ belongs to the class S_r (resp. S_l) if $f = G \cup d$ where G is a certain set in $\bar{\mathbf{F}}_\Delta$ and d is a vertical segment that contains the right (resp. left) side of the rectangle $P(G)$ with endpoints that are at least at a distance of $b - a$ from $P(G)$.

Let f_1 (resp. f_2) denote the restriction of a function $f \in \bar{\mathbf{F}}_\Delta$ to the closed interval $[a, (a + b)/2]$ (resp. $[(a + b)/2, b]$). We say that f belongs to the class T_r (resp. T_l) provided

(a) $f \in S_r$ (resp. $f \in S_l$);
(b) f_1 (resp. f_2) belongs to S_l (resp. S_r).

We set $S = S_r \cup S_l$ and $T = T_r \cup T_l$.

Lemma 5.1. If $f \in \bar{\mathbf{F}}_\Delta$ and $E(\mathbf{M}_1, \Delta, 1; f) = b - a$, then $f \in S$. If $f \notin S$, then $r(\Delta; f, l) < b - a$ holds for one of the diagonals l of the rectangle $P(f)$.

Proof. The second part is obvious, from which the first follows.

Lemma 5.2. Suppose $f \in \bar{\mathbf{F}}_\Delta$ with $E(\mathbf{M}_1, \Delta, 1; f) = b - a$ and $E(\mathbf{M}_2, \Delta, 1; f) = (b - a)/2$. Then $f \in T$.

Proof. By Lemma 5.1, we have $f \in S$. Without loss of generality, let $f \in S_l$. We set

$$f_0 = \{(x,y) : y \in f(x),\ (b + a)/2 < x \le b\},$$

$$f_1 = \{(x,y) : y \in f(x),\ a \le x \le (b + a)/2\},$$

$$m_i = \min\{y : (x,y) \in f_i\}, \qquad M_i = \max\{y : (x,y) \in f_i\},\ i = 0, 1.$$

Since $f \in S_l$, we have $m_1 < m_0 \le M_0 < M_1$.

Suppose that $f \notin T$. Then $f_0 \notin S_r$. On the other hand, by construction, $f_0 \notin S_l$, so that $f_0 \notin S$. By Lemma 5.1, $r(\Delta; f_0, l) < (b - a)/2$ for one of the diagonals l of the rectangle $P(f_0)$, say, $l = [((a + b)/2, M_0), (b, m_0)]$.

Consider the polygonal path $L = [(a, m_1), (a + (b - a)/4, M_1), ((a + b)/2, M_0), (b, m_0)]$. Since $m_1 < m_0 \le M_0 < M_1$, we have $L \subset P(f_0) \cup P(f_1)$, and L has two regions of monotonicity. Since $f_1 \in S_l$, it is easy to see that all the points of the polygonal path $L \cap P(f_1)$, except for the point $A = ((a + b)/2, M_0)$, lie in the open $(b - a)/2$-neighborhood of the set f_1. Also, A and all of $L \cap P(f_0)$ lie in the open $(b - a)/2$-neighborhood of the set f_0. As a result, the entire path L lies in the open $(b - a)/2$-neighborhood of the set f. On the other hand the rectangle $P(f_1)$ lies in the open $(b - a)/2$-neighborhood of $L \cap P(f_0)$, because $r(f_0, L \cap P(f_0)) < (b - a)/2$. Thus, f and L lie in open Hausdorff $(b - a)/2$-neighborhoods of each other, whence $r(\Delta; f, L) < (b - a)/2$. This contradicts $E(\mathbf{M}_2, \Delta, 1; f) = (b - a)/2$, completing the proof of the lemma.

Proof of Theorem 5.3. Let us denote by f_1 and f_2 subsets of the set f that consist of all $(x,y) \in f$ for which $x \in [a, a + (b - a)/3]$ and $x \in [a + 2(b - a)/3, b]$, respectively. Also, denote by f_0 the closure of the set of all $(x,y) \in f$ for which $x \in (a + (b - a)/3, a +$

$2(b - a)/3)$. Obviously, f_0 is the closure of $f \setminus (f_1 \cup f_2)$. If $E(M_1, \Delta, 1; f) = b - a$ and $E(M_2, \Delta, 1; f) = (b - a)/2$, then $f \in T$ by Lemma 5.2. For definiteness, suppose $f \in T_r$. Then $f_1 \in S_l$ and $f_2 \in S_r$. From the definition of f_0 it is easy to see that $r(f_0, P(f_0)) < (b - a)/3$. We now write for $i = 0, 1, 2$, $m_i = \min \{y : (x,y) \in f_i\}$ and $M_i = \max \{y : (x,y) \in f_i\}$. We introduce a five component polygonal path, whose compositon depends on the relative sizes of M_0 and M_1: if $M_0 \geq M_1$, then $L = [(a,m_1), (a + (b - a)/3, M_1), (a + (b - a)/2, M_0), (a + 2(b - a)/3, m_0), (a + 5(b - a)/6, m_2), (b, M_2)]$, and if $M_0 < M_1$, then $L = [(a,m_1), (a + (b - a)/6, M_1), (a + (b - a)/3, M_0), (a + 2(b - a)/3, m_0), (a + 5(b - a)/6, m_2), (b, M_2)]$.

As in the proof of Lemma 5.2, the Hausdorff distance between f_0 and $L \cap P(f_0)$ is less than the width of the rectangle $P(f_0)$, i.e., $r(f_0, L \cap P(f_0)) < (b - a)/3$. Using arguments similar to those employed in the proof of Lemma 5.2, we see that the set $f_1 \cup f_0$ and that part of L whose projection on the horizontal axis is the interval $[a, a + 2(b - a)/3]$ lie in open Hausdorff $(b - a)/3$-neighborhoods of one another, and $f_0 \cup f_2$ and that part of L whose projection on the horizontal axis is the interval $[a + (b - a)/3, b]$ also lie in open Hausdorff $(b - a)/3$-neighborhoods of one another. Hence, $r(\Delta; f, L) < (b - a)/3$, and Theorem 5.3 is proved.

In §4.3, we obtained an exact estimate of the approximation of $f \in \overline{F}_\Delta$, $\Delta = [a, b]$, by spline functions with equidistant knots, namely

$$E(\overline{S}_{k,n}, \Delta, 1; f) \leq \frac{(k + 3)(b - a)}{2n}.$$

In the present context, it is natural to ask whether there exists $f \in \overline{F}_\Delta$ with preassigned values $E(\overline{S}_{k,n}, \Delta, 1; f)$. More precisely, if $\langle A_n \rangle_1^\infty$ is a sequence of numbers for which $0 \leq A_{n+1} \leq A_n \leq (k + 3)(b - a)/2n$, $n = 1, 2, 3, \ldots$, does there exist $f \in \overline{F}_\Delta$ for which $E(\overline{S}_{k,n}, \Delta, 1; f) = A_n$ for each n? We shall show that the answer to this question is negative for $k = 1$. First, we establish

Lemma 5.3. For every positive integer n we have the inequality

$$E(\overline{S}_{1,n}, \Delta, 1; \overline{S}_{1,n+1}) \leq \frac{3(b - a)}{2n}. \tag{5.5}$$

Proof. Let $\varphi \in \bar{S}_{1,n+1}$ be arbitrary. Write $m_i = \varphi(x_{i,n+1})$ where $x_{i,n+1} = a + i(b - a)/(n + 1)$ for $i = 0, 1, 2, \ldots, n + 1$. We define $\psi \in \bar{S}_{1,n}$ at the knots $x_{i,n} = a + i(b - a)/n$ as follows. For $i = 0, 1, 2, \ldots, q = [(n - 1)/2]$, set $\psi(x_{i,n}) = \varphi(x_{i,n+1}) = m_i$. For $i = q + 1, q + 2, \ldots, n$, the definition of $\psi(x_{i,n})$ depends on which of m_q and m_{q+1} is larger:

(a) If $m_q \le m_{q+1}$, we set $\psi(x_{q+1,n}) = \max\{m_{q+1}, m_{q+2}\}$, $\psi(x_{q+2,n}) = \min\{m_{q+2}, m_{q+3}\}$, $\psi(x_{q+3,n}) = \max\{m_{q+3}, m_{q+4}\}$, and so forth.

(b) If $m_q > m_{q+1}$, we set $\psi(x_{q+1,n}) = \min\{m_{q+1}, m_{q+2}\}$, $\psi(x_{q+2,n}) = \max\{m_{q+2}, m_{q+3}\}$, $\psi(x_{q+3,n}) = \min\{m_{q+3}, m_{q+4}\}$, and so forth.

Keeping in mind that $\varphi \in \bar{S}_{1,n+1}$ and $\psi \in \bar{S}_{1,n}$, it is not hard to see that

$$r(\Delta, 1; \varphi, \psi) \le \frac{b - a}{n} + d(b - a), \tag{5.6}$$

where $d = \max\{q/n - q/(n + 1), (q + 1)/(n + 1) - q/n\}$. When n is odd, say $n = 2m + 1$, we have $q = m = (n - 1)/2$ and $d = \max\{(n - 1)/2n(n + 1), 1/2n\} = 1/2n$. For $n = 2m$, we have $q = m - 1 = (n - 2)/2$ and $d = (n - 2)/2n(n + 1) < 1/2n$. Hence, it follows from (5.6) that $r(\Delta, 1; \varphi, \psi) \le 3(b - a)/2n$. Since $\varphi \in \bar{S}_{1,n+1}$ was arbitrary, the lemma is proved.

Theorem 5.4. There exists a nonincreasing sequence of reals $\langle A_n \rangle_1^\infty$ with $0 \le A_n \le 2(b - a)/n$ such that no function $f \in \bar{F}_\Delta$ satsifies $E(\bar{S}_{1,n}, \Delta, 1; f) = A_n$, $n = 1, 2, \ldots$.

Proof. Take such a sequence such that for a certain fixed integer p, we have $A_p = 2(b - a)/p$ and $A_{p+1} < (b - a)/2p$. Assume now that there does exist $f_0 \in \bar{F}_\Delta$ for which $E(\bar{S}_{1,p}, \Delta, 1; f) = A_p$ and $E(\bar{S}_{1,p+1}, \Delta, 1; f) = A_{p+1}$. Choose $\psi \in \bar{S}_{1,p+1}$ such that $r(\Delta; f_0, \psi) = A_{p+1}$, and, by Lemma 5.3, choose $\varphi \in \bar{S}_{1,p}$ satisfying $r(\Delta; \psi, \varphi) \le 3(b - a)/2p$. Then

$$A_p = \frac{2(b - a)}{p} \le r(\Delta; f_0, \varphi) \le r(\Delta; f_0, \psi) + r(\Delta; \psi, \varphi)$$

$$\le A_{p+1} + \frac{3(b - a)}{2p} < \frac{2(b - a)}{p} .$$

This contradicts the definition of A_p, so that no such function f_0 can exist.

The method of proof of Theorem 5.4, resting on Lemma 5.3, cannot be adapted to prove Theorem 5.3 (on approximation by piecewise monotone functions) because $E(\mathbf{M}_n, \Delta, 1; \mathbf{M}_{n+1}) = E(\mathbf{M}_n, \Delta, 1; \overline{\mathbf{F}}_\Delta)$. It is not known, however, whether this method can be used to obtain Theorem 5.2. Suppose we could compute $E(H_n, \Delta, 1; H_{n+1})$. If it turns out $E(H_n, \Delta, 1; H_{n+1}) < E(H_n, \Delta, 1; \overline{\mathbf{F}}_\Delta)$, then Theorem 5.2 follows as a corollary of this inequality.

In conclusion, we note that the converse problem for the approximation of spline functions in $\overline{S}_{k,n}$ for $k = 2, 3, 4, \ldots$ is still unsolved. If $k = 0$, i.e., in the case of approximation by step functions, the problem has an elementary (negative) solution as in Theorem 5.4.

§ 5.2. Converse theorems for the approximation by algebraic and trigonometric polynomials

The order of best Hausdorff approximation by algebraic and trigonometric polynomials can be used to establish certain properties of the approximated function. The conditions on the order of best approximation, which ensure the continuity of the approximated function, are characteristic of Hausdorff approximation. As we mentioned earlier, the constants in front of the order of approximation, as well as the order itself, play an essential role.

We begin by establishing a few preliminary results for the estimate of the modulus of continuity of a function by its best Hausdorff approximations. For this purpose, we invoke Theorem 2.3, which stated that for every two segment functions f and g, defined on a finite or infinite segment Δ, the inequality

$$r(\Delta, \alpha; f, g) \leq \|f - g\| \leq r(\Delta, \alpha; f, g) + \omega(g; \alpha r(\Delta, \alpha; f, g)) \tag{5.7}$$

holds, where $\|f - g\| = \sup\{|f(x) - g(x)| : x \in \Delta\}$ is the uniform distance between f and g.

Lemma 5.4. Let $<\varphi_p>_0^\infty$ be a sequence of differentaible functions, defined on the closed interval Δ for which

$$\varphi'_0(x) \equiv 0 \quad \text{and} \quad \|\varphi'_p - \varphi'_{p-1}\| \leq c_p\|\varphi_p - \varphi_{p-1}\|, \qquad p = 1, 2, 3, \ldots . \tag{5.8}$$

Then the following inequality is valid:

(5.9) $$\|\varphi_p'\| \le \alpha^{-1}\left(-1 + \prod_{i=1}^{p}(1 + \alpha c_i\, r(\Delta, \alpha; \varphi_i, \varphi_{i-1}))\right).$$

Proof . Using inequalities of (5.7) and (5.8) and the properties of the modulus of continuity, we obtain

$$\|\varphi_p'\| \le \|\varphi_p' - \varphi_{p-1}'\| + \|\varphi_{p-1}'\| \le c_p\|\varphi_p - \varphi_{p-1}\| + \|\varphi_{p-1}'\|$$

$$\le c_p r(\alpha; \varphi_p, \varphi_{p-1})) + c_p\omega(\varphi_{p-1}; \alpha r(\alpha; \varphi_p, \varphi_{p-1})) + \|\varphi_{p-1}'\|$$

$$\le c_p r(\alpha; \varphi_p, \varphi_{p-1})) + (1 + \alpha c_p r(\alpha; \varphi_p, \varphi_{p-1}))\|\varphi_{p-1}'\|,$$

or

$$1 + \alpha\|\varphi_p'\| \le (1 + \alpha c_p r(\alpha; \varphi_p, \varphi_{p-1}))(1 + \alpha\|\varphi_{p-1}'\|).$$

In view of (5.8) and $\|\varphi_0'\| = 0$, the last inequality yields (5.9). The lemma is proved.

5.2.1. THE TRIGONOMETRIC CASE

Let $\tau_n \in T_n$ be a trigonometric polynomial of best approximation for $f \in \bar{F}_{2\pi}$. From (5.7) we have

$$E(T_n, \alpha; f) + \omega(\tau_n; \alpha E(T_n, \alpha; f)) \ge \|f - \tau_n\| \ge E_n^T(f) ,$$

where $E_n^T(f)$ is the best uniform approximation of the function f by trigonometric polynomials of nth order. Also, using the left inequality in (5.7), we obtain

(5.10) $$E(T_n, \alpha; f) \le E_n^T(f) \le E(T_n, \alpha; f) + \omega(\tau_n; \alpha E(T_n, \alpha; f)).$$

Using Bernstein's inequality for trigonometric polynomials, namely $\|\tau'\| \le n\|\tau\|$, for $\tau \in T_n$, and Lemma 5.4, one can estimate $\omega(\tau_n; \delta)$:

(5.11) $$\omega(\tau_n; \alpha E(T_n, \alpha; f)) \le \alpha E(T_n, \alpha; f)\, \|\tau_n'\|$$

$$\le E(T_n, \alpha; f)\left(-1 + \prod_{i=1}^{k}(1 + \alpha n_i\, r(\alpha; \tau_{n_i}, \tau_{n_{i-1}}))\right),$$

where n_i are positive integers with $0 = n_0 < n_1 < n_2 < n_3 < \cdots < n_k = n$, and τ_{n_i} is a trigonometric polynomial of best Hausdorff approximation for the function f of order n_i, $i = 0, 1, 2, \ldots, k$. It is directly seen that

$$(5.12) \qquad \prod_{i=1}^{k} (1 + \alpha n_i\, r(\alpha; \tau_{n_i}, \tau_{n_{i-1}})) \le$$

$$(1 + \alpha n_1 a(0))(1 + \alpha n_k a(n_k)) \prod_{i=1}^{k-1} (1 + \alpha(n_{i+1} - n_i)a(n_i)),$$

where $a(m) = E(T_m, \alpha; f)$. It is obvious that

$$(5.13) \qquad 1 + \alpha n_k a(n_k) \le \left(1 + \frac{\alpha n_k a(n_k)}{n_k - n_{k-1}}\right)^{n_k - n_{k-1}},$$

$$(5.14) \qquad 1 + \alpha(n_{i+1} - n_i)a(n_i) \le \left(1 + \frac{\alpha a(n_i)(n_{i+1} - n_i)}{n_i - n_{i-1}}\right)^{n_i - n_{i-1}},$$

where $n_0 = 0$.

We claim that for every positive integer n, we can choose a finite sequence of positive integers $n_0 = 0 < n_1 < n_2 < \cdots < n_k = n$ such that

$$(5.15) \qquad \max\left\{n_1, \frac{n_2 + n_1}{n_1 - n_0}, \frac{n_3 + n_2}{n_2 - n_1}, \ldots, \frac{n_k + n_{k-1}}{n_{k-1} - n_{k-2}}, \frac{n_k}{n_k - n_{k-1}}\right\} \le (1 + \sqrt{2})^2.$$

Indeed, it suffices to let k be the largest positive integer for which $(1 + \sqrt{2})^k - 3 < n$, and then to set $n_i = [(1 + \sqrt{2})^{i+1}] - 3$, $i = 1, 2, 3, \ldots, k - 1$. Then $n_1 = 2 < (1 + \sqrt{2})^2$, and

$$\frac{n_{i+1} + n_i}{n_i - n_{i-1}} \le \frac{(1 + \sqrt{2})^{i+2} + (1 + \sqrt{2})^{i+1} - 6}{(1 + \sqrt{2})^{i+1} - (1 + \sqrt{2})^i - 1}$$

$$\le (1 + \sqrt{2})^2 \frac{((1 + \sqrt{2})^i + (1 - \sqrt{2}))((1 + \sqrt{2})^i - 1)}{(\sqrt{2}(1 + \sqrt{2})^i - 1)}$$

$$= (1 + \sqrt{2})^2.$$

From (5.12) - (5.15), we obtain

$$\prod_{i=1}^{k} (1 + \alpha n_i\, r(\alpha; \tau_{n_i}, \tau_{n_{i-1}})) \le \prod_{i=0}^{k} (1 + \alpha(1 + \sqrt{2})^2\, a(n_i))^{n_i - n_{i-1}}$$

$$\le \prod_{i=0}^{n} (1 + \alpha(1 + \sqrt{2})^2\, E(T_i, \alpha; f)).$$

Combining the last inequality with (5.10) and (5.11), we obtain

Lemma 5.5. For every 2π-periodic function f, the inequality

$$E(T_n, \alpha; f) \le E_n^T(f) \le E(T_n, \alpha; f) \prod_{i=0}^{n} (1 + \alpha(1 + \sqrt{2})^2\, E(T_i, \alpha; f)) \tag{5.16}$$

holds for the best uniform approximations and best Hausdorff approximations by trigonometric polynomials.

From (5.16), we see that if the series $\sum_{i=0}^{\infty} E(T_i, \alpha; f)$ is convergent, then $E(T_n, \alpha; f)$ and $E_n^T(f)$ have the same order. This inequality also yields an inequality of Dolzenko and Sevastianov [1]:

$$E(T_n, \alpha; f) \le E_n^T(f) \le E(T_n, \alpha; f) \exp\left(\alpha(1 + \sqrt{2})^2 \sum_{i=0}^{n} E(T_i, \alpha; f)\right), \tag{5.17}$$

which is useful in applications.

We now seek conditions on the order of best Hausdorff approximation by trigonometric polynomials that guarantee continuity of the approximated function. Obviously, for the function $f_0(x) \equiv [-1, 1]$, we have $r(\alpha; f_0, \sin nx) < \pi/\alpha n$, and, consequently, $E(T_n, \alpha; f_0) \le \pi/\alpha n$. This shows that the order of approximation $E(T_n, \alpha; f) = O(n^{-1})$ does not alone ensure continuity of the function f. But it follows from (5.17) that if $E(T_n, \alpha; f) \le c/\alpha n$, then $E_n^T(f) \le c\alpha^{-1} n^{-1} n^{c(3 + 2\sqrt{2})}$, and $E_n^T(f)$ tends to zero for $c < (3 + 2\sqrt{2})^{-1}$ so that f has to be single valued and continuous. It is interesting to find the maximal constant that ensures the continuity of the approximated function. E. Dolzenko and E. Sevastianov [1] showed that if $c < \pi/3$, then the approximated function is continuous. The final result is the following.

Theorem 5.5. Let $f \in \bar{F}_{2\pi}$. If

$$(5.18) \qquad \liminf_{n\to\infty} nE(T_n, \alpha; f) < \frac{\pi}{2\alpha},$$

then f is single valued and hence continuous. The constant $\pi/2\alpha$ on the right-hand side of (5.18) is exact, i.e., there is a multivalued function $g \in \bar{F}_{2\pi}$ where

$$(5.19) \qquad \liminf_{n\to\infty} nE(T_n, \alpha; g) = \frac{\pi}{2\alpha} .$$

The proof of Theorem 5.5 requires some preliminary lemmas.

Lemma 5.6. Let A_n^+ and A_n^- denote the following point sets in the plane:

$$A_n^+ = \{(x,y) : |x| < \pi/2n, \ |\sin nx| < y \le 1\},$$

$$A_n^- = \{(x,y) : |x| < \pi/2n, \ -1 \le y < -|\sin nx|\}.$$

For every polynomial $\tau_n \in T_n$ with $\|\tau_n\| < 1$, and for every two points $x_1, x_2 \in (-\pi/2n, \pi/2n)$, the following conditions are mutually exclusive: $(x_1, \tau_n(x_1)) \in A_n^+$, and $(x_2, \tau_n(x_2)) \in A_n^-$.

Proof. Let us assume to the contrary that there exists such a polynomial $\tau_n \in T_n$. Then there exists a constant $\lambda > 1$ such that either $\tau_n(x) - \lambda\sin(nx)$ or $\tau_n(x) + \lambda\sin(nx)$ has more than $2n$ zeros in the interval $(-\pi, \pi)$, which is impossible for polynomials of nth order. The lemma is proved.

Lemma 5.7. Let $\langle a(n)\rangle_1^\infty$ be a sequence of reals such that $a(n) \ge n^{-1/2}$ and $\lim_{n\to\infty} a_n = 0$, and let $\langle b(n)\rangle_1^\infty$ be another sequence such that $0 < b(n) \le b$. Then there exists a sequence of trigonometric polynomials $\langle q(n)\rangle_1^\infty$, where $q(n)$ is of order no higher than $n^{3/4}$ and a positive constant c such that

$$(5.20) \qquad |q(n; x)| \le b(n) \text{ for all } x;$$

$$|q(n; x) - b(n)| \le cn^{-1/2} \text{ for } |x| \le a(n);$$

$$|q(n; x)| \leq cn^{-1/2} \quad \text{for} \quad 2a(n) \leq |x| < \pi.$$

Proof. The conclusion of the lemma follows directly from the results of §4.1.2.2 on the universal estimate for best Hausdorff approximation of all bounded 2π-periodic functions, and this estimate is of order $m^{-1}\ln m$ for the approximation by trigonometric polynomials of order m. We shall only note that $m^{-1}\ln m = (3/4)n^{-3/4}\ln n < 8n^{-1/2}$ for $m = n^{3/4}$.

Proof of Theorem 5.5. According to the statement of the theorem, we may assume for the given $f \in \bar{F}_{2\pi}$ that there exists an increasing sequence of indices $\langle n_i \rangle_1^\infty$ such that

$$\lim_{i\to\infty} n_i E(n_i) < \pi/2\alpha, \tag{5.21}$$

where $E(n) = E(T_n, \alpha; f)$. We assume that f is not single-valued at a certain point x_0, i.e., $S(f; x_0) - I(f; x_0) = 2h > 0$. Without loss of generality, we shall assume $x_0 = 0$ and $S(f; x_0) = -I(f; x_0) = h$. Since $f \in \bar{F}_{2\pi}$ and $S(f; x)$ (resp. $I(f; x)$) is upper (resp. lower) semicontinuous, there exists an even, continuous, bounded function φ that is monotonically increasing for $x \geq 0$ such that $\varphi(0) = h$ and $-\varphi(x) \leq I(f; x) \leq S(f; x) \leq \varphi(x)$ for all x.

Let $\tau(n)$ be a trigonometric polynomial of best approximation of f from T_n, i.e., $r(\alpha; \tau(n), f) = E(n) = E(T_n, \alpha; f)$. Define numbers $a(n_i)$ and $b(n_i)$ by

$$a(n_i) = \max\{n_i^{-1/2}, E(n_i)\}, \qquad b(n_i) = \frac{1}{\varphi(2a(n_i)) + E(n_i)}. \tag{5.22}$$

The polynomial of order at most $n_i^{3/4}$ which satisfies the conditions (5.20) of Lemma 5.7 for the numbers $a(n_i)$ and $b(n_i)$ prescribed by (5.22), will be denoted by $q(n_i)$. Then the trigonometric polynomial $p(n_i) = \tau(n_i)q(n_i)$ will be of order at most $n_i + n_i^{3/4}$, and this trigonometric polynomial will satisfy the following conditions:

(a) $\|p(n_i)\| < 1$ for sufficiently large i;

(b) there exist points $x_i', x_i'' \in [-\alpha E(n_i), \alpha E(n_i)]$ such that $1 - p(n_i; x_i') \leq \beta(n_i)$ and $1 + p(n_i; x_i'') \leq \beta(n_i)$, where $\beta(n_i) = 1 - (h - E(n_i))(b(n_i) - cn_i^{-1/2})$.

Since φ is continuous and $\varphi(0) = h$, we have $\lim_{i\to\infty} b(n_i) = 1/h$; so, $\lim_{i\to\infty} \beta(n_i) = 0$. By Lemma 5.6, and conditions (a) and (b) above, for i so large that $n_i^{-1/2} < \pi/2 - \alpha E(n_i)$, it follows that

$$\beta(n_i) \geq 1 - |\sin (n_i + n_i^{3/4})x_i'| \geq 1 - \sin (\alpha E(n_i)(n_i + n_i^{3/4})). \tag{5.23}$$

But according to the hypotheses of the theorem,

$$\lim_{i\to\infty} \alpha E(n_i)(n_i + n_i^{3/4}) = \lim_{i\to\infty} \alpha n_i E(n_i) < \pi/2,$$

which contradicts (5.23), because $\lim_{i\to\infty} \beta(n_i) = 0$. This completes the proof of the first part of the theorem.

Now we shall construct a function $g \in \mathbf{F}_{2\pi}$ for which condition (5.19) holds, where g is multivalued (discontinuous) at the point $x = 0$. This function will be of the form

$$g(0) = 0, \quad g(x) = \sum_{i=0}^{\infty} p(n_i; x) \qquad \text{for } x \neq 0, \tag{5.24}$$

where $p(n_i)$, $i = 0, 1, 2, \ldots$, are odd trigonometric polynomials of degree n_0, $n_1 = m_1 + k_1$, $n_2 = m_2 + k_2, \ldots$, satisfying the following conditions:

(1) m_i and k_i are positive integers for which $n_0 < m_1$, $m_i < m_{i+1}$, and $k_i < k_{i+1}$, $i = 1, 2, 3, \ldots$;

(2*) $49 \sum_{i=\nu}^{\infty} \frac{n_i}{k_{i+1}} \leq n_\nu^{-2}$, $\nu = 0, 1, 2, \ldots$;

(3*) $\lim_{i\to\infty} \frac{n_i}{m_i} = 1$;

(4) $p(n_0; x) = P(n_0; x) = \sin (n_0 x)$ and $z_0 = \pi/2n_0$;

(5) $P(n_\nu; x) = \sum_{i=0}^{\nu} p(n_i; x)$, $P(n_\nu; \pi/2m_\nu) \geq 1$, and if $z_\nu = \inf \{x : P(n_\nu; x) = 1 \text{ for } x \in [0, \pi/2m_\nu]\}$, then $z_\nu = \pi/2m_\nu$ and $z_{\nu-1} > z_\nu > 0$ for $\nu = 1, 2, 3, \ldots$;

(6) whenever $0 < x \leq z \leq z_\nu$, we have $\frac{x}{z} P(n_\nu; z) \leq P(n_\nu; x) \leq 1$;

(7) for $x \in [z_\nu, z_{\nu-1}]$, $\nu = 1, 2, 3, \ldots$

$$\frac{2x}{z_{\nu-1}} - 1 - \frac{49n_{\nu-1}}{k_\nu} \leq P(n_\nu; x) \leq 1 + \frac{49n_{\nu-1}}{k_\nu} ;$$

(8) $|p(n_\nu; x)| < \frac{49n_{\nu-1}}{k_\nu}$ for $z_{\nu-1} \leq |x| \leq \pi$, $\nu = 1, 2, 3, \ldots$;

(9) $\|P(n_\nu)\| \leq 1 + 49 \sum_{i=1}^{\nu-1} \frac{n_i}{k_{i+1}} \leq 2$, $\nu = 1, 2, 3, \ldots$;

and consequently $\|P'(n_\nu)\| \leq 2n_\nu$, $\nu = 0, 1, 2, \ldots$ where the norm is the usual supremum norm over the entire line.

It should be noted that conditions (2*) and (3*) follow from the conditions

(2) $\frac{49n_i}{k_{i+1}} \leq 2^{p-i-1} n_p^{-2}$, $p = 0, 1, 2, \ldots$, $i = 0, 1, 2, \ldots$;

(3) $m_i \geq k_i^2$, $i = 1, 2, 3, \ldots$.

The sequence of polynomials $\langle p(n_i)\rangle_0^\infty$ and the sequences of positive integers $\langle m_i\rangle_1^\infty$ and $\langle k_i\rangle_1^\infty$ satisfying the condtions (1) - (9) will be constructed by induction.

Let n_0 be an arbitrary positive integer and set

$$p(n_0; x) = P(n_0; x) = \sin n_0 x, \quad \text{and } z_0 = \frac{\pi}{2n_0} .$$

Choose for k_1 a positive integer such that $49n_0/k_1 \leq 2^{-1} n_0^{-2}$. Let $g(n_0)$ be this 2π-periodic function, specified on $[-\pi, \pi]$ by

$$g(n_0; x) = \begin{cases} 1 - |P(n_0; x)| & \text{if } |x| \leq z_0 \\ 0 & \text{if } z_0 < |x| \leq \pi \end{cases} .$$

Since $g(n_0)$ is an even function, satisfying a Lipschitz condition with constant n_0, $0 < g(n_0; x) \leq 1$, and $g(n_0; 0) = 1$, it follows from Jackson's theorem that there exists a trigonometric polynomial $q(k_1)$, of order no higher than k_1, for which

(5.25) $$|g(n_0; x) - q(k_1; x)| \leq \frac{25n_0}{k_1}, \qquad -\infty < x < \infty,$$

and $q(k_1; 0) > 1$.

We now choose a positive integer m_1 such that $m_1 > n_0$ and $m_1 > k_1^2$, $q(k_1; x) \geq 1$ for $x \in [0, \pi/2m_1]$, and such that the function $q(k_1; x) \sin m_1 x$ is convex on the segment $[0, \pi/2m_1]$. All this is possible for m_1 sufficiently large. We set $p(n_1; x) = q(k_1; x) \sin m_1 x$ and $P(n_1; x) = P(n_0; x) + p(n_1; x)$. Obviously, both $p(n_1)$ and $P(n_1)$ are odd trigonometric polynomials of order no higher than $n_1 = k_1 + m_1$. Since $P(n_0; x) \geq 0$ for $x \in [0, z_0]$ and $m_1 > n_0$, condition (5) above is fulfilled. Condition (6) follows from the convexity of $p(n_1; x)$ and $P(n_0; x)$ on $[0, z_1]$.

We now verify condition (7) for $P(n_1)$. From (5.25) we have for $x \in [z_1, z_2]$ that

$$\begin{aligned} P(n_1; x) &= P(n_0; x) + q(k_1; x) \sin m_1 x \\ &= P(n_0; x) + (1 - |P(n_0; x)|) \sin m_1 x + (q(k_1; x) - g(n_0; x)) \sin m_1 x \\ &\leq P(n_0; x) + 1 - P(n_0; x) + |g(n_0; x) - q(k_1; x)| \leq 1 + \frac{25 n_0}{k_1}. \end{aligned}$$

For the other inequality in (7), using the convexity of $P(n_0; x)$ on $[0, z_0]$ and (5.25), we obtain

$$\begin{aligned} P(n_1; x) &= P(n_0; x) + (1 - |P(n_0; x)|) \sin m_1 x + (q(k_1; x) - g(n_0; x)) \sin m_1 x \\ &\geq P(n_0; x) - 1 + P(n_0; x) - |g(n_0; x) - q(k_1; x)| \\ &\geq \frac{2x}{z_0} - 1 + \frac{25 n_0}{k_1}. \end{aligned}$$

Conditions (8) and (9) are also verified directly using (5.25), so that (1) - (9) hold for the index 1.

Suppose now that $p(n_i)$, m_i, and k_i have been chosen for $i = 1, 2, 3, \ldots, \nu$, satisfying conditions (1) - (9). We now describe $p(n_{\nu+1})$, $m_{\nu+1}$, and $k_{\nu+1}$. First, $k_{\nu+1}$ is chosen so that the inequalities $49 n_\nu / k_{\nu+1} \leq 2^{p-\nu-1} n_p^{-2}$ hold for $p = 0, 1, 2, \ldots, \nu$, i.e., so that condition (2) holds. We now construct a 2π-periodic function $g(n_\nu)$:

$$g(n_\nu; x) = \begin{cases} 1 - |P(n_\nu; x)| & \text{if } |x| < z_\nu \\ 0 & \text{if } z_\nu < |x| \leq \pi \end{cases}.$$

Since $g(n_\nu)$ is an even function, satisfying a Lipschitz condition with constant $2n_\nu$, $0 \le g(n_\nu; x) \le 1$, and $g(n_\nu; 0) = 1$, it follows from Jackson's theorem that there exists a trigonometric polynomial $q(k_{\nu+1})$, of order no higher than $k_{\nu+1}$, for which

$$(5.26) \qquad |g(n_\nu; x) - q(k_{\nu+1}; x)| \le \frac{49n_\nu}{k_{\nu+1}}, \qquad -\infty < x < \infty,$$

and $q(k_{\nu+1}; 0) > 1$.

The positive integer $m_{\nu+1}$ is to be chosen so that $m_{\nu+1} \ge m_1 > k_{\nu+1}^2$, $\pi/2m_{\nu+1} < z_\nu$, $q(k_{\nu+1}; x) \ge 1$ for $x \in [0, \pi/2m_{\nu+1}]$, and such that $q(k_{\nu+1}; x) \sin m_{\nu+1}x$ is convex on the segment $[0, \pi/2m_{\nu+1}]$. Let us set $p(n_{\nu+1}; x) = q(k_{\nu+1}; x) \sin m_{\nu+1}x$ and $P(\nu+1; x) = P(n_\nu; x) + p(n_{\nu+1}; x)$. Obviously, both $p(n_{\nu+1})$ and $P(n_{\nu+1})$ are odd trigonometric polynomials of order no higher than $n_{\nu+1} = k_{\nu+1} + m_{\nu+1}$. Since $P(n_\nu; x) \ge 0$ for $x \in [0, z_\nu]$ and $\pi/2m_{\nu+1} < z_\nu$, condition (5) above is fulfilled. Condition (6) follows from the convexity of $p(n_{\nu+1}; x)$ and $P(n_{\nu+1}; x)$ on $[0, z_{\nu+1}]$. Condition (7) is verified as in the case $\nu = 1$, as are conditions (8) and (9), using (5.26) instead of (5.25). We have now shown that the sequences $\langle p(n_i)\rangle_0^\infty$, $\langle m_i\rangle_1^\infty$ and $\langle k_i\rangle_1^\infty$ exist.

We claim that for the function g defined by (5.24), the inequality

$$(5.27) \qquad \limsup_{i\to\infty} n_i\, r(\alpha; g, P(n_i)) \le \frac{\pi}{2\alpha}$$

holds, and that g is discontinuous at $x = 0$. From (2*) and (8) we have

$$(5.28) \qquad g(z_\nu) = P(n_\nu; z_\nu) + \sum\nolimits_{i=\nu+1}^{\infty} p(n_i; z_\nu)$$

$$\ge 1 - \sum\nolimits_{i=\nu+1}^{\infty} |p(n_i; z_\nu)| \ge 1 - 49 \sum\nolimits_{i=\nu+1}^{\infty} \frac{n_i}{k_{i+1}} \ge 1 - n_{\nu+1}^{-2}.$$

Consequently, $\limsup_{\nu\to\infty} g(z_\nu) \ge 1$. But since g is an odd function, it now follows that g is discontinuous at $x = 0$.

We now estimate the Hausdorff distance between g and $P(n_\nu)$. First, we establish these inequalities:

$$(5.29) \qquad |g(x) - P(n_\nu; x)| \le n_\nu^{-2} \quad \text{for } z_\nu \le |x| \le \pi,$$

$$(5.30) \qquad \frac{2x}{z_i} - 1 - n_i^{-2} \le g(x) \le 1 + n_i^{-2} \quad \text{for } x \in [z_{i+1}, z_i], \text{ where } i \ge \nu.$$

Indeed, (2*) and (8) jointly yield (5.29), whereas for $x \in [z_{i+1}, z_i]$, we obtain from (2*), (7), and (8)

$$g(x) = P(n_{i+1}; x) + \sum_{\nu=i+2}^{\infty} p(n_\nu; x)$$

$$\leq 1 + 49 \sum_{\nu=i}^{\infty} \frac{n_\nu}{k_{\nu+1}} \leq 1 + n_i^{-2} ,$$

and similarly,

$$g(x) = P(n_{i+1}; x) + \sum_{\nu=i+2}^{\infty} p(n_\nu; x)$$

$$\geq \frac{2x}{z_i} - 1 - \frac{49 n_i}{k_{i+1}} - 49 \sum_{\nu=i+1}^{\infty} \frac{n_\nu}{k_{\nu+1}} \geq \frac{2x}{z_i} - 1 - n_i^{-2} .$$

This proves (5.30).

Keeping in mind that $P(n_\nu; x)$ is an odd function, $z_\nu \leq \pi/2m_\nu$, $|P(n_\nu; x)| \leq 1$ for $|x| \leq z_\nu$ and that $P(n_\nu; z_\nu) = 1$, we obtain from (5.28) - (5.30) that

$$r(\alpha; g, P(n_\nu)) \leq \frac{\pi}{2\alpha m_\nu} + 2n_\nu^{-2} . \tag{5.31}$$

The inequality (5.27) follows from (5.31), i.e., $E(T_{n_k}, \alpha; g) \leq \pi/2\alpha n_k$ holds. But since g is discontinuous, we see that

$$\lim_{k\to\infty} n_k E(T_{n_k}, \alpha; g) = \frac{\pi}{2\alpha} .$$

This completes the proof of the theorem.

The estimate for the modulus of continuity of a given function f through its best Hausdorff approximations by trigonometric polynomials can be obtained as was done for the inequalities of (5.16) and (5.17). To see this, let $f \in \mathbf{F}_{2\pi}$ be arbitrary, and let $p_n \in T_n$ be one of the trigonometric polynomials of best Hausdorff approximation for f. By using (5.7), we obtain for $\delta = E(T_n, \alpha; f)$ that

$$\omega(f; \delta) < \omega(f - p_n; \delta) + \omega(p_n; \delta) \leq 2\|f - p_n\| + \omega(p_n; \delta) \leq 2\delta + 3\omega(p_n; \delta),$$

or

(5.32) $$\omega(f;\alpha E(T_n,\alpha;f)) \le 3E(T_n,\alpha;f)\prod_{i=0}^{n}(1+\alpha(1+\sqrt{2})^2 E(T_i,\alpha;f)) .$$

From (5.32) we can obtain an analogue of the inequality of E. Dolzenko and E. Sevastianov [1]: for every $f \in \mathbf{F}_{2\pi}$,

(5.33) $$\omega(f;\alpha E(T_n,\alpha;f)) \le 3E(T_n,\alpha;f)\exp\left(\alpha(1+\sqrt{2})^2\sum_{i=0}^{n} E(T_i,\alpha;f)\right) .$$

Using either (5.32) or (5.33) one can obtain

Theorem 5.6. (E. Dolzenko and E. Sevastianov [1]). If $E(T_n,\alpha;f) \le \varepsilon/\alpha n$, then for sufficiently small positive ε, f satisfies a Lipschitz-Hölder condition of degree $\beta(\varepsilon)$, where $\beta(\varepsilon) \to 1$ as $\varepsilon \to 0$.

5.2.2. THE ALGEBRAIC CASE

Converse theorems for best Hausdorff approximation by algebraic polynomials can be obtained as was done in the trigonometric case, but the flavor of the analysis here has been influenced by Nikol'skii's effect, which is closely connected with the endpoints of the interval on which the functions are approximated.

First we obtain an analogue of (5.16) in the algebraic case. Let $\Delta = [a, b]$, and let $p_n \in H_n$ be an algebraic polynomial of best Hausdorff approximation for a function $f \in \overline{F}_\Delta$. Then (5.7) yields

(5.34) $$E(H_n,\Delta,\alpha;f) \le E_n(\Delta;f) \le E(H_n,\Delta,\alpha;f) + \omega(p_n;\alpha E(H_n,\Delta,\alpha;f)),$$

where $E_n(\Delta;f)$ is the best uniform approximation of the function f by algebraic polynomials of degree no higher than n on Δ. Applying Lemma 5.4 and A. Markov's inequality (I. Natanson [1, p.174]),

$$\|p_n'\| \le \frac{2n^2\|p_n\|}{b-a},$$

where $\|\cdot\|$ is the uniform norm with respect to the domain Δ, we obtain

$$\omega(p_n;\delta) \le \delta\|p_n'\| \le \frac{\delta}{b-a}\left(-1+\prod_{i=1}^{k}\left(1+2\alpha n_i^2\, r(\Delta,\alpha;p_{n_i},p_{n_{i-1}})\right)\right), \tag{5.35}$$

where $0 = n_0 < n_1 < n_2 < \cdots < n_k$, and p_{n_i} is an algebraic polynomial of best Hausdorff approximation for f of degree at most n_i. It is directly seen that

$$\prod_{i=1}^{k}\left(1+\lambda n_i^2\, r(\Delta,\alpha;p_{n_i},p_{n_{i-1}})\right) \le (1+\lambda n_1^2 a(0))(1+\lambda n_k^2 a(n_k))\prod_{i=1}^{k-1}\left(1+\lambda(n_{i+1}^2+n_i^2)a(n_i)\right), \tag{5.36}$$

where for each positive integer m, $a(m) = E(H_m, \Delta, \alpha; f)$. We choose a positive integer k such that $2^{k-1} < n \le 2^k$, and we set $n_i = 2^i$. Then $(n_{i+1}^2 + n_i^2)/n_i(n_i - n_{i-1}) = 10$ for $i = 2, 3, \ldots, k-1$, and keeping in mind (5.13) and (5.14), we obtain from (5.36) that

$$\prod_{i=1}^{k}\left(1+\lambda n_i^2\, r(\Delta,\alpha;p_{n_i},p_{n_{i-1}})\right) < \prod_{i=0}^{n}\left(1+10\lambda a(i)\right).$$

Therefore, from (5.34) and (5.35), we finally obtain this result.

Lemma 5.8. For every $f \in \bar{F}_\Delta$, $\Delta = [a, b]$, the following inequality for best uniform approximation and best Hausdorff approximation by algebraic polynomials is valid:

$$E(H_n,\Delta,\alpha;f) \le E_n(\Delta;f) \le E(H_n,\Delta,\alpha;f)\left(1+\frac{8\alpha E(H_0,\Delta,\alpha;f)}{b-a}\right)\prod_{i=1}^{n}\left(1+\frac{20\alpha i E(H_i,\Delta,\alpha;f)}{b-a}\right). \tag{5.37}$$

From (5.37) we get the following inequality :

$$E(H_n,\Delta,\alpha;f) \le E_n(\Delta;f) \le \tag{5.38}$$

$$E(H_n, \Delta, \alpha; f) \exp\left(\frac{2\alpha}{b - a}\left(4E(H_0, \Delta, \alpha; f) + \sum_{i=1}^{n} 10i\, E(H_i, \Delta, \alpha; f)\right)\right).$$

Actually, the constant 10 can be improved to 9.42, or to be exact, $3\mu + 1$, where μ is the positive root of the equation $2t^3 - 3t^2 - 1 = 0$. We also obtain from Lemma 5.8 that if the series $\sum_{i=1}^{\infty} iE(H_i, \Delta, \alpha; f)$ converges, then $E(H_n, \Delta, \alpha; f)$ and $E_n(\Delta; f)$ have the same order.

To ensure continuity of the approximated function f on the entire closed interval $\Delta = [a, b]$, the rate of decrease of $E(H_n, \Delta, \alpha; f)$ to zero must be not less than n^{-2}. Here, too, the constant in front of the order plays an essential role. Suppose $E(H_n, \Delta, \alpha; f) \leq c(b - a)/\alpha n^2$. Then it follows from (5.38) that $E_n(\Delta; f) = O(n^{-2+20c})$, and hence if $c < 1/10$, then f must be continuous on the entire segment $\Delta = [a, b]$. It is of interest to find the best estimate for the constant c with the above property. The following result, which we state without proof, can be obtained using the method of proof of Theorem 5.5.

Theorem 5.7. (P. Petrusev and Sp. Tasev [1]). Let $f \in \bar{F}_\Delta$, $\Delta = [a, b]$. If

$$\liminf_{n\to\infty} n^2 E(H_n, \Delta, \alpha; f) < \frac{\pi^2(b - a)}{2\alpha},$$

then f is single valued (continuous) at every point of the segment [a, b]. The constant $\pi^2(b - a)/2\alpha$ on the right-hand side is exact, i.e., there exists $f_0 \in \bar{F}_\Delta$, multivalued at the endpoints of Δ, for which

$$\liminf_{n\to\infty} n^2 E(H_n, \Delta, \alpha; f) = \frac{\pi^2(b - a)}{2\alpha}.$$

Continuity at an interior point of the segment is ensured if the order of approximation $E(H_n, \Delta, \alpha; f)$ equals n^{-1}, and the constant in front of the order reflects the distance between the point and the endpoints of the interval. The notion of Hausdorff difference, introduced in §4.1.2.7 for the purpose of producing an analogue of Nikolskii's theorem for the estimate of best Hausdorff approximation by algebraic polynomials, arises again in this context.

Theorem 5.8. (P. Petrusev and Sp. Tasev [1]). Let $f \in \overline{F}_\Delta$, $\Delta = [-1, 1]$. If there exists a sequence of algebraic polynomials $\langle p_n \rangle_1^\infty$, where $p_n \in H_n$, such that for each $x \in (-1, 1)$ we have

$$\liminf_{n\to\infty} n(1 - x^2)^{-1/2} |f(x) - p_n(x)|_\alpha < \pi/2\alpha, \tag{5.39}$$

then f is single valued at every $x \in (-1, 1)$. The constant $\pi/2\alpha$ on the right-hand side of (5.39) cannot be improved.

Without going into the details of the proof of the Theorem 5.8, we shall note that it is similar to the proof of Theorem 5.5, but the extremal function used here is the Chebyshev polynomial $T_n(x) = \cos(n \arccos x)$, instead of $\sin x$.

In order to restate Theorem 5.8 in terms of best Hausdorff approximations, we consider the Hausdorff distance with weight $\sqrt{(x - a)(b - x)}$ on $[a, b]$, which amounts to a weight of $\sqrt{1 - x^2}$ for $[a, b] = [-1, 1]$. Applying Bernstien's inequality $n\|p_n\| \geq \|\sqrt{1 - x^2}\, p_n'\|$, valid for every polynomial $p_n \in H_n$, and making use of Lemma 5.4, one can obtain analogues of (5.16) and (5.17).

§ 5.3. Converse theorems for approximation by spline functions

We shall consider here only spline functions with equidistant knots, as in §4.3. Let us recall that by $\overline{S}_{k,n}$ we mean the set of all spline functions of order (k,n) on [a, b] with knots $x_i = a + i(b - a)/n$, $i = 0, 1, 2, \ldots, n$, where k, n are positive integers. Each function $\varphi \in \overline{S}_{k,n}$ belongs to $C^{(k-1)}_{[a, b]}$, i.e., it has a continuous (k - 1)st derivative on [a, b]), and it coincides with an algebraic polynomial of degree at most k on every segment $\Delta_i = [x_{i-1}, x_i]$, $i = 1, 2, 3, \ldots, n$. The set of all step functions on [a, b] will be denoted by $\overline{S}_{0,n}$, but for these functions we choose the knots differently: $x_0 = a$, $x_1 = a + h/2$, $x_2 = x_1 + h, \ldots, x_{n-1} = x_{n-2} + h$, $x_n = b = x_{n-1} + h/2$, where $h = (b - a)/(n - 1)$. We may view $\varphi \in \overline{S}_{0,n}$ as a segment function whose value at the knot x_i is given by

$$\varphi(x_i) = \lim_{x\to x_i - 0} \varphi(x) \vee \lim_{x\to x_i + 0} \varphi(x),$$

$i = 1, 2, 3, \ldots, n - 1.$

In order to identify the order of best Hausdorff approximation by spline functions that ensures single valuedness and continuity of the approximated segment function, let us consider perhaps the simplest multivalued function $f_1(x) = [-1, 1]$ for $x \in [a, b]$.

Lemma 5.9. For $f_M(x) = [-M, M]$, $x \in \Delta = [a, b]$, we have

$$E(\overline{S}_{0,n}, \Delta, \alpha; f_M) \leq \frac{b - a}{2\alpha(n - 1)} \qquad n = 2, 3, 4, \ldots, \tag{5.40}$$

and equality in (5.40) holds for $n \geq (b - a)/2\alpha M$. For larger k, we have

$$E(\overline{S}_{k,n}, \Delta, \alpha; f_M) \leq \frac{b - a}{\alpha n}, \qquad k \text{ odd and } n = 1, 2, 3, \ldots, \tag{5.41}$$

$$E(\overline{S}_{k,n}, \Delta, \alpha; f_M) \leq \frac{3(b - a)}{2\alpha n}, \qquad k \text{ even and } n = 1, 2, 3, \ldots. \tag{5.42}$$

Proof . The inequality (5.40) is established by the approximation of f_M by the step function $s(x)$ defined as follows: $s(a) = -M$, $s(x) = (-1)^i M$ for $x \in (x_{i-1}, x_i)$, $i = 1, 2, 3, \ldots, n$, $s(b) = (-1)^n M$, and $s(x_i) = [-M, M]$ for $i = 1, 2, 3, \ldots, n - 1$.

For (5.41) and (5.42), it is sufficient to establish the existence of a spline function in $\overline{S}_{k,n}$, which has equal maxima and minima, attained for values of the argument that are equally spaced at a distance of $2(b - a)/n$ from one another. Separate constructions will be provided for even and odd k.

For odd k, $k = 2m + 1$, we construct an odd polynomial $P(x) = a_0x - a_1x^3 - \cdots - a_mx^{2m+1}$ satisfying the following conditions: $p(1) = 1$ and $p^{(j)}(1) = 0$ for $j = 1, 3, 5, \ldots, 2m - 1$. These conditions are compatible and uniquely determine the polynomial P. It is easy to see that the function

$$S_M(x) = Mp((-1)^j(2n(x - a)/(b - a) - 2j - 1))$$

for $x \in [x_j, x_{j+1}]$, $j = 0, 1, 2, \ldots, n - 1$ is a spline function in $\overline{S}_{k,n}$. Moreover, it is directly verified that $r(\Delta, \alpha; f_M, S_M) < (b - a)/\alpha n$.

For even k, $k = 2m$, we consider the even polynomial $Q(x) = 1 + b_1x^2 + b_2x^4 + \cdots + b_mx^{2m}$, satisfying the conditions $Q^{(2j)}(1) = 0$, for $j = 0, 1, 2, \ldots, m - 1$. These conditions are compatible and they define the polynomial Q uniquely. It is not hard to see that

$$\theta_M(x) = (-1)^j MQ(2n(x - a - (b - a)/2n)/(b - a) - 2j)$$

for $x \in [x_j, x_{j+1}]$, $j = 0, 1, 2, \ldots, n - 1$, is a spline function in $\bar{S}_{k,n}$, and that $r(\Delta, \alpha; f_M, \theta_M) < 3(b - a)/2\alpha n$. This proves the lemma.

We note that if for even positive k, we choose our knots as was done in the case $k = 0$, then on the right-hand side of (5.42), we can put $(b - a)/\alpha(n - 1)$ in place of $3(b - a)/2\alpha n$.

With respect to converse theorems, we first look at best Hausdorff approximations by step functions, i.e., by functions in $\bar{S}_{0,n}$. This requires some new constructions.

For $f \in \bar{\mathbf{F}}_\Delta$ and $x_0 \in \Delta$, we say that x_0 is a **max-point** of f if

$$S(f; x_0) > \limsup_{\delta \to 0} \{y : y \in f(x), \ |x - x_0| < \delta, \ x \neq x_0\}.$$

Similarly, we say that x_0 is a **min-point** of f if

$$I(f; x_0) < \liminf_{\delta \to 0} \{y : y \in f(x), \ |x - x_0| < \delta, \ x \neq x_0\}.$$

Segment functions in $\bar{\mathbf{F}}_\Delta$ that are completed graphs of single-valued functions are easily characterized in terms of these concepts.

Lemma 5.10. A necessary and sufficient condition for a function $f \in \bar{\mathbf{F}}_\Delta$ to be the completed graph of a single-valued function g, i.e., $f(x) = F(g; x)$, is the following: there does not exist a point $x_0 \in \Delta$ that is simultaneously a max-point and a min-point of f.

Proof. Necessity of the condition is obvious. For sufficiency, we construct our single valued function g as follows. If x is a max-point of f, or x is rational and not a min-point of f, we write $g(x) = S(f; x)$. Otherwise, we set $g(x) = I(f; x)$. By assumption, g is well-defined, and it can be shown that $F(g) = f$.

We define now a functional C_0 on $\bar{\mathbf{F}}_\Delta$ by

$$C_0(f) = \limsup_{n \to \infty} \frac{\alpha n \, E(\bar{S}_{0,n}, \Delta, \alpha; f)}{b - a}.$$

According to the estimate obtained in §4.3, for each $f \in \bar{\mathbf{F}}_\Delta$, the inequality $C_0(f) \leq 3/2$ holds.

Theorem 5.9. A necessary and sufficient condition for the function $f \in \bar{\mathbf{F}}_\Delta$ to be the completed graph of a single-valued (bounded) function is that $C_0(f) < 3/2$.

Proof. Let us assume that $C_0(f) < 3/2$, but that f is not a completed graph of a single-valued function. By Lemma 5.10, there exists a point $x_0 \in \Delta$ such that x_0 is simultaneously a max-point and a min-point of f. This means that there exist two positive numbers $h > 0$, $\delta > 0$, such that

$$S(f; x_0) > h + \sup\{y : y \in f(x),\ |x - x_0| < \delta,\ x \neq x_0\},$$

$$I(f; x_0) < -h + \inf\{y : y \in f(x),\ |x - x_0| < \delta,\ x \neq x_0\}.$$

Since for every real number v, the inequality $|v - p/q| < q^{-2}$ has infinitely many integer solutions p, q (see for instance, J. Kassels [1, p. 6]), then for infinitely many values of n the point x_0 will be at a distance at most n^{-2} from the middle of the closed interval that joins two knots. Then for infinitely many values of n, the inequality

$$E(\bar{S}_{0,n}, \Delta, \alpha; f) \geq \frac{3(b-a)}{2\alpha n} - \frac{1}{\alpha n^2}$$

will hold (see Figure 5.1). But from this, it follows that $C_0(f) \geq 3/2$, which contradicts the hypothesis. The theorem is proved.

Theorem 5.10. Suppose $f \in \bar{\mathbf{F}}_\Delta$, and $C_0(f) < 1$. Then f has neither a max-point nor a min-point.

Proof. Let us suppose that f has a max-point or a min point at ξ_0. Then, as in the proof of the preceding theorem, keeping in mind that for infinitely many values of n, the point ξ_0 will be at a distance of at most n^{-2} from a certain knot, we come to the conclusion that $C_0(f) \geq 1$. This proves the theorem.

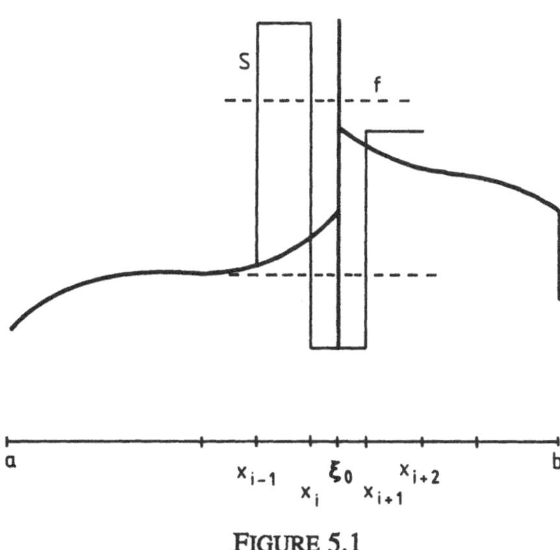

FIGURE 5.1

Theorem 5.11. A necessary and sufficient condition for the function $f \in \overline{\mathbf{F}}_\Delta$ to be single valued (and continuous) is that $C_0(f) < 1/2$.

Proof. Let us assume that the condition $C_0(f) < 1/2$ holds, and f is multivalued at a certain point $x_0 \in \Delta$, i.e. $S(f; x_0) > I(f; x_0)$. As in the proof of Theorem 5.9, keeping in mind that for infinitely many values of n the point x_0 will be at a distance of at most n^{-2} from the middle of a closed interval joining two knots, if x_0 is an interior point of the segment Δ, we come to the conclusion that $C_0(f) \geq 1/2$. If x_0 coincides with one of the endpoints of Δ, then obviously $C_0(f) \geq 1/2$ holds again. This establishes sufficiency.

Necessity follows from (5.40) of Lemma 5.9, which shows that $C_0(f_M) = 1/2$. But the function f_M is multivalued. This completes the proof.

Theorem 5.12. Let $f \in C_\Delta$. Then a necessary and sufficient condition for f to satisfy a Lipschitz condition with constant M ($f \in \text{Lip}_M 1$) is that $C_0(f) \leq \alpha M/(1 + \alpha M)$.

Proof. Since $f \in \text{Lip}_M 1$, we have $|f(x - h) - f(x)| \leq Mh$ whenever $x, x - h \in \Delta, h > 0$, then for $\psi \in \overline{S}_{0,n}$ satisfying $\psi(x) = f((x_{i-1} + x_i)/2)$ for $x \in (x_{i-1}, x_i)$, $i = 1, 2, 3, \ldots, n$, we have

$$r(\Delta, \alpha; f, \psi) \le \frac{Mh}{1 + \alpha M},$$

where $h = (b - a)/(n - 1)$. Hence, $C_0(f) \le \alpha M/(1 + \alpha M)$. This proves necessity.

Suppose sufficiency fails, i.e., the inequality is satisfied but $f \notin \mathrm{Lip}_M 1$. Then we may assume without loss of generailty that for some $x_0 \in \Delta$,

$$\sup \left\{ \frac{|f(\xi_0 - h) - f(\xi_0)|}{h} : \xi_0 - h \in \Delta,\ h > 0 \right\} > M. \tag{5.43}$$

But for infinitely many values of n, the point x_0 will be at a distance of at most n^{-2} from the middle of a closed interval joining two knots. Then it follows (see Figure 5.2) from (5.43) that

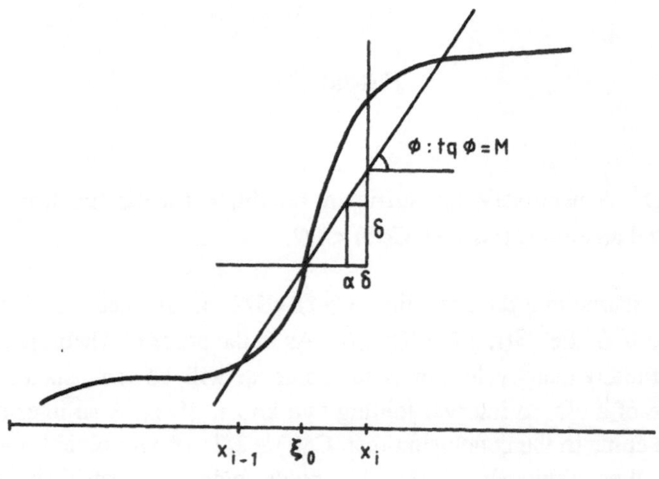

FIGURE 5.2

$$C_0(f) = \limsup_{n\to\infty} \frac{\alpha n\, E(\bar{S}_{0,n}, \Delta, \alpha; f)}{b - a}$$

$$> \limsup_{n\to\infty} \frac{\alpha n M((b - a)/n - n^{-2})}{b - a} \cdot \frac{1}{1 + \alpha M}$$

$$= \frac{\alpha M}{1 + \alpha M},$$

which contradicts the given condition. The theorem is proved.

Corollary 5.1. If $f \in \bar{F}_\Delta$ and $E(\bar{S}_{0,n}, \Delta, \alpha; f) = o(n^{-1})$, then f is a horizontal segment, i.e., $f(x) = c$, where c is real number.

From Theorem 5.9 we obtain

Corollary 5.2. If $f \in \bar{F}_\Delta$ is not the completed graph of a single valued function defined on Δ, then $C_0(f) = 3/2$.

Theorems 5.9 through 5.12 give the complete spectrum of characterizations of functions by their best Hausdorff approximations by step functions with equidistance knots. Here, we approximate by means of discontinuous functions, and the best approximations may be used to characterize functions that are single valued, continuous, or Lipschitz. It is natural to expect characterizations of approximated functions up to a Lipschitz condition for their derivatives of kth order, provided we consider the best Hausdorff approximations by spline functions in $\bar{S}_{k,n}$.

Our goal now is to obtain a converse theorem for best Hausdorff approximations by spline functions in $\bar{S}_{k,n}$ for $k \geq 1$ that implies single valuedness of the approximated function.

Lemma 5.11. If $\varphi, \psi \in \bar{S}_{k,n}$, $k \geq 1$, then

$$(5.44) \qquad \|\varphi' - \psi'\| \leq \frac{4nk^2 \|\varphi - \psi\|}{b - a} .$$

Proof. Clearly, $\varphi - \psi \in \bar{S}_{k,2n}$. Applying A. Markov's inequality with respect to each segment between two knots of the spline function in $\bar{S}_{k,2n}$, we obtain (5.44) directly.

Having (5.44) at our disposal, we can use Lemma 5.4 to establish inequalities of the type (5.16) for spline-functions in $\bar{S}_{k,n}$. This procedure is delicate because the sequence $a_n = E(\bar{S}_{k,n}, \Delta, \alpha; f)$ is not necessarilly monotonically decreasing. However a subsequence whose indices form a geometric progression will be monotonically decreasing.

Lemma 5.12. Let $f \in \bar{F}_\Delta$ and write $a_k(n) = E(\bar{S}_{k,n}, \Delta, \alpha; f)$. Then

$$\omega(f; \alpha a_k(2^m)) \le a_k(2^m) \prod_{i=0}^{m} (1 + \alpha k^2(b - a)^{-1} 2^{i+2} (a_k(2^i) + a_k(2^{i-1}))) ,$$

where $a_k(2^{-1}) \le \max \{|y| : y \in f(x),\ x \in \Delta\}$.

Proof. According to the inequality (5.7), if $s_k(n) \in \bar{S}_{k,n}$ is a spline function of best Hausdorff approximation for f, then

$$\text{(5.45)} \qquad \omega(f; \alpha a_k(n)) \le \omega(f - s_k(n); \alpha a_k(n)) + \omega(s_k(n); \alpha a_k(n))$$

$$\le 2\|f - s_k(n)\| + \omega(s_k(n); \alpha a_k(n))$$

$$\le 2a_k(n) + 3\omega(s_k(n); \alpha a_k(n)).$$

Applying Lemma 5.4 and Lemma 5.11, we obtain

$$\omega(s_k(2^m); \alpha a_k(2^m)) \le \alpha a_k(2^m)\|s_k'(2^m)\|$$

$$\le a_k(2^m)(-1 + \prod_{i=0}^{m} (1 + 4k^2 2^i \alpha (b - a)^{-1} (a_k(2^i) + a_k(2^{i-1})))),$$

where $a_k(2^{-1})$ is the best approximation of f by a constant, i.e, $a_k(2^{-1}) \le M(f) = \sup \{|y| : y \in f(x),\ x \in \Delta\}$. The conclusion of the lemma follows from (5.45).

Corollary 5.3. Suppose $f \in \bar{F}_\Delta$ satisfies the following inequalities:

$$a_k(n) = E(\bar{S}_{k,n}, \Delta, \alpha; f) \le \frac{c(b - a)}{\alpha n}, \qquad n = 1, 2, 3, \ldots.$$

Then

$$\text{(5.46)} \qquad \omega(f; c(b - a)2^{-m}) \le c(b - a)\alpha^{-1}(1 + 4\alpha k^2(a_k(1) + M(f)))/(b - a)(\tfrac{1}{2} + 6ck^2),$$

where $M(f) = \sup \{|y| : y \in f(x),\ x \in \Delta\}$.

We next introduce a functional C_k on $\overline{\mathbf{F}}_\Delta$, $\Delta = [a, b]$, defined by

$$C_k(f) = \limsup_{n\to\infty} \frac{\alpha n\, E(\overline{S}_{k,n}, \Delta, \alpha; f)}{b - a}.$$

The results of §4.3 show that for every $f \in \overline{\mathbf{F}}_\Delta$ we have

$$C_k(f) \le \frac{k-3}{2}, \qquad k = 1, 2, 3, \ldots . \tag{5.47}$$

From (5.46), we obtain the following statement:

Theorem 5.13. If $f \in \overline{\mathbf{F}}_\Delta$ and $C_k(f) < 1/12k^2$, then f is singlevalued on Δ, i.e., $f \in C_\Delta$.

By C_k^0 we denote the largest constant such that if $C_k(f) < C_k^0$ holds, then f is single valued (and continuous). According to Lemma 5.9 and Theorem 5.13, we have the inequalities

$$\frac{1}{12k^2} \le C_k^0 \le 1 \quad \text{for } k \text{ odd},$$

$$\frac{1}{12k^2} \le C_k^0 \le 3/2 \quad \text{for } k \text{ even}.$$

The exact values of C_k^0 for $k = 1, 2, 3, \ldots$, are unknown. As shown by Theorem 5.11, we have $C_0^0 = 1/2$.

§ 5.4. Converse theorems for approximation by rational and partially monotone functions

We now consider converse theorems associated with best Hausdorff approximation by rational functions. In this case, no order of approximation ensures single valuedness (continutiy) of the approximated function. For example, for the function

$$f_0(x) = \begin{cases} 0 & \text{if } x \in [-1, 1],\ x \ne 0 \\ [0, 1] & \text{if } x = 0 \end{cases},$$

we have $E(R_2, \Delta, 1; f_0) = 0$, since $q_\varepsilon(x) = \varepsilon^3/(\varepsilon + x^2) \in R_2$ for each $\varepsilon > 0$ and $r(\Delta, 1; f_0, q_\varepsilon) < \varepsilon$.

K. Lungu [1] demonstrated that $E(R_2, \Delta, 1; f) = o(n^{-1})$ ensures continuity of f almost everywhere on Δ. This result is a corollary of Dolzenko's bound [2] for the area $\sigma(L_n,\varepsilon)$ of an ε-neighborhood of the part of an algebraic curve L_n of nth order that lies in the unit square: $\sigma(L_n,\varepsilon) \leq c(n\varepsilon - (n\varepsilon)^2)$, where c is a constant. To see this, suppose $E(R_2, \Delta, 1; f) = o(n^{-1})$. Then the area of the graph of f is zero, since for every n, this graph is inside the ε-neighborhood of the graph of a rational function of best approximation of order no higher than n for $\varepsilon = \sqrt{2E(R_2, \Delta, 1; f)}$. Hence the intersection of this graph with each vertical line has zero length, i.e., it is a point.

We shall obtain a generalization of Lungu's theorem [1] for the best Hausdorff approximation by piecewise monotone functions.

Theorem 5.14. (E. Dolzenko and E. Sevastianov [1]). If $f \in \bar{F}_\Delta$, $\Delta = [a, b]$, and $E(M_n, \Delta, \alpha; f) = o(n^{-1})$, then f is single valued and continuous almost everywhere.

Proof. Let $M = M(f) = \sup \{|y| : y \in f(x),\ x \in \Delta\}$, and let $\varphi_n \in M_n$ be a piecewise monotone function for which

$$r(\Delta, \alpha; f, \varphi_n) = E(M_n, \Delta, \alpha; f). \tag{5.48}$$

Without loss of generality, we may assume that $M(\varphi_n) = \sup \{|y| : y \in \varphi_n(x),\ x \in \Delta\} \leq M$. Since $\varphi_n \in M_n$, it has total variation $V_a^b \varphi_n \leq nM$. Then the area $\sigma(\varphi_n,\varepsilon)$ of the ε-neighborhood of φ_n with respect to Hausdorff distance with parameter α will satisfy the inequality $\sigma(\varphi_n,\varepsilon) \leq (1 - \alpha)\varepsilon nM - (b - a)\varepsilon$. If we take $\varepsilon = E(M_n, \Delta, \alpha; f) = o(n^{-1})$, then according to (5.48), the graph of f will be contained in this ε-neighborhood of φ_n. As a result,

$$\int_a^b (S(f; x) - I(f; x))\, dx \leq (1 + \alpha)\varepsilon nM + (b - a)\varepsilon \to 0,$$

so that $\int_a^b S(f; x)\, dx = \int_a^b I(f; x)\, dx$. This means that f is Riemann integrable and is thus continuous almost everywhere on Δ. The theorem is proved.

E. Dolzenko and E. Sevastianov [1] studied in detail the relationship between the measure of the set of points of discontinuity of f and the rate at which $E(M_n, \Delta, \alpha; f)$

decreases with n. We state the next two results without proof.

Theorem 5.15. (E. Dolzenko and E. Sevastianov [1]). Let mes P(f) denote the measure of the set of points of nonsingle-valuedness of $f \in \bar{F}_\Delta$. The following bound holds: $\text{mes}\, P(f) \le \liminf_{n\to\infty} nE(M_n, \Delta, 1; f)$.

Theorem 5.16. (E. Dolzenko [1]). Suppose $f \in \bar{F}_\Delta$, and for some positive integer k, the series $\sum_{n=1}^{\infty} (E(R_n, \Delta, \alpha; f))^{1/k}$ is convergent. Then the function f has a differential of order k at almost every point $x \in \Delta$. The result cannot be improved.

§ 5.5. Converse theorems for approximation by positive linear operators

In the third chapter, we looked at a series of problems involving the approximation of functions by positive linear operators with respect to Hausdorff distance. Here we obtain some converse theorems in this direction. First, we prove a few auxiliary results, in the spirit of Lemma 5.5.

Lemma 5.13 (V. Hristov (unpublished)). Let $B \subset A_\Delta$ and let $\langle L_n \rangle_0^\infty$ be a sequence of linear operators defined on B with values in B which commute with each other, i.e., $L_k(L_n(f)) = L_n(L_k(f))$ for every $f \in B$. Suppose for each n and f, $L_n'(f)$ exists. If $\langle c_n \rangle_0^\infty$ is a real sequence such that

$$r(f, L_n(f)) = r(\Delta, \alpha; f, L_n(f)) \le c_{n-1}^{-1},$$

and $\|L_n'(f)\| \le c_n \|L_n(f)\|$, then

$$1 + \alpha \|L_k'(f)\| \le \frac{(1 + \alpha \|L_{k-1}'(f)\|)(1 + \alpha c_k\, r(f, L_{k-1}(f)))}{1 - \alpha c_{k-1}\, r(f, L_k(f))}, \tag{5.49}$$

where $\|\cdot\|$ denotes, as usual, the uniform norm on Δ.

Proof. According to the hypotheses of the lemma and (5.7), we have

$$\|L_k'(f)\| \le \|L_k'(f) - L_{k-1}'(f)\| + \|L_{k-1}'(f)\|$$

$$= \|(L_k(f - L_{k-1}(f)) - L_{k-1}(f - L_k(f)))'\| + \|L_{k-1}'(f)\|$$

$$\le \|(L_k'(f - L_{k-1}(f))\| + \|L_{k-1}'(f - L_k(f))\| + \|L_{k-1}'(f)\|$$

$$\le c_k\|f - L_{k-1}(f)\| + c_{k-1}\,\|f - L_k(f)\| + \|L_{k-1}'(f)\|$$

$$\le\; c_k\, r(f, L_{k-1}(f)) + c_k\, \omega(L_{k-1}(f); \alpha r(f, L_{k-1}(f))) + c_{k-1}\, r(f, L_k(f)) + c_{k-1}\, \omega(L_k(f); \alpha r(f, L_k(f))) + \|L_{k-1}'(f)\|$$

$$\le\; c_k\, r(f, L_{k-1}(f)) + \alpha c_k r(f, L_{k-1}(f))\|L_{k-1}'(f)\| + c_{k-1}\, r(f, L_k(f)) + \alpha c_{k-1} r(f, L_k(f))\|L_k'(f)\| + \|L_{k-1}'(f)\|.$$

Hence,

$$\alpha^{-1} + \|L_k'(f)\| - c_{k-1}\, r(f, L_k(f)) - \alpha\, c_{k-1}\, r(f, L_k(f))\, \|L_k'(f)\|$$

$$\le \alpha^{-1} + \|L_{k-1}'(f)\| + c_k\, r(f, L_{k-1}(f)) + \alpha\, c_k\, r(f, L_{k-1}(f))\, \|L_{k-1}'(f)\|\,.$$

As the last inequality is equivalent to (5.49), the lemma is proved.

Theorem 5.17. Let $\langle L_n\rangle_0^\infty$ be a sequence of linear operators, satisfying the hypotheses of Lemma 5.13. If $L_0'(f) \equiv 0$, then for an arbitrary positive integer n the inequality

$$\omega(f; \alpha r(f, L_n(f))) \le 3\, r(f, L_n(f)) \prod_{k=1}^{n} \frac{1 + \alpha c_k\, r(f, L_{k-1}(f))}{1 - \alpha c_{k-1}\, r(f, L_k(f))} \tag{5.50}$$

holds, where $r(f, g) = r(\Delta, \alpha; f, g)$.

Proof. According to (5.7), we have

$$\omega(f; \delta) \le \omega(f - L_n(f); \delta) + \omega(L_n(f); \delta) \le 2\,\|f - L_n(f)\| + \omega(L_n(f); \delta)$$

$$\le\; 2r(f, L_n(f)) + 2\omega(L_n(f); \alpha r(f, L_n(f))) + \omega(L_n(f); \delta).$$

Setting $\delta = \alpha r(f, L_n(f))$, we obtain

$$\omega(f; \alpha r(f, L_n(f)) \le 2r(f, L_n(f)) + 3\alpha\|L_n'(f)\|\, r(f, L_n(f))$$

$$\le 3r(f, L_n(f))(1 + \alpha\|L_n'(f)\|).$$

Since $L_0'(f) \equiv 0$, from Lemma 5.13 and the last inequality, we obtain (5.50).

The inequality of (5.50) might be used find conditions on the rate of decrease of $r(f, L_n(f))$ to zero that would ensure continuity of f.

Theorem 5.18. (V. Hristov (unpublished)). Suppose for the sequence $\langle L_n\rangle_0^\infty$ of linear operators, we have for some fixed $\lambda > 0$ and $k = 0, 1, 2, \ldots$

$$\|L_k'(f)\| \le ck^\lambda \|f\|.$$

If for a certain function f,

$$\limsup_{n\to\infty} \alpha c n^\lambda r(\Delta, \alpha, f, L_n(f)) < 1 \tag{5.51}$$

is valid, then f is continuous on Δ.

Proof. We consider the auxiliary sequence of linear operators $\langle V_n\rangle_0^\infty$, where $V_0 = L_0$, and $V_n = L_{q^n}$ for $n \ge 1$, where $q \ge 2$ is a certain positive integer. Condition (5.51) means that for a certain integer m and all $n > m$, the inequality $r(f, V_n(f)) < \mu/\alpha c q^{\lambda n}$ holds for some $\mu \in (0, 1)$. Let $n > m$; we apply Theorem 5.17 to the finite sequence $V_0, V_m, V_{m+1}, \ldots, V_n$ to obtain

$$\omega(f; \alpha r(f, V_n(f))) \le 3r(f, V_n(f))\, A \prod_{k=m-1}^{n} \frac{1 + \mu q^\lambda}{1 - \mu q^{-\lambda}}, \tag{5.52}$$

where the constant $A = (1 + \alpha c_m\, r(f, V_0(f)))/(1 - \alpha c_0 r(f, L_m(f)))$ is independent of n. We write (5.52) in the form

$$\omega(f; \alpha r(f, V_n(f))) \le 3A\, \mu c^{-1} q^{-\lambda n} \left(\frac{1 + \mu q^\lambda}{1 - \mu q^{-\lambda}}\right)^{n-m} \tag{5.53}$$

$$= 3A\,\mu c^{-1} q^{-\lambda m} \left(\frac{1 + \mu q^{\lambda}}{q^{\lambda} - \mu}\right)^{n-m} .$$

Since $\mu < 1$, for sufficiently large q, we have $(1 + \mu q^{\lambda})/(q^{\lambda} - \mu) < 1$. From (5.53), we see that $\lim_{n \to \infty} \omega(f; \alpha r(f, V_n(f))) = 0$, i.e., f is continuous. The proof is complete.

To apply 5.19, we utilize estimates of the type of Bernastien's inequality for the corresponding operators. It is easy to obtain such inequalitites for integral operators with bell-shaped, differentiable, symmetric kernels.

Lemma 5.14. Let

$$L(f; x) = \int_{-\pi}^{\pi} f(x + t)\, K(t)\, dt,$$

where $K(t)$ is an even, differentiable, nonnegative, 2π-periodic bell-shaped function, i.e., $K(t)$ is monotonically increasing on $[-\pi,0]$ and monotonically decreasing on $[0, \pi]$. Then we have

$$\|L'(f)\| \le 2K(0)\, \|f\|. \tag{5.54}$$

Proof.

$$\|L'(f; x)\| = \| \int_{-\pi}^{\pi} f(t)\, K'(t)\, dt \| = \| \int_{-\pi}^{\pi} f(x + t)\, K'(t)\, dt \|$$

$$\le \|f\| \left(\int_{-\pi}^{0} K'(t)\, dt - \int_{0}^{\pi} K'(t)\, dt \right) = 2\|f\| (K(0) - K(\pi)) \le 2K(0)\, \|f\|.$$

Obviously, the inequality (5.54) is valid for the bell-shaped kernels on $(-\infty, \infty)$, too; a typical example is the **Gauss-Weierstrass operator**

$$W_n(f; x) = \left(\frac{n}{\pi}\right)^{1/2} \int_{-\infty}^{\infty} f(t)\, e^{-n(t - x)^2}\, dt .$$

By Lemma 5.14, the inequality $\|W_n'(f)\| \le 2(n/\pi)^{1/2} \|f\|$ holds. Applying Theorem 5.18, we obtain

Corollary 5.4. Let f be a function satisfying

(5.55) $$\limsup_{n\to\infty} \alpha n^{1/2}\, r(\alpha; f, W_n(f)) < \frac{\pi^{1/2}}{2} .$$

Then f is continuous on $(-\infty, \infty)$.

On the other hand, applying Lemma 5.14 to the **Abel-Poisson operator**

$$P_\rho(f; x) = \frac{1}{2\pi} \int_{-\pi}^{\pi} f(x - t) \frac{1 - \rho^2}{1 - 2\rho\cos t - \rho^2} dt$$

we obtain $\|P'_\rho(f)\| \le \dfrac{2}{\pi(1-\rho)}\|f\|$. By Theorem 5.18, we have

Corollary 5.5. Let f be a 2π-periodic function satisfying

(5.56) $$\limsup_{\rho\to 1-0} \frac{\alpha}{1-\rho}\, r(\alpha; f, P_\rho(f)) < \frac{\pi}{2} ,$$

then f is continuous on $[-\pi, \pi]$.

Lemma 5.14 can also be applied to the Valleé-Poussin operators

$$V_n(f; x) = \frac{(n!)^2}{2\pi(2n)!} \int_{-\pi}^{\pi} f(x - t)\, (2\cos \tfrac{t}{2})^{2n}\, dt$$

for which we have

$$\|V_n'(f)\| \le \frac{2^{2n+1}(n!)^2\, \|f\|}{2\pi(2n)!} \le \left((\tfrac{n}{\pi})^{1/2} + O(1)\right)\|f\|.$$

Applying Theorem 5.18 once again, we obtain

Corollary 5.6. Let f be a 2π-periodic function satisfying

(5.57) $$\limsup_{n\to\infty} \alpha n^{1/2}\, r(\alpha; f, V_n(f)) < \pi^{1/2} .$$

Then f is continuous for all x.

The inequalities (5.55) - (5.57) give sufficient conditions for continuity of the approximated function f. The problem of finding the best constants on the right-hand side guaranteeing continuity remains open.

Since approximation in Hausdorff distance is sideways rather than vertical, points of discontinuity of the second kind do not seem to present additional difficulties for approximation than the points of discontinuity of the first kind. Thus, finding conditions that ensure the lack of points of discontinuity of the first kind of the approximated function is of interest. V. Veselinov [18] was the first to consider such problems. Veselinov's results show that, if on the right-hand side of the inequalities of (5.55)-(5.57), we put an arbitrarily large constant, then the resulting conditions ensure the lack of discontinuities of the first kind of the approximated function f. It is not known to what extent these results are final. For example, we know of no function that has discontinuities of the second kind and which is approximated by positive linear operators at such a rate that discontinuities of the first kind are precluded.

Remarks . The first converse theorems on best Hausdorff approximation by algebraic polynomials were published by Bl. Sendov [2]. More recent converse theorems on best Hausdorff approximation are presented in the paper of E. Dolzenko [1].

Chapter 6

ε-Entropy, ε-Capacity and Widths

The concepts of ε-entropy and ε-capacity of function spaces were introduced by A. Kolmogorov [1] to characterize massiveness of these spaces. Let us provide the relevant terminology.

Let X be a metric space and let A be a nonempty subset of X. A collection of subsets Γ of X is called an **ε-covering** of A if the diameter d(U) of each element U $\in \Gamma$ is at most 2ε, and $A \subset \cup \{U : U \in \Gamma\}$. A subset V of X is called **ε-discrete** if the distance between distinct points of V is greater than ε. By Zorn's lemma, each subset A of X has a maximal ε-discrete subset V, and the balls of radius ε whose centers run over V form an ε-covering of A. The set A is called **totally bounded** provided it has a finite ε-covering for each $\varepsilon > 0$.

Suppose now that A is totally bounded. We introduce the following notation:

$N_\varepsilon(A)$ = the minimal number of elements in an ε-covering of A;

$M_\varepsilon(A)$ = the maximal number of elements in an ε-discrete subset of A;

$H_\varepsilon(A) = \log_2 N_\varepsilon(A)$;

$C_\varepsilon(A) = \log_2 M_\varepsilon(A)$.

We call $H_\varepsilon(A)$ the **ε-entropy** of A, and $C_\varepsilon(A)$ the **ε-capacity** of A. The following theorem holds.

Theorem 6.1. (A. Kolmogorov and V. Tihomirov [1]). Let A be a totally bounded subset of a metric space X. The following inequalities hold:

$$M_{2\varepsilon}(A) \leq N_\varepsilon(A) \leq M_\varepsilon(A).$$

As a result, $C_{2\varepsilon}(A) \leq H_\varepsilon(A) \leq C_\varepsilon(A)$.

§ 6.1. ε-entropy and ε-capacity of the set F_Δ^M

Let $\Delta = [a, b]$ and let M be a positive constant. We have denoted by F_Δ^M those segment functions $f \in F_\Delta$ for which

$$\max \{|y| : y \in f(x),\ x \in \Delta\} \le M,$$

Thus, if $f \in F_\Delta^M$, then its graph is a compact point set in the plane lying between the horizontal lines $y = -M$ and $y = M$, and whose projection on the x-axis is $[a, b]$. Moroeover, each vertical section of the graph of f is convex. Results of Chapter 2 show that F_Δ^M is a compact subset of F_Δ, equipped with the Hausdorff metric. Our main task here will be to compute the ε-entropy and ε-capacity of F_Δ^M.

Consider the rectangle $D = [a, b] \times [-M, M]$, and the two sequences of numbers

$$a = x_0 < x_1 < x_2 < \cdots < x_p = b, \tag{6.1}$$

$$-M = y_0 < y_1 < y_2 < \cdots < y_q = M.$$

The rectangle D is divided into pq closed elementary rectangles by the two families of parallel lines $x = x_i$, $i = 0, 1, \ldots, p$ and $y = y_j$, $j = 0, 1, 2, \ldots, q$. We denote that rectangle whose lower left corner is (x_i, y_j) by $\{x_i; y_j\}$.

Definition 6.1. We say that a set K of elementary rectangles from such a partition of D forms a **corridor** if the following conditions hold:

(a) If $\{x_i; y_j\}$ and $\{x_i; y_k\}$ are in K, then so is $\{x_i; y_l\}$ whenever $j < l < k$;

(b) For each $i = 0, 1, 2, \ldots, p - 2$ there exists a $j \in \{0, 1, 2, \ldots, q - 1\}$ such that both $\{x_i; y_j\}$ and $\{x_{i+1}; y_j\}$ belong to K.

Notice that the union of the rectangles in a corridor is itself an element of F_Δ^M; thus we may view a corridor as an element of F_Δ^M. Denote by $K_{p,q}$ the set of all corridors of rectangles for the above partition of D. Clearly, $K_{p,q}$ is a finite subset of F_Δ^M; we write $k_{p,q}$ for the number of elements in $K_{p,q}$.

Definition 6.2. Let $f \in \mathbf{F}_\Delta^M$. We say that f belongs to the corridor $K \in K_{p,q}$ if every point of f belongs to some rectangle in K, and every rectangle which has a point in common with f belongs to K.

It is easy to see that for every $f \in \mathbf{F}_\Delta^M$ there exists one and only one corridor K in $K_{p,q}$ to which f belongs. For each $\varepsilon > 0$ and $\alpha > 0$ we define a positive integer p as follows: $p = -[-(b - a)/2\alpha\varepsilon]$. Set $t = (b - a - 2\alpha\varepsilon(p - 2))/2$; we claim that $\alpha\varepsilon < t \leq 2\alpha\varepsilon$. First, if $(b - a)/2\alpha\varepsilon$ is an integer, then $t = 2\alpha\varepsilon$. Otherwise,

$$t = \frac{b-a}{2} - ([(b-a)/2\alpha\varepsilon] - 1)\alpha\varepsilon > \frac{b-a}{2} - \frac{b-a}{2} + \alpha\varepsilon = \alpha\varepsilon,$$

$$t = \frac{b-a}{2} - ([(b-a)/2\alpha\varepsilon] - 1)\alpha\varepsilon < \frac{b-a}{2} - \frac{b-a}{2} + 2\alpha\varepsilon = 2\alpha\varepsilon.$$

Let $q = [-M/\varepsilon]$ and $h = M - (q - 2)\varepsilon$. We now introduce the following two systems of numbers:

$$x_0 = a,\ x_1 = a + t,\ x_2 = x_1 + 2\alpha\varepsilon, \ldots, \quad x_{p-1} = x_{p-2} + 2\alpha\varepsilon,\ x_p = x_{p-1} + t = b; \tag{6.2}$$

$$y_0 = -M,\ y_1 = -M + h,\ y_2 = y_1 + 2\varepsilon, \ldots, \quad y_{q-1} = y_{q-2} + 2\varepsilon,\ y_q = y_{q-1} + h = M.$$

The set of corridors obtained under the induced partition of D will be called the ε-**covering corridors** with parameter α. We denote these by $K'_{p,q}$.

Lemma 6.1. If $f, g \in \mathbf{F}_\Delta^M$ and both belong to the same corridor K of $K'_{p,q}$, then

$$r(\alpha; f, g) \leq 2\varepsilon. \tag{6.3}$$

Proof. Let $A = (x,y)$ be an arbitrary point of f. The point A belongs to a certain rectangle $\{x_i; y_j\}$ in K, and hence there exists a point $B_A \in g$ with $B_A \in \{x_i; y_j\}$. Thus,

$$\min_{B \in g} \rho_\alpha(A, B) = \min_{(\xi,\eta) \in g} \{\alpha^{-1}|x - \xi|, |y - \eta|\}$$

$$\leq \rho_\alpha(A, B_A) = \max\{\alpha^{-1}2\alpha\varepsilon, 2\varepsilon\} = 2\varepsilon.$$

Consequently,

$$\max_{A\in f} \min_{B\in g} \rho_\alpha(A, B) \leq 2\varepsilon.$$

In the same way we obtain

$$\max_{A\in g} \min_{B\in f} \rho_\alpha(A, B) \leq 2\varepsilon.$$

Together these yield (6.3).

Now let $\delta = (b - a)/2(p - 1) - \alpha\varepsilon$, $\lambda = M/(q - 1) - \varepsilon$, and $\gamma = \min\{\delta, \lambda\}$. It is easy to see that $\delta > 0$, $\lambda > 0$, and hence $\gamma > 0$. We introduce a second pair of systems of numbers in $[a, b]$ and $[-M, M]$:

(6.4)
$$\xi_0 = a,\ \xi_1 = a + \alpha\varepsilon + \mu,\ \xi_2 = \xi_1 + 2(\alpha\varepsilon + \gamma), \ldots,$$
$$\xi_{p-1} = \xi_{p-2} + 2(\alpha\varepsilon + \gamma),\ \xi_p = \xi_{p-1} + \alpha\varepsilon + \mu = b;$$

$$\eta_0 = -M,\ \eta_1 = -M + \varepsilon + \nu,\ \eta_2 = \eta_1 + 2(\varepsilon + \gamma), \ldots,$$
$$\eta_{q-1} = \eta_{q-2} + 2(\varepsilon + \gamma),\ \eta_q = \eta_{q-1} + \varepsilon + \nu = M,$$

where $\mu = (b - a)/2 - (p - 2)(\alpha\varepsilon + \gamma) - \alpha\varepsilon$ and $\nu = M - (q - 2)(\varepsilon + \gamma) - \varepsilon$. It is directly verified that $\mu > \gamma$ and $\nu > \gamma$.

The set of corridors, obtained by the partition of D induced by the systems of (6.4), will be denoted by $K''_{p,q}$; these will be call the **ε-discrete corridors.** Obviously, the number of corridors in both $K'_{p,q}$ and $K''_{p,q}$ is just $k_{p,q}$. Let $K \in K''_{p,q}$ be fixed. The **γ-center** of K is this subset G of K:

$$G = \{A \in K : \min\{\rho_\alpha(A,B) : B \in \partial K \backslash \partial D \geq \varepsilon + \gamma/2\}.$$

Here, the symbol ∂ represents the boundary operator. It follows from the definition of ε-discrete corridor that every corridor in $K''_{p,q}$ has a nonempty γ-center and that the γ-center belongs to F^M_Δ.

We may also define the γ-center of a given rectangle $d_{ij} = \{\xi_i; \eta_j\}$ of an ε-discrete corridor of D. Specifically, the γ-center of d_{ij} is the set of all points $A \in d_{ij}$ for which

$$\min\{\rho_\alpha(A,B) : B \in \partial d_{ij} \backslash \partial D \geq \varepsilon + \gamma/2\}.$$

If $d_{ij} \subset K \in K''_{p,q}$, then the γ-center of d_{ij} is contained in the γ-center of K, too.

We now show that the centers of the ε-discrete corridors form a 2ε-discrete set with respect to Hausdorff distance.

Lemma 6.2. Suppose $K_1, K_2 \in K''_{p,q}$ are distinct corridors, and G_1, G_2 are their respective γ-centers. Then

$$r(\alpha; G_1, G_2) > 2\varepsilon. \tag{6.5}$$

Proof . Since K_1 and K_2 are distinct, we may assume there exists a rectangle $d_{ij} = \{\xi_i; \eta_j\}$ that belongs to K_1 but not to K_2. For each point A of the γ-center of d_{ij} and for each point B of the γ-center of K_2 we have $\rho_\alpha(A,B) > 2\varepsilon$. From the last inequality and Lemma 2.1, we get (6.5). The lemma is proved.

Lemma 6.3. The set $\mathbf{F}^M_\Delta$ is totally bounded with respect to Hausdorff distance with parameter $\alpha > 0$. For this distance, we have

$$H_\varepsilon(\mathbf{F}^M_\Delta) = C_{2\varepsilon}(\mathbf{F}^M_\Delta) = \log_2 k_{p,q}.$$

Proof. Identifying corridors $K \in K'_{p,q}$ with the set of all elements $f \in \mathbf{F}^M_\Delta$ that belong to K, we see that the diameter of such a set is at most 2ε by Lemma 6.1. This means that the set of all corridors in $K'_{p,q}$ is a finite ε-covering of the set $\mathbf{F}^M_\Delta$. Since the number of elements in $K'_{p,q}$ is $k_{p,q}$, we have

$$H_\varepsilon(\mathbf{F}^M_\Delta) = \log_2 N_\varepsilon(\mathbf{F}^M_\Delta) \leq \log_2 k_{p,q}. \tag{6.6}$$

On the other hand, by Lemma 6.2, we can construct $k_{p,q}$ elements of $\mathbf{F}^M_\Delta$ such that the distance between every two of them exceeds 2ε. Consequently,

$$C_{2\varepsilon}(\mathbf{F}^M_\Delta) = \log_2 M_{2\varepsilon}(\mathbf{F}^M_\Delta) \geq \log_2 k_{p,q}. \tag{6.7}$$

Combining (6.6) and (6.7), we get $C_{2\varepsilon}(\mathbf{F}^M_\Delta) \geq H_\varepsilon(\mathbf{F}^M_\Delta)$. But by Theorem 6.1, the reverse inequality is valid, and hence

$$H_\varepsilon(\mathbf{F}^M_\Delta) = C_{2\varepsilon}(\mathbf{F}^M_\Delta) = \log_2 k_{p,q}.$$

This completes the proof.

Lemma 6.3 reduces the computation of the ε-entropy and ε-capacity of the set $\mathbf{F}^M_\Delta$ to the calculation of the number of (p,q)-corridors $k_{p,q}$. Obviously, $k_{p,q}$ depends only on the positive integers p and q, and not on the placement of the knots. Thus, we need only consider corridors in the rectangle $D_{p,q} = [0, p] \times [0, q]$, using the straight lines $x = i$, $i = 0, 1, 2, \ldots, p$, and $y = j$, $j = 0, 1, 2, \ldots, q$. With respect to this partition, the vertices of the subrectangles are all integral lattice points.

The calculation of the number $k_{p,q}$ of (p,q)-corridors in $D_{p,q}$ is quite complicated. First, we produce rough upper and lower bounds for $k_{p,q}$, obtained in an elementary way, which allows us to determine the asymptotic behavior of $\log_2 k_{p,q}$.

Lemma 6.4. For positive integers p and q, the following inequalities are valid:

$$2^{-p}\binom{q+1}{2}^p < k_{p,q} \le \binom{q+1}{2}^p . \tag{6.8}$$

Proof. From the definition of a corridor, we have

$$k_{p,q} \le k_{s,q}k_{p-s,q}, \quad s = 1, 2, 3, \ldots, p-1, \tag{6.9}$$

$$k_{1,q} = \binom{q+1}{2}. \tag{6.10}$$

From (6.9) and (6.10), we obtain the right-hand side of the inequality of (6.8):

$$k_{p,q} \le k_{1,q}k_{p-1,q} \le \cdots \le k_{1,q}^p = \binom{q+1}{2}^p .$$

Denote by $k_{p,q}(i,j)$ the number of corridors in $K_{p,q}$ which have squares between the lines $y = i$ and $y = q - j$ in the rightmost vertical strip (see Figure 6.1). We shall say that these corridors have an end (i,j). Clearly, $k_{p,q}(i,j)$ makes sense for $i + j \le q - 1$, and it follows directly from the definition of a corridor that

$$k_{p,q} = k_{p+1,q}(0,0). \tag{6.11}$$

By symmetry, the number of corridors in $K_{p,q}$ that have squares between the horizontal lines $y = i$ and $y = q - j$ in the leftmost vertical strip is also $k_{p,q}(i,j)$.

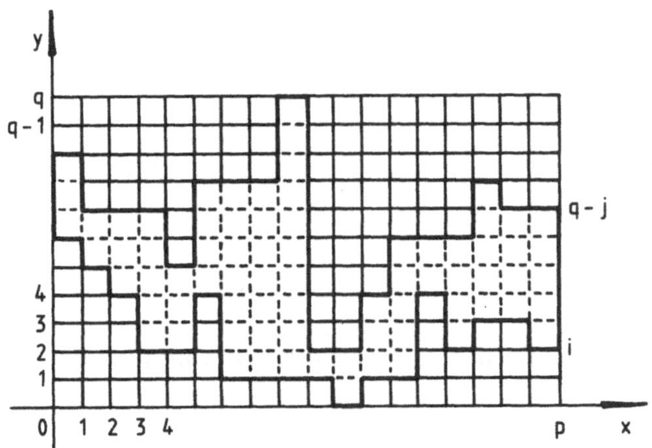

FIGURE 6.1

It is clear that the concatenation of the corridor K_1 in the rectangle $[0, s] \times [0, q]$ with the corridor K_2 in the rectangle $[s, p] \times [0, q]$ forms a corridor in $[0, p] \times [0, q]$ if and only if K_1 has an end (i,j) and K_2 has a beginning (m,n), where the open intervals (i,j) and (m,n) have nonempty intersection. From this, we have

$$k_{p,q} \geq \left(\sum_{i,j} k_{1,q}(i,j)\right)^p \geq (([q/2]+1)(q-[q/2]))^p \geq 2^{-p}\left(\frac{q+1}{2}\right)^p,$$

for $0 \leq i \leq [q/2]$ and $0 \leq j \leq q - [q/2] - 1$. The lemma is proved.

If we take $p = -[-(b-a)/2\alpha\varepsilon]$ and $q = -[-M/\varepsilon]$, then Lemma 6.3 and (6.8) yield

Theorem 6.2. (Bl. Sendov and B. Penkov [1,2]). The asymptotic values of ε-entropy and ε-capacity of the set $\mathbf{F}^M_\Delta$ with respect to Hausdorff distance with parameter α are given by

$$H_\varepsilon(\mathbf{F}^M_\Delta) = C_{2\varepsilon}(\mathbf{F}^M_\Delta) \sim \frac{b-a}{\alpha\varepsilon} \log_2 \frac{1}{\varepsilon}. \tag{6.12}$$

Corollary 6.1. For the set C_Δ^M of continuous functions of uniform norm at most M on Δ, we have

$$H_\varepsilon(C_\Delta^M) = C_{2\varepsilon}(C_\Delta^M) \sim \frac{b-a}{2\alpha\varepsilon}\log_2\frac{1}{\varepsilon}$$

with respect to the Hausdorff distance with parameter α.

Proof. In Chapter 2, we showed that C_Δ^M is a dense subset of $\mathbf{F}_\Delta^M$. The result now follows from (6.12).

6.2. The number of (p,q)- corridors

If we use recurrence relations, we obtain the following formulas, which hold for all positive integers q:

$$K_{1,q} = \binom{q+1}{2},$$

$$K_{2,q} = \frac{1}{3}\binom{q+1}{2}(q^2-q-1),$$

$$K_{3,q} = \frac{1}{90}\binom{q+1}{2}(11q^4-22q^3-28q^2-17q-12),$$

$$K_{4,q} = \frac{1}{180}\binom{q+1}{2}(8q^6-24q^5-41q^4-42q^3-35q^2-18q-12)$$

$$K_{5,q} = \frac{1}{113400}\binom{q+1}{2}(1837q^8-\cdots).$$

In general, one can show (H. Hitov [1]) that for a fixed integer p, the value of $k_{p,q}$ is a polynomial in the variable q of degree 2p. For fixed q, we can actually compute $k_{p,q}$ as a function of p. For instance, for $q = 2$ we have

$$k_{p,q} = \frac{1}{2}((1+\sqrt{2})^{p+1} - (1-\sqrt{2})^{p+1}).$$

In order to express the asymptotic value of $k_{p,q}$ for large p and q, it is convenient to introduce numbers $\lambda_{p,q}$ defined by

$$k_{p,q} = (\lambda_{p,q} \binom{q+1}{2})^p .$$

According to (6.8) we have

$$\frac{1}{2} \le \lambda_{p,q} \le 1. \tag{6.13}$$

L. Bondarenko [1] has obtained a sharper inequality than (6.13) for large p and q, namely,

$$0.7272\ldots = \frac{8}{11} < \lambda_{p,q} < \frac{1837}{2520} = 0.728968\ldots .$$

In addition, A. Panov [1,2,4] found the exact asymptotic value of $\lambda_{p,q}$ for p and q tending to infinity:

$$\lim_{p,q\to\infty} \lambda_{p,q} = \lambda = 0\,.728850\ldots,$$

where λ is the maximal positive root of the equation ${}_1F_1(-1/2 - 1/8\lambda, 1/2, 1/2\lambda) = 0$, where ${}_1F_1$ is a degenerate hyperbolic function. More preceisely, we shall prove

Theorem 6.3 (A. Panov [4]). For every positive integer q, there exist numbers λ_q and c_q such that as $p \to \infty$,

$$k_{p,q} \sim c_q (\lambda_q \binom{q+1}{2})^p .$$

Moreover there exist numbers λ and c such that $\lim_{q\to\infty} \lambda_q = \lim_{p,q\to\infty} \lambda_q = \lambda$ and $\lim_{q\to\infty} c_q = c$. The number λ is the positive root of maximum modulus of the equation ${}_1F_1(-1/2 - 1/8\lambda, 1/2, 1/2\lambda) = 0$, where ${}_1F_1$ is a degenerate hyperbolic function.

The proof of this theorem is involved, and we shall need a number of preliminary lemmas. To each closed integral subinterval $[i, j]$ of $[0, q]$, we associated the point $(x,y) = (i/q, 1 - j/q)$ in the plane. The points (x,y) corresponding to all such closed subintervals will uniformly fill out a triangle τ, enclosed by the coordinate axes and the line $x + y = 1$. We denote this set of points by τ_q; obviously, τ_q consists of $\binom{q+1}{2}$ points. The following fact is geometrically obvious.

Lemma 6.5. Suppose $(x,y) \in \tau_q$ corresponds to the integral closed interval $[i, j]$. Then the points $(u,v) \in \tau$ corresponding to the integral closed intervals $[m, n]$ which intersect $[i, j]$ in an interval of positive length satisfy the condition

$$\sigma(1 - x - v)\sigma(1 - y - u) = 1, \tag{6.14}$$

where $\sigma(t) = 0$ for $t \leq 0$ and $\sigma(t) = 1$ for $t > 0$.

The set of points (u,v) of the triangle τ for which equality holds in (6.14) will be denoted by $\tau(x,y)$, and $\tau_q(x,y) = \tau_q \cap \tau(x,y)$.

We now introduce an operator T_q defined on the set of functions with (finite) domain τ_q, that will be of importance in the sequel:

$$T_q(f; x, y) = \binom{q+1}{2}^{-1} \sum_{(u,v) \in \tau_q(x,y)} f(u,v) \tag{6.15}$$

Note that the number of points in τ_q is equal to $\binom{q+1}{2}$, and these points form a rectangular lattice with step $1/q$ in the triangle τ, of area $1/2$. Therefore, if we consider an integral operator T acting on functions $f(x,y)$ defined on $\tau(x,y)$, described by

$$T(f; x, y) = 2 \iint_{\tau(x,y)} f(u,v)\, du\, dv, \tag{6.16}$$

then the operator of (6.15) is generated by the operator of (6.16) upon replacing the integral by a corresponding Riemann sum. Using (6.14), the operator T can be written as

$$T(f; x, y) = 2 \iint_{\tau} \sigma(1 - x - v)\sigma(1 - y - u)\, f(u, v)\, du\, dv,$$

or

$$T(f; x, y) = 2 \int_0^x \int_0^{1-x} f(u, v)\, du\, dv + 2 \int_x^{1-y} \int_0^{1-u} f(u, v)\, dv\, du\ . \tag{6.17}$$

We now define a scalar product $\langle f, g\rangle_q$ with induced norm $\|f\|_q$ on the set of functions with domain τ_q:

$$(6.18)\qquad \langle f, g\rangle_q = \binom{q+1}{2}^{-1} \sum_{(x,y)\in\tau_q} f(x,y)g(x,y)\,, \qquad \|f\|_q = \sqrt{\langle f, f\rangle_q}\,.$$

Lemma 6.6. The operator T_q is symmetric with respect to the scalar product of (6.18), i.e., $\langle T_q(f), g\rangle_q = \langle f, T_q(g)\rangle_q$.

Proof. We have from (6.15) and (6.18) that

$$(6.19)\qquad \langle T_q(f), g\rangle_q = \binom{q+1}{2}^{-1} \sum_{(x,y)\in\tau_q}\ \sum_{(u,v)\in\tau_q(x,y)} f(u,v)g(x,y).$$

By Lemma 6.5, the point (u,v) belongs to $\tau_q(x,y)$ if and only if the segments corresponding to the points (u,v) and (x,y) intersect in a segment of positive length. Thus, for arbitrary (u,v) and (x,y) in τ_q, the condition $(u,v)\in\tau_q(x,y)$ implies $(x,y)\in\tau_q(u,v)$. Consequently, the summation on the right-hand side of (6.19) can be written as

$$\sum_{(x,y)\in\tau_q}\ \sum_{(u,v)\in\tau_q(x,y)} f(u,v)g(x,y) = \sum_{(x,y)\in\tau_q(u,v),\ (u,v)\in\tau_q(x,y)} f(u,v)g(x,y)\ ,$$

where f and g participate symmetrically. As a result, $\langle T_q(f), g\rangle_q = \langle f, T_q(g)\rangle_q$.

Lemma 6.7. The operator $T_q^2 = T_q \circ T_q$ is positive, i.e., it transforms every nonnegative function with domain τ_q that is not identically zero into a strictly positive one.

Proof. If the function f with domain τ_q is positive at a certain point $(x,y)\in\tau_q$, then by (6.15), $T_q(f;\,0,\,0) > 0$ because $\tau_q(0,0) = \tau_q$. On the other hand, if $f(0,0) > 0$, then $T_q(f;\,x,\,y) > 0$ for every point $(x,y)\in\tau_q$, because for each such (x,y), we have $(0,0)\in\tau_q(x,y)$. The lemma is proved.

The numbers $k_{p,q}$ can be expressed through the operator T_q, as we next show.

Lemma 6.8. Let e_q be the function identically equal to unity on τ_q, and let $(x,y)\in\tau_q$ correspond to the segment $[i,j]$. Then

$$(6.20)\qquad \alpha_{p,q}(x,y) = k_{p,q}(i,j) = \binom{q+1}{2}^{p-1} T_q^{p-1}(e_q;\,x,\,y).$$

Thus, according to (6.11), the total number of (p,q)-corridors is given by

$$k_{p,q} = \alpha_{p+1,q}(0,0) = \binom{q+1}{2}^p T_q^p(e_q; 0, 0). \tag{6.21}$$

Proof. Since there is a one-to-one correspondence between each (1,q)-corridor and the integral subinterval [i, j] of the interval [0, q], with which this corridor ends (and begins), we have

$$\alpha_{1,q}(x,y) = k_{1,q}(i,j) = e_q(x,y) = 1, \tag{6.22}$$

where the point $(x,y) \in \tau_q$ correspond to the segment [i, j].

To obtain all the (p,q)-corridors with end [i, j], we take all (p - 1,q)-corridors whose ends are segments that intersect [i, j] in a segment of positive length, and adjoin to each of these corridors a (1,q)-corridor on the right corresponding to [i, j]. By Lemma 6.5, the points corresponding to the segments that intersect [i, j] in a segment of positive length fill out $\tau_q(x,y)$. Thus, we obtain for the number of (p,q)-corridors with end [i, j]

$$k_{p,q}(i,j) = \alpha_{p,q}(x,y) = \sum_{(u,v)\in\tau_q(x,y)} \alpha_{p-1,q}(u,v),$$

so that by (6.15),

$$\alpha_{p,q}(x,y) = \binom{q+1}{2} T_q(\alpha_{p-1,q}; x, y).$$

Combining this equality and (6.22), we obtain (6.20) by induction. The lemma is proved.

We see from (6.21) that the number $k_{p,q}$ can be computed from the value of the p-th iteration of the operator T_q of the unit function e_q at the point (0,0).

Along with the operator T_q we also consider the integral operator T of (6.16), which is symmetric as well, since it has a symmetric kernel. Moreover, T^2 is positive, which is proved as in Lemma 6.7. By Perron's Theorem on positive operators and its generalization to integral operators, in view of Lemma 6.6, Lemma 6.7 and the last remark, we obtain

Corollary 6.2. The operators T_q and T have unique characteristic numbers λ_q and λ with greatest absolute value. These characteristic numbers are positive and relatively prime. The eigenfunctions φ_q and φ corresponding to the characteristic numbers λ_q and λ are positive, i.e., φ_q is positive on τ_q and φ is positive almost everywhere on τ.

We shall assume that these positive eigenfunctions are normalized, i.e., $\|\varphi_q\|_q = 1$ and $\|\varphi\| = 1$, where the last norm is induced by a scalar product defined by analogy with (6.18):

$$<f, g> = 2 \iint_\tau f(x,y)\, g(x,y)\, dx\, dy.$$

It follows from (6.17) that the eigenfunction φ of the operator T is infinitely differentiable.

Proof of Theorem 6.3. Denote by μ_q and μ the second in magnitude (after λ_q and λ) moduli of the characteristic numbers of the operaors T_q and T. Since T_q and T are symmetric operators, the numbers μ_q and μ are equal to the norms of these operators on the orthogonal complements of the eigenfunctions φ_q and φ, respectively. Because of the maximality of λ_q and λ (see Corllary 6.2) we have

(6.23) $$\lambda_q > \mu_q \geq 0, \quad \lambda > \mu \geq 0.$$

Since the operators T_q have been obtained from operator T by replacing the integral in (6.16) by the integral sums of (6.15) over a lattice with step 1/q, this substitution may be viewed as a numerical method for solving integral equations (see, for example, L. Kantorovic and G. Akilov [1]). Using a result about the convergence of these numerical methods, we can assert that

(6.24) $$\lim_{q\to\infty} \lambda_q = \lambda, \quad \lim_{q\to\infty} \mu_q = \mu,$$

$$\lim_{q\to\infty} \|\varphi_q - \varphi\| = 0,$$

where φ_q is regarded as the restriction of φ to the set τ_q.

Now we shall need the following inequality for the functions defined on τ_q:

(6.25) $$|T_q(f; x, y)| \leq \|f\|_q, \qquad (x,y) \in \tau_q.$$

This follows from

$$|T_q(f;x,y)| = \binom{q+1}{2}^{-1} \Big| \sum_{(u,v)\in\tau_q(x,y)} f(u,v) \Big| \le \binom{q+1}{2}^{-1} \sum_{(u,v)\in\tau_q} |f(u,v)|,$$

after applying the Schwarz inequality:

$$\sum_{(u,v)\in\tau_q} |f(u,v)| \le \left(\sum_{(u,v)\in\tau_q} |f(u,v)| \right)^{1/2} \left(\sum_{(u,v)\in\tau_q} 1 \right)^{1/2}$$

$$= \binom{q+1}{2}^{1/2} \left(\sum_{(u,v)\in\tau_q} |f(u,v)| \right)^{1/2} = \binom{q+1}{2}^{1/2} \|f\|_q .$$

It follows from (6.25) that the eigenfunctions φ_q uniformly converge to φ, i.e.,

(6.26) $$\lim_{q\to\infty} \{|\varphi_q(x,y) - \varphi(x,y)| : (x,y) \in \tau_q\} = 0 .$$

Indeed, according to (6.25) and (6.24)

$$|T_q(\varphi_q - \varphi; x, y)| \le \|\varphi_q - \varphi\| \to 0 \text{ as } q \to \infty,$$

and for large q the action of T_q on the functions φ_q and φ tends to multiplication by $\lambda > 0$. Thus, (6.26) follows from the last inequality; in particular, $\lim_{q\to\infty} \varphi_q(0,0) = \varphi(0,0)$.

We now take orthognonal decompositions of the unit functions e_q and e :

(6.27) $$e_q = b_q\,\varphi_q - \psi_q, \quad e = b\varphi - \psi$$

on τ_q and τ respectively, where $b_q = \langle e_q, \varphi_q\rangle_q$ and $b = \langle e, \varphi\rangle$, and ψ_q and ψ belong to the orthogonal complements of φ_p and φ, respectively. Since φ_q and φ are positive, we have $b_q > 0$ and $b > 0$. According to (6.26), we have

$$\lim_{q\to\infty} b_q = b, \quad \lim_{q\to\infty} \|\psi_q\|_q = \|\psi\|.$$

Applying the operator T_q^p in the left equality of (6.27), we obtain

$$T_q^p(e_q) = b_q T_q^p(\varphi_q) - T_q^p(\psi_q) = b_q \lambda_q^p \varphi_q - T_q^p(\psi_q).$$

According to (6.25), we have

(6.28) $$|T_q^p(e_q; 0, 0) - b_q \lambda_q^p \varphi_q(0,0)| \le \|T_q^{p-1}(\psi_q)\|_q \le \mu_q^{p-1} \|\psi_q\|_q ,$$

where μ_q is the norm of T_q on the orthogonal complement of φ_q.

By (6.23), $\lambda_q > \mu_q > 0$. For a fixed q, letting p tend to infinity in (6.28), we have

$$T_q^p(e_q; 0, 0) = b_q \lambda_q^p \varphi_q(0,0) + o(\lambda_q^p) \qquad \text{as } p \to \infty.$$

Substituting $c_q = b_q \varphi_q(0,0) > 0$ and keeping in mind (6.21), we obtain

$$k_{p,q} \sim c_q \left(\lambda_q \binom{q+1}{2}\right)^p ,$$

which establishes the first part of the theorem.

Define $\lambda_{p,q}$ by $\lambda_{p,q} = (T_q^p(e_q; 0, 0))^{1/p}$. Now letting p and q in (6.28) tend independently and simultaneously to infinity, and using $\lambda_q \to \lambda$, $\mu_q \to \mu$, $c_q \to c = b\varphi(0,0) > 0$, $\|\psi_q\|_q \to \|\psi\|$, and $\lambda > \mu > 0$, we see that the numbers $\lambda_{p,q}$ have as limit λ which by (6.21) proves the second assertion of the theorem.

It remains to show that λ is the positive root of maximum modulus of the equation ${}_1F_1(-1/2 - 1/8\lambda, 1/2, 1/2\lambda) = 0$. First, we examine how T acts on functions of the form

(6.29) $$f(x,y) = \psi(x) + \psi(y),$$

where ψ is defined on [0, 1]. From (6.17) we obtain

(6.30) $$\begin{aligned} T(f; x, y) &= 2(1-x)\int_0^x \psi(u)\,du + 2\int_x^{1-y} (1-u)\psi(u)\,du \\ &\quad + 2(1-y)\int_0^y \psi(u)\,du + 2\int_y^{1-x} (1-u)\psi(u)\,du \\ &= A(\psi; x) + A(\psi; y), \end{aligned}$$

where A is an operator acting on functions defined on [0, 1].

The unit function e, defined on τ, is naturally represented in the form (6.29). From (6.30), the same is true for its iterations. But it follows from (6.27) that the iterations of e for an appropriate normalization will converge to the eigenfunction φ of T. As a result, φ is also representable in the form (6.29). From this, we see that the characteristic numbers with maximal modulus of the operators T and A coincide.

Thus, in order to find λ, it is sufficient to solve the integral equation

$$\lambda\psi(x) - A(\psi; x) \equiv \lambda\psi(x) - 2(1-x)\int_0^x \psi(u)\,du - 2\int_x^{1-x} (1-u)\psi(u)\,du = 0.$$

Differentiating twice yields the equation

$$\lambda\psi''(x) + 2\psi(x) + 2\psi(1-x) - 2\psi'(1-x) = 0 \tag{6.31}$$

with boundary conditions

$$\psi(0) - \psi(1) = 0, \qquad \psi'(0) = 0. \tag{6.32}$$

To solve the differential equation, let us represent the function ψ on the interval [0, 1] as a sum of a symmetric function g and an antisymmetric function h, i.e.,

$$\psi(x) = g(x) + h(x),$$

$$g(1-x) = g(x),\ h(1-x) = -h(x), \qquad 0 \le x \le 1.$$

Representing ψ in this way in (6.31), we obtain

$$\lambda g''(x) + 2xg'(x) + 2g(x) + \lambda h''(x) - 2xh'(x) = 0, \tag{6.33}$$

where the boundary conditions (6.33) are transformed into

$$g(0) = 0, \qquad g'(0) + h'(0) = 0. \tag{6.34}$$

Separating the symmetric and antisymmetric parts on [0, 1] in (6.33), we obtain the system

$$\lambda g''(x) + 2(x - 1/2)g'(x) + 4g(x) = h'(x),$$

$$\lambda h''(x) - 2(x - 1/2)h'(x) = - g'(x).$$

Eliminating h, we obtain this differential equation for g:

(6.35) $$\lambda^2 g'''(x) + (6\lambda + 1 - 4(x - 1/2)^2)g'(x) - 8(x - 1/2)\, g(x) = 0,$$

with boundary conditions

(6.36) $$g(0) = 0, \qquad g''(0) = 0.$$

The equation (6.35) is an equation in total differentials, and in view of (6.36), its integration gives

$$\lambda^2 g''(x) + (6\lambda + 1 - 4(x - 1/2)^2)g(x) = 0.$$

The last equation coincides with equation 2273 of E. Kamke's handbook [1]. From the symmetry of the function g, we obtain

$$g(x) = \exp(-\lambda^{-1}(x - 1/2)^2)\, {}_1F_1(-1/2 - 1/8\lambda, 1/2, \lambda^{-1}(x - 1/2)^2),$$

and the condition $g(0) = 0$ leads to the equation ${}_1F_1(-1/2 - 1/8\lambda, 1/2, \lambda/2) = 0$. Thus, the proof of the theorem is completed.

Note that the most effective method of computing the constants λ and c in Theorem 6.3 give the formulas suggested by by A. Panov in [4] :

$$\lambda = \lim_{p\to\infty} \frac{A^p(e; 0, 0)}{A^{p-1}(e; 0, 0)}, \qquad c = \lim_{p\to\infty} \lambda^{-p} A^p(e; 0, 0),$$

where e is the unit function on [0, 1]. In this way, we find that $\lambda = 0.7288500\ldots$ and $c = 1.260077\ldots$. With the help of Theorem 6.3, we can sharpen Theorem 6.2.

Corollary 6.3. The relation

$$H_\varepsilon(F_\Delta^M) = C_{2\varepsilon}(F_\Delta^M) = \frac{b - a}{\alpha\varepsilon} \log_2 \varepsilon^{-1} + \log_2 (M\sqrt{\lambda/2}) - o(\varepsilon^{-1})$$

holds for the ε-entropy and the ε-capacity of the set F_Δ^M, $\Delta = [a, b]$, with respect to the Hausdorff distance with parameter α.

§ 6.3. Labyrinths

Let Z_D denote the set of all the connected compact subsets of a rectangle D in the plane, and let Γ_D denote the set of all continuous curves that are contained in the rectangle D. To compute the ε-entropy and the ε-capacity of these sets with respect to Hausdorff distance, we utilize other configurations of squares based on the integer lattice, in the spirit of corridors, which have been named labyrinths (Bl. Sendov, S. Dimiev, and B. Penkov [1]).

Definition 6.3. Let $D_{p,q} = [0, p] \times [0, q]$, partitioned into squares by the lines $x = i$, $i = 0, 1, \ldots, p$ and $y = j$, $j = 0, 1, 2, \ldots, q$. A set L of squares in $D_{p,q}$ is called a **(p,q)-labyrinth** if it cannot be represented as $L = L' \cup L''$ where L' and L" are nonempty sets of squares in $D_{p,q}$ where $L' \cap L''$ consists of at most finitely many points.

A set of squares that forms a (p,q)-labyrinth is shown in Figure 6.2. The set of all (p,q)-labyrinths will be denoted by $L_{p,q}$ and their number by $l_{p,q}$. It is easy to obtain the following upper and lower bounds for $l_{p,q}$:

(6.37) $$2^{(2/3)(p-2)(q-1)} \le l_{p,q} \le 2^{pq}.$$

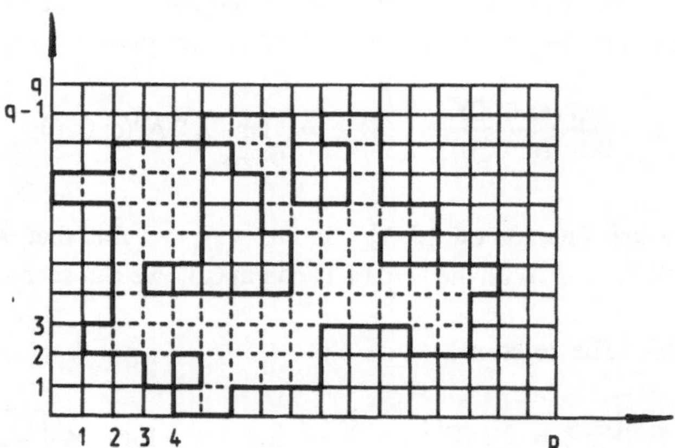

FIGURE 6.2

Indeed, the right-hand side of (6.37) is trivial, since the set of all possible collections of squares which belong to $D_{p,q}$ has 2^{pq} elements. To obtain the left-hand side of (6.37), we choose a number s divisible by 3 satisfying $p - 2 \leq s \leq p$. Consider the comb-shaped labyrinth $L^* \in L_{s,q}$ displayed in Figure 6.3.

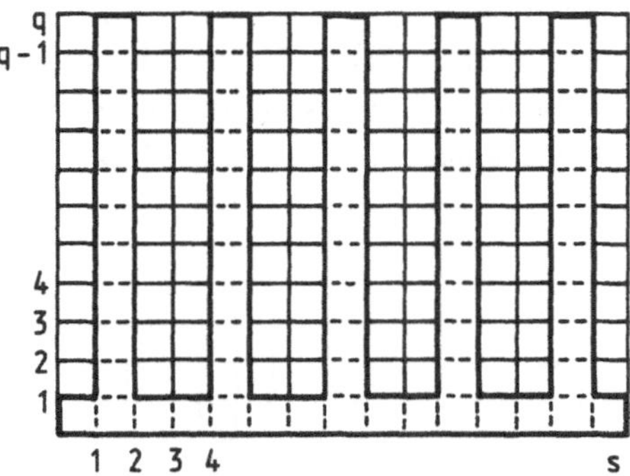

FIGURE 6.3

Let M be the set of squares in $D_{p,q}$ not in L^*. Evidently, the number of squares in M is $\frac{2}{3}s(q-1)$. On the other hand, the union of every subset of squares within M with L^* forms a labyrinth in $L_{s,q}$, and thus in $L_{p,q}$, since $L_{s,q} \subset L_{p,q}$. Hence the number of labyrinths in $L_{p,q}$ is not less than the total number of all the subsets of squares within M, i.e.,

$$2^{(2/3)(p-2)(q-1)} \leq 2^{(2/3)s(q-1)}.$$

This establishes the left inequality in (6.37).

A nontrivial upper bound for $l_{p,q}$ can be found by using a wider set of subsets of squares than those in the set $L_{p,q}$.

Theorem 6.4. (A. Panov [3]). The inequality

$$l_{p,q} \leq 496^{(p+2)(q+2)/9} + pq, \qquad 496^{1/9} = 1.992\ldots \tag{6.38}$$

holds for arbitrary positive integers p and q.

Proof. We say that a set M consisting of elementary squares does not contain an isolated square provided we cannot write M as $M' \cup m$ where m is an elementary square, M' is a set of elementary squares, and $M' \cap m$ contains at most finitely many points. Let $M_{p,q}$ denote the collection of all subsets of squares in $D_{p,q}$ that have no isolated squares. Clearly, all (p,q)-labyrinths except for those consisting of an elementary square alone belong to $M_{p,q}$, so that $l_{p,q} \leq m_{p,q} + pq$, where $m_{p,q}$ is the number of elements in $M_{p,q}$.

For the given p and q, choose numbers s and t both divisible by 3 for which $D_{p,q} \subset D_{s,t} \subset D_{p+2,q+2}$. Divide the rectangle $D_{s,t}$ into 3 by 3 squares (there are st/9 such squares). The number of all subsets of squares in the 3 by 3 square is $2^9 = 512$. Intersecting a 3 by 3 square by an arbitrary element of $M_{s,t}$, out of these 512 subsets, we cannot obtain the $2^4 = 16$ subsets consisting of the central square plus some of the four vertex squares of the 3 by 3 square (see Figure 6.4 (a)).

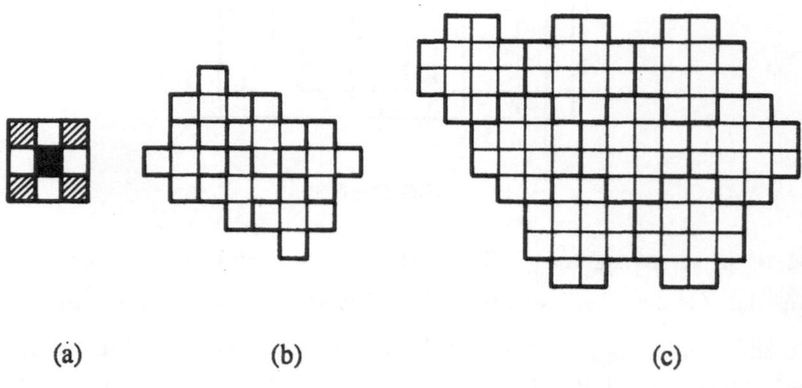

(a) (b) (c)

FIGURE 6.4

Thus, intersecting an element of $M_{s,t}$ with an arbitrary 3 by 3 square, one has at most 512 - 16 = 496 possibilities. Since the number of 3 by 3 squares that cover $D_{s,t}$ is st/9, we have $m_{s,t} \leq 496^{st/9}$, whence

$$l_{p,q} \leq m_{p,q} + pq \leq m_{s,t} + pq \leq 496^{(p+2)(q+2)/9} + pq.$$

This completes the proof of the theorem.

A Panov [3] demonstrated a way in which the estimate (6.38) could be improved. Instead of 3 by 3 squares, the rectangle $D_{p,q}$ could be covered by other configurations of elementary squares, such as crosses, as represented in Figure 6.4(b). Using such coverings and essentially the same logic as in the proof of Theorem 6.4, we obtain a sharper estimate, namely, the number $496^{1/9} = 1.992\ldots$ in (6.38) can be replaced by $31^{1/5} = 1.987\ldots$. Using the coverings in Figure 6.4(c), we obtain an even better estimate: $3279^{1/12} = 1.963\ldots$.

Corollary 6.4. The following inequality for the number of (p,q)-labyrinths is valid:

$$\lim_{p,q\to\infty} (l_{p,q})^{1/pq} < 1.963\ldots \quad (6.39)$$

Define numbers $\theta_{p,q}$ by

$$\theta_{p,q} = \frac{1}{pq}\log_2 l_{p,q}.$$

It follows from (6.37) and (6.39) that for large p and q, we have

$$\frac{2}{3} = 0.666\ldots \le \theta_{p,q} \le 0.973\ldots . \quad (6.40)$$

Later we shall show that the numbers $\theta_{p,q}$ tend to a certain number θ as p and q tend independently to infinity. Clearly, $2/3 \le \theta \le 0.973\ldots$; although it is possible to slightly improve these bounds, a method of computing θ with arbitrary precision is yet unavailable.

In the table on the following page, we present the values of $l_{p,q}$ and $\theta_{p,q}$ for small p and q, keeping in mind that $l_{p,q} = l_{q,p}$ and $\theta_{p,q} = \theta_{q,p}$. The empty set of squares is included as a labyrinth, too.

Definition 6.4. A labyrinth $L \in L_{p,q}$ is called a **(p,q)'-labyrinth** if its projection on the x-axis is of length p. The number of (p,q)'-labyrinths will be denoted by $l'_{p,q}$.

Since the segment [0, p] has $p - i + 1$ integral subsegments of length i and the empty labyrinth is included in $L_{p,q}$, we see that

$$l_{p,q} = 1 + \sum_{i=1}^{p} (p - i + 1)\, l'_{i,q} . \quad (6.41)$$

It follows directly from (6.41) that $l_{p,1} = p(p + 1)/2 + 1$. In order to compute $l_{p,2}$ we introduce the following notation: a_p (resp. b_p) is the number of (p,2)'-labyrinths that have one (resp. two) elementary squares in the right-most strip of $D_{p,2}$. The following recurrence relations are valid:

Values of $l_{p,q}$

	p = 1	p = 2	p = 3	p = 4	p = 5	p = 6
q = 1	2					
q = 2	4	14				
q = 3	7	41	219			
q = 4	11	109	1127	11507		
q = 5	16	276	5727	116167	2301878	
q = 6	22	682	28993	1168587	45280510	1732082742

Values of $\theta_{p,q}$

	p = 1	p = 2	p = 3	p = 4	p = 5	p = 6
q = 1	1.000					
q = 2	1.000	0.952				
q = 3	0.936	0.893	0.864			
q = 4	0.865	0.846	0.845	0.843		
q = 5	0.800	0.811	0.832	0.841	0.845	
q = 6	0.743	0.784	0.824	0.840	0.848	0.8525

TABLE

$$(6.42) \qquad a_{i+1} = a_i + b_i, \qquad b_{i+1} = 2a_i + b_i .$$

Moreover, $l'_{p,q} = 2a_p + b_p$. Solving the system of recurrence equations (6.42) with initial conditions $a_1 = b_1 = 1$, $a_2 = 2$ and $b_2 = 3$, we obtain

$$a_p = \frac{(1+\sqrt{2})^p - (1-\sqrt{2})^p}{2\sqrt{2}}, \qquad b_p = \frac{(1+\sqrt{2})^p + (1-\sqrt{2})^p}{2},$$

$$l'_{p,q} = \frac{(1+\sqrt{2})^{p+1} + (1-\sqrt{2})^{p+1}}{2},$$

because $1+\sqrt{2}$ and $1-\sqrt{2}$ are roots of the characteristic equation

$$\begin{vmatrix} 1-\lambda & 1 \\ 2 & 1-\lambda \end{vmatrix} = \lambda^2 - 2\lambda - 1 = 0$$

of the system (6.42). Using the identity

$$(6.43) \qquad \sum_{i=1}^{p} (p-i+1)\lambda^i = -p-1-\frac{p+2}{\lambda-1}+\frac{\lambda^{p+2}-1}{(\lambda-1)^2},$$

we obtain form (6.41) that

$$l_{p,2} = \frac{(1+\sqrt{2})^{p+3} + (1-\sqrt{2})^{p+3}}{4} - 2p - \frac{5}{2},$$

or

$$l_{p,2} \sim c_2 2^{\theta_2 2p},$$

where $c_2 = (1+\sqrt{2})^3 = 14.071\ldots$ and $\theta_2 = 0.6357\ldots$.

A similar procedure applies for $l_{p,3}$. However, in this situation, we obtain a system of recurrence equations of the sixth order, with characteristic equation

$$(6.44) \qquad \lambda^6 - 7\lambda^5 + 11\lambda^4 - 6\lambda^3 - \lambda^2 + 7\lambda - 1 = 0$$

which has as greatest positive root $\lambda_1 = 5.05662\ldots$. Hence,

$$l_{p,3} \sim c_3\lambda_1^p = c_3 2^{\theta_3 3p},$$

where $\theta_3 = 0.77939\ldots$. Note that in computing $l'_{p,3}$, we have to count not only (p,q)'-labyrinths that end the given pattern of squares, but also such sets of squares that would turn into (p+1,q)'-labyrinths after an addition on the right of a suitable set of squares in the last vertical strip.

Applying this method, it can be shown that

$$(6.45) \qquad l_{p,4} \sim c_4 2^{\theta_4 4p}, \qquad l_{p,5} \sim c_5 2^{\theta_5 5p}$$

where $\theta_4 = 0.83243\ldots$ and $\theta_5 = 0.85916\ldots$.

For general q one can obtain the following result.

Theorem 6.5. (A. Panov [3]). To each $q > 1$ there correspond numbers $c_q > 1$ and $\theta_q > 0$ such that the number $l_{p,q}$ of (p,q)-labyrinths satisfies

$$(6.46) \qquad l_{p,q} \sim c_q 2^{\theta_q qp} \quad \text{as } p \to \infty.$$

The following statement holds true as well; the special case $p = q$ was considered independently by J. Mycilski [1].

Theorem 6.6. (A. Panov [3]). For given $q \geq 2$ and $\varepsilon > 0$, there exist numbers s_0 and t_0 such that for $s > s_0$ and $t > t_0$, we have

$$(6.47) \qquad l_{s,t} > (2^{\theta_q} - \varepsilon)^{st} = (\mu_q - \varepsilon)^{st}.$$

Moreover, a limiting value θ of θ_q exists, and we have

$$(6.48) \qquad \theta = \lim_{q\to\infty} \theta_q = \lim_{p,q\to\infty} \theta_{p,q}.$$

Finally, the inequality $\theta_q \leq \theta$ holds for each q.

Proof. Write $\mu_q = 2^{\theta_q}$, and choose s_0 such that for every $s > s_0$, we have

$$(6.49) \qquad l'_{s-1,q} > (\mu_q - \frac{\varepsilon}{2})^{sq}.$$

Such an s_0 exists by virtue of (6.41), (6.43), and Theorem 6.5. For an arbitrary t, we will divide the rectangle $D_{s,t}$ as shown in Figure 6.5. In this partition, we create $m =$

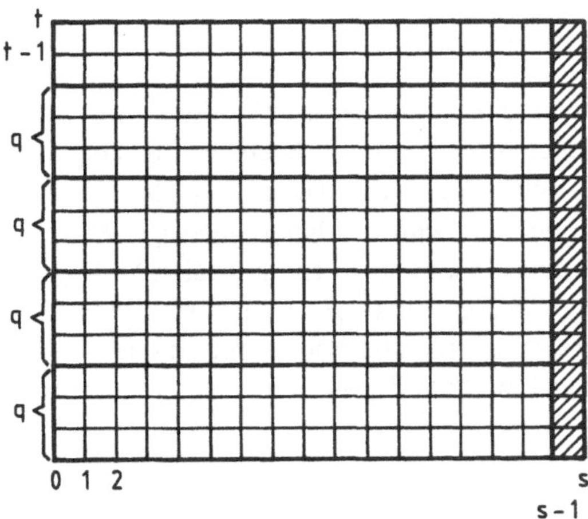

FIGURE 6.5

$[t/q]$ rectangles $D_{s-1,q}$ and a vertical rectangle $D_{1,t}$ on the right. Consider those (s,t)-labyrinths that contain $D_{1,t}$ whose intersection with each rectangle $D_{s-1,q}$ gives a $(s-1,q)'$-labyrinth. The number of these is at least $(l'_{s-1,q})^m$. According to (6.49), from this we obtain

$$l_{s,t} \geq (l'_{s-1,q})^m > (\mu_q - \frac{\varepsilon}{2})^{sqm} = ((\mu_q - \frac{\varepsilon}{2})^{qm/t})^{st}. \tag{6.50}$$

But $qm/t = q[t/q]/t > 1 - 9/t$, and hence we can choose t_0 such that the inequality $(\mu_q - \frac{\varepsilon}{2})^{qm/t} \geq \mu_q - \varepsilon$ holds for $t > t_0$. This proves (6.47).

Taking in (6.47) the root of degree st and letting s tend to infinity, we obtain $\mu_t \geq \mu_q - \varepsilon$ for $s > s_0$ as well. From here we obtain the existence of the limit $\mu = \lim_{q\to\infty} \mu_q$, and the inequalities $\mu_q \leq 2$, $\mu_q \leq \mu$. Therefore, since $\mu_q = 2^{\theta_q}$, we have $\theta = \lim_{q\to\infty} \theta_q$, where $\mu = 2^\theta$ and $\theta_q \leq \theta$.

From (6.50), it also follows that

$$\liminf_{p,q\to\infty} (l_{p,q})^{1/pq} \geq \mu = 2^\theta. \tag{6.51}$$

To complete the proof of (6.48), we will show that $\limsup_{p,q\to\infty} (l_{p,q})^{1/pq} \leq \mu = 2^{\theta}$. For given q and $\varepsilon > 0$, choose s_0 such that for $s > s_0$, we have

$$(6.52) \qquad l_{s,q} \leq (\mu_q + \frac{\varepsilon}{2})^{sq}.$$

This is possible by Theorem 6.5. For arbitrary t, divide the rectangle $D_{s,t}$ by horizontal lines into $[t/(q-2)]$ rectangles $D_{s,q-2}$. Each such rectangle can be embedded into a rectangle $D_{s,q}$ by adding to $D_{s,q-2}$ a rectangle $D_{s,1}$ above and below. Let $L \in L_{s,t}$ be arbitrary. The intersection of L with an arbitrary rectangle $D_{s,q-2} \subset D_{s,t}$ is either a $(s,q-2)$-labyrinth and then, according to the inclusion $D_{s,q-2} \subset D_{s,q}$, it is a (s,q)-labyrinth, or $L \cap D_{s,q-2}$ becomes a (s,q)-labyrinth after the addition to $(L \cap D_{s,q-2}) \subset D_{s,q}$ all of the squares of the top and bottom rectangles $D_{s,1}$. Since the number of rectangles $D_{s,q-2}$ is equal to $[t/(q-2)] = n$, obviously we have $l_{s,t} \leq (l_{s,q})^{n+1}$. According to (6.52), we have

$$l_{s,t} \leq (l_{s,q})^{n+1} \leq ((\mu_q + \frac{\varepsilon}{2})^{sq})^{n+1}.$$

Selecting s_0 in such a way that $(\mu_q + \frac{\varepsilon}{2})^{n+1} \leq (\mu_q + \frac{\varepsilon}{2})^{s/(q-2)}$ holds for $s > s_0$, we see that $l_{s,t} \leq ((\mu_q + \varepsilon)^{q/(q-2)})^{st}$ holds for $s > s_0$ and $t > t_0$. Thus, we obtain

$$\limsup_{s,t\to\infty} (l_{s,t})^{1/st} \leq \mu.$$

From this last inequality and (6.51), it follows that $\lim_{p,q\to\infty} (l_{p,q})^{1/pq} = \mu$, fully establishing (6.48). The theorem is proved.

Corollary 6.5. For the number θ determine by the relation

$$(6.53) \qquad \log_2 l_{p,q} \sim \theta pq,$$

one has

$$(6.54) \qquad 0.859\ldots < \theta < 0.973\ldots.$$

We again note that although the estimate (6.54) can be improved, there is no known efficient method for computing θ with arbitrary precision.

6.3.1. PASSAGES IN LABYRINTHS

We consider an additional question with respect to labyrinths $L_{p,q}$ in the rectangle $D_{p,q} = \{(x,y) : 0 \leq x \leq p, 0 \leq y \leq q\}$, where p and q are positive integers. Let $k \geq 2$; a sequence of elementary squares $P_1, P_2, \ldots, P_k$ contained in $D_{p,q}$ such that P_i and P_{i+1} have a side in common for $i = 1, 2, 3, \ldots, k - 1$ will be called a **path of length k - 1** joining P_1 and P_k. An argument parallel to a standard connectness argument from general topology shows that each two elementary squares in a labyrinth can be joined by a path totally within the labyrinth.

Definition 6.5. A **passage** L in a labyrinth is a path that includes all of the elementary squares of the labyrinth and only these squares.

It follows from the path connectedness of a labyrinth that each has a passage; a typical passage may go through a given square several times. A passage within a labyrinth L is called minimal if its length l is minimal among all passages through the labyrinth. We next obtain an upper bound for l.

Theorem 6.7. (Bl. Sendov and T. Boyanov [1]). Let L be a (p,q)-labyrinth, and let l be the length of its minimal passage. We have

$$l \leq \frac{4}{3}pq + \frac{2}{3}(p + q - 1) . \tag{6.55}$$

Moreover, the constant 4/3 in front of pq is exact.

Proof. Let $L \in L_{p,q}$ be arbitrary consisting of at least two squares. We call a path containing all peripheral squares of L and beginning and ending at the same elementary square the circumference E(L) of L. We now describe the construction of E(L). Let P_1 be an elementary square in L with a corner at minimal distance from the boundary of the rectangle $D_{p,q}$. A side s_1 containing this corner cannot be a side of any other square of L. By s_2 we denote the first side of P_1 that belongs to another square P_2 of L that follows s_1 with respect to the order defined by a clockwise rotation in the plane. This square P_2 will be second square in the circumference path. Proceeding from s_2, we define P_3 and s_3, and so on. This process will end at P_1. Each of the sides following s_2 and preceeding s_3, following s_3 and preceeding s_4, and so on, together with s_1, belongs to only one square in L. They form a closed polygonal curve that will be called the contour Γ(L) of the labyrinth L. We denote by e(L) the length of the circumference E(L) and by γ(L) the number of sides in Γ(L).

Let us produce a new labyrinth L', adding to L a square P' that touches $\Gamma(L)$ on i consecutive sides, $1 \le i \le 4$. We obtain increments Δe and $\Delta\gamma$, for the length of the circumference and the number of sides in the contour, given by

(6.56) $$\Delta e = -2(i - 1) + 2 \quad \text{and} \quad \Delta\gamma = 4 - 2i, \quad i = 1, 2, 3, 4.$$

If L has only two elementary squares, we have $e(L) = 2$ and $\gamma(L) = 6$. Using induction and (6.56), we obtain this equality valid for any labyrinth L:

(6.57) $$e(L) = \gamma(L) - 4.$$

Now let L be an arbitrary (p,q)-labyrinth. The set $L \setminus E(L)$, if it is nonempty, can be decomposed into several labyrinths without points in common. If L' is one of them, then $\Gamma(L')$ has a side s' that belongs to a square of E(L) but at the same time, $\Gamma(L) \cap \Gamma(L') = \emptyset$. We need two supplementary steps in order to join E(L) and E(L') into one path by double-passing through s'. We would not count the side s' as a side of $\Gamma(L')$ anymore, because we passed through it. For the above reason, we must modify (6.57) in order to apply it to L'. It is necessary to add two extra steps to E(L) and to subtract one side of $\Gamma(L)$. Now the relation becomes

(6.58) $$e(L') = \gamma(L') - 1.$$

Note that (6.58) is true for the newly constructed path. The consequtive addition of circumferences of the remaining labyrinths to the obtained path (after the exclusion of already associated circumferences) leads to a path in L that is a passage. Keeping in mind the construction of this passage, its length is equal to the number of sides crossed in passing from one square into another, where no side of a square is passed more than twice. If we denote by e^* the number of all the noncrossed sides of L and by l^* the length of this passage, then

(6.59) $$e^* + \frac{1}{2} l^* \le p(q + 1) + q(p + 1) = 2pq + p + q,$$

where at the right hand side is the number of all sides in $D_{p,q}$. But on the other hand, e^* and l^* satisfy (6.58). Hence, we obtain (6.55) from (6.58) and (6.59).

problem of the minimal passage within labyrinths formed by regular hexagons is not solved. It turns out that in this case, the method of proof of Theorem 6.7 is inapplicable.

We mention in closing that the problems discussed in this section are of interest from the point of view of combinatorial analysis and graph theory.

6.4. ε-entropy and ε-capacity of bounded sets of connected compact sets

Let Z_D denote the set of all compact connected subsets of the rectangle $D = \{(x,y) : a \le x \le b \text{ and } c \le y \le d\}$. Hausdorff distance in Z_D is defined in the usual way: for $F, G \in Z_D$ we have

$$r(F, G) = \max \{ \max_{(x,y)\in F} \inf_{(\xi,\eta)\in G} \max \{|x - \xi|, |y - \eta|\}, \max_{(x,y)\in G} \inf_{(\xi,\eta)\in F} \max \{|x - \xi|, |y - \eta|\} \}. \tag{6.60}$$

The set Z_D equipped with this distance is a compact metric space. This follows from the fact that the compact subsets of a compact metric space equipped with Hausdorff distance form a compact metric space (see, e.g., Hausdorff [1], Castaing and Valadier [1] or Nadler [1]), and a compact set is connected if and only if it is ε-connected for each $\varepsilon > 0$, a property that is preserved under limits with respect to Hausdorff distance. Here, of course, the underlying metric inducing r is the box metric $\rho((x,y), (\xi,\eta)) = \max \{|x - \xi|, |y - \eta|\}$, where closed balls are squares. This remains, for our purposes, particularly convenient. But any other Minkowski distance, e.g., the usual Euclidean distance $\rho_0((x,y), (\xi,\eta)) = ((x - \xi)^2 + (y - \eta)^2)^{1/2}$, can be used to induce r with analagous results, since any two Minkowski metrics are equivalent.

Starting with two number systems

$$a = x_0 < x_1 < x_2 < \cdots < x_p = b, \tag{6.61}$$

$$c = y_0 < y_1 < y_2 < \cdots < y_q = d,$$

we obtain a partition of D into pq elementary rectangles. Each rectangle will be considered as a closed point set, and we denote the rectangle with left lower corner (x_i,y_j) by $\{x_i; y_j\}$. The notion of (p,q)-labyrinth presented in §6.3 can be extended to the present context in the natural way. A set L of elementary rectangles in $D_{p,q}$ is called a **(p,q)-labyrinth** if it cannot be represented as $L = L' \cup L''$ where L' and L'' are nonempty sets of rectangles $\{x_i; y_j\}$ in D and $L' \cap L''$ consists of at most finitely many points. We denote the set of these labyrinths by $L^D_{p,q}$. Obviously the number of such rectangular labyrinths depends only on the positive integers p and q and not on the shapes of the rectangles and must be $l_{p,q}$ as well.

Definition 6.6. Let $F \in Z_D$. We say that F belongs to the labyrinth $L \in L^D_{p,q}$ if every point of F belongs to some elementrary rectangle $\{x_i; y_j\}$ of L, and every elementary rectangle which meets F belongs to L.

Denote by ε_1 the maximum of $\{x_i - x_{i-1} : i = 1, 2, \ldots, p\}$ and by ε_2 the maximum of $\{y_j - y_{j-1} : j = 1, 2, \ldots, q\}$. With $\varepsilon = \max\{\varepsilon_1, \varepsilon_2\}$, we have the following result whose proof is similar to the proof of Lemma 6.1.

Lemma 6.9. If $F, G \in Z_D$ belong to the labyrinth $L \in L^D_{p,q}$, then $r(F,G) \le \varepsilon$.

Reasoning for labyrinths as with corridors, we also obtain

Lemma 6.10. The set Z_D is totally bounded with respect to the Hausdorff distance, and for this distance, we have

$$H_\varepsilon(Z_D) = C_{2\varepsilon}(Z_D) = \log_2 l_{p,q},$$

where $p = -[-(b - a)/\varepsilon]$ and $q = -[-(c - d)/\varepsilon]$.

From Lemma 6.10 and Corollary 6.5, we finally obtain

Theorem 6.8. The asymptotic value of ε-entropy and ε-capacity of the set Z_D with respect to Hausdorff distance is

$$H_\varepsilon(Z_D) = C_{2\varepsilon}(Z_D) \sim \frac{\theta(b - a)(d - c)}{\varepsilon^2}$$

where $0.859\ldots < \theta < 0.973\ldots$.

Since the ε-entropy of all subsets of D is equal to $(b - a)(d - c)/\varepsilon^2$, the number θ indicates the massiveness of the set of all connected subsets with respect to all the subsets of a given rectangle in the plane. It is not hard to prove that θ does not depend on the choice of distance in the plane generating Hausdorff distance in Z_D.

§ 6.5. Widths

Let X be a metric linear space with distance $\rho(x,y)$, $x, y \in X$, and let F and G be two subsets of X. Then

$$\delta(F; G) = \sup_{x\in F} \inf_{y\in G} \rho(x,y)$$

is called the **excess** of F over G or the **one-sided Hausdorff distance** from F to G (see §4.1). We can consider G as a set that approximates F with respect to this one-sided Hausdorff distance, and since finite dimensional subspaces are widely used in approximation theory as a means of approximation, it is natural to seek the best linear subspace of the space X of a given dimension for the approximation of the subset F.

Definition 6.7. (A. Kolmogorov [3]). Let X be a metric linear space and let $F \subset X$. Let L_n be the set of all linear subspaces of the space X of dimension at most n. The number

$$d_n(F) = \inf \{ \delta(F; L) : L \in L_n \}$$

is called the **nth-width** of the set F. If for a certain $L^* \in L_n$ the equality $d_n(F) = \delta(F; L^*)$ holds, then L^* is called extremal for F among all the linear subspaces of dimension at most n.

Width as just defined is often called Kolmogorov width. The general method for estimation of widths from above and below in Banach spaces was developed by V. Tihomirov [1,2]. Our aim here will be to compute widths of linear function spaces metrized by Hausdorff distance- spaces that are not Banach spaces. As such, it is impossible to apply the general methods of V. Tihomirov.

6.5.1. WIDTHS OF THE SET OF BOUNDED REAL FUNCTIONS

In this section we shall compute the widths of the set of all bounded functions defined on a segment. We shall confine ourselves to single-valued functions only, since the definition of width itself is formulated in the context of linear spaces, and the set of segment functions is not a linear space.

In the sequel, A_Δ will represent the set of single valued bounded functions on the segment $\Delta = [a, b]$ (our original definition in Chapter 1 allowed for unbounded functions). The main result we are after here is the following:

Theorem 6.9. If A_Δ is metrized by Hausdorff distance with parameter α, then

$$d_n(A_\Delta) = \begin{cases} 3(b - a)/2\alpha n & \text{if } n = 3k \\ 3(b - a)/2\alpha(n - 1) & \text{if } n = 3k + 1 \\ 3(b - a)/2\alpha(n - 1/2) & \text{if } n = 3k + 2 \end{cases},$$

for $n = 1, 2, 3, \ldots,$ where k is an integer.

Since the set C_Δ of continuous functions on the segment Δ is dense in A_Δ with respect to Hausdorff distance, from Theorem 6.9 we obtain

Corollary 6.6. (T. Boyanov and V. Popov [1]). The nth-width of C_Δ is given by

$$d_n(C_\Delta) = \begin{cases} 3(b - a)/2\alpha n & \text{if } n = 3k \\ 3(b - a)/2\alpha(n - 1) & \text{if } n = 3k + 1 \\ 3(b - a)/2\alpha(n - 1/2) & \text{if } n = 3k + 2 \end{cases},$$

with respect to Hausdorff distance with parameter α.

It should be noted that the asymptotic behavior $d_n(C_\Delta) \sim 3(b - a)/2\alpha n$ was discovered by Bl. Sendov and B. Penkov [3,4].

The proof of Theorem 6.9 will be broken into two parts. First, we find upper bounds for the widths $d_n(A_\Delta)$, and then we shall show that these are also lower bounds. Following T. Boyanov and V. Popov [1], we first obtain some preliminary facts. We shall make use of functions $\eta(h; x)$ $(h > 0)$ and $\delta(x)$ given by

$$\eta(h; x) = \begin{cases} 1 & \text{if } |x| < h \\ 1/2 & \text{if } |x| = h \\ 0 & \text{if } |x| > h \end{cases}, \tag{6.62}$$

and

$$\delta(x) = \begin{cases} 1 & \text{if } x = 0 \\ 0 & \text{if } x \neq 0 \end{cases}.$$

Let $a = x_0 < x_1 < x_2 < \cdots < x_n = b$, and for $i = 1, 2, 3, \ldots, n$, let $h_i = (x_i - x_{i-1})/2$ and $\xi_i = (x_i + x_{i-1})/2$. We have $x \in \Delta_i = [x_{i-1}, x_i]$ if and only if $|x - \xi_i| \le h_i$. Write $h = \max\{h_i : 1 \le i \le n\}$, and divide the index set $\{1, 2, \ldots, n\}$ into two parts N_1 and N_2 as follows:

$$N_1 = \{i : h_i > h/2\}, \qquad N_2 = \{i : h_i \le h/2\},$$

Let k be the number of indices in N_1; then N_2 contains $l = n - k$ indices.

Lemma 6.11. For k, l and h as introduced above, we have

$$d_{3k+2l}(A_\Delta) \le \frac{h}{\alpha}.$$

Proof. Using (6.62), we define these functions on $\Delta = [a, b]$:

$$\eta_1(x) = \begin{cases} 1 & \text{if } a \le x \le \xi_1 \\ \eta(h_1; x - \xi_1) & \text{if } \xi_1 \le x \le b \end{cases},$$

$$\eta_i(x) = \eta(h_i; x - \xi_i), \quad i = 2, 3, 4, \ldots, n-1,$$

$$\eta_n(x) = \begin{cases} \eta(h_n; x - \xi_n) & \text{if } a \le x \le \xi_n \\ 1 & \text{if } \xi_n \le x \le b \end{cases}.$$

Fix ε with $0 < \varepsilon < \min\{h_i : 1 \le i \le n\}$. We define functions $\delta_i^-(x)$ and $\delta_j^+(x)$ by

$$\delta_i^-(x) = \delta(x - \xi_i + \varepsilon), \qquad i = 1, 2, 3, \ldots, n,$$

$$\delta_j^+(x) = \delta(x - \xi_j - \varepsilon), \qquad j \in N_1.$$

Finally, for $n = 1$, we set $\eta_1(x) \equiv 1$.

The following properties of these functions follow directly from their definition:

(6.63) $$0 \le \eta_i(x) \le 1, \ 0 \le \delta_i^-(x) \le 1, \ 0 \le \delta_j^+(x) \le 1, \quad i = 1, 2, \ldots, n, \quad j \in N_1;$$

(6.64) $$\eta_1(x) = 1 \text{ for } x \in [a, x_1) \text{ and } \eta_1(x) = 0 \text{ for } x \in (x_1, 1];$$

$$\eta_i(x) = 1 \text{ for } |x - \xi_i| < h_i \text{ and } \eta_i(x) = 0 \text{ for } |x - \xi_i| > h_i, \quad 2 \le i \le n-1;$$

$$\eta_n(x) = 1 \text{ for } x \in (x_{n-1}, b] \text{ and } \eta_1(x) = 0 \text{ for } x \in [a, x_{n-1});$$

$$\delta_i^-(\xi_i - \varepsilon) = 1 \text{ and } \delta_i^-(x) = 0 \text{ for } x \ne \xi_i - \varepsilon, \quad i = 1, 2, 3, \ldots, n; \tag{6.65}$$

$$\delta_j^+(\xi_j + \varepsilon) = 1 \text{ and } \delta_j^-(x) = 0 \text{ for } x \ne \xi_j + \varepsilon, \qquad j \in N_1. \tag{6.66}$$

Consider the $2n + k = 3k + 21$ dimensional linear subspace L_ε of A_Δ spanned by the functions $\eta_1, \ldots, \eta_n, \delta_1^-, \ldots, \delta_n^-$ and δ_j^+, $j \in N_1$. The lemma will be proved if we can show for each $\varepsilon > 0$

$$\delta(A_\Delta; L_\varepsilon) \le \frac{h + \varepsilon}{\alpha}. \tag{6.67}$$

To prove (6.67), we show that for each $f \in A_\Delta$ there exists a function $g \in L_\varepsilon$ such that

$$r(\alpha; f, g) \le \frac{h + \varepsilon}{\alpha}. \tag{6.68}$$

To specify g for a given $f \in A_\Delta$, let us write as usual

$$m_i = \inf\{f(x) : x \in \Delta_i\}, \quad M_i = \sup\{f(x) : x \in \Delta_i\}$$

for $\Delta_i = [x_{i-1}, x_i]$, $i = 1, 2, \ldots, n$. Our function g is given by

$$g(x) = \sum_{i \in N_1} \left(f(\xi_i)\eta_i(x) + (m_i - f(\xi_i))\delta_i^-(x) + (M_i - f(\xi_i))\delta_i^+(x)\right) + \sum_{i \in N_2} \left(M_i\eta_i(x) + (m_i - M_i)\delta_i^-(x)\right). \tag{6.69}$$

The completed graph of g for arguments interior to the segment Δ_i for $i \in N_1$ appears in Figure 6.7 (a), and for $i \in N_2$ in Figure 6.7 (b). Keeping in mind that $h/2 < x_i - x_{i-1} \le h$ for $i \in N_1$ and $x_i - x_{i-1} \le h/2$ for $i \in N_2$, it can be seen from these figures that the

To prove that (6.55) is exact with respect to the constant 4/3, we count the number of steps required for a passage of the labyrinth of the type presented in Figure 6.6 (a). As it is shown in Figure 6.6 (b), for a passage of 6 squares in $D_{p,q}$, usually 8 steps are involved, i.e., 4/3 steps per square on the average, so that the length of such a labyrinth for large p and q will be aymptotically equal to $\frac{4}{3}pq$. The theorem is proved.

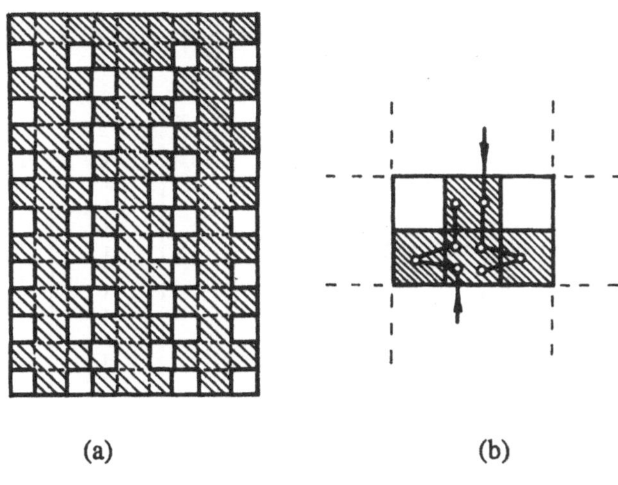

(a) (b)

FIGURE 6.6

It should be noted that the passage constructed in the proof of Theorem 6.7 is closed, i.e., its initial and terminal squares coincide.

An analagous problem can be considered for labyrinths formed by equilateral triangles within a given equilateral triangle T. Suppose the length of each side of T is a positive integer m. The triangle T can be divided into equal triangles with sides of length one by lines parallel to the sides of T, and these smaller triangles can be used to form labyrinths. If we employ arguments similar to those used for squares, the following upper bound holds for the length l of the minimal passage within a labyrinth in T for the given partition :

$$l \leq \frac{3}{2}m^2 + \frac{3}{2}m - 1.$$

The constant $\frac{3}{2}$ in front of m^2 is exact, too.

The problem for the minimal passage within a labyrinth is not solved in the most general case, when the labyrinths are formed by arbitrary polygons. In particular, the

Hausdorff distance between (the completed graphs of) f and g satisfies (6.68). This completes the proof.

Let us now set $x_i = a + i(b - a)/n$, $i = 0, 1, 2, \ldots, n$. In this case, for each index i, we have $h_i = (b - a)/2n$ and the set N_2 is empty, i.e., $k = 0$. From Lemma 6.13, we get this estimate of $d_n(A_\Delta)$:

Corollary 6.7. For each positive integer n we have $d_{3n}(A_\Delta) \leq \dfrac{b-a}{2\alpha n}$.

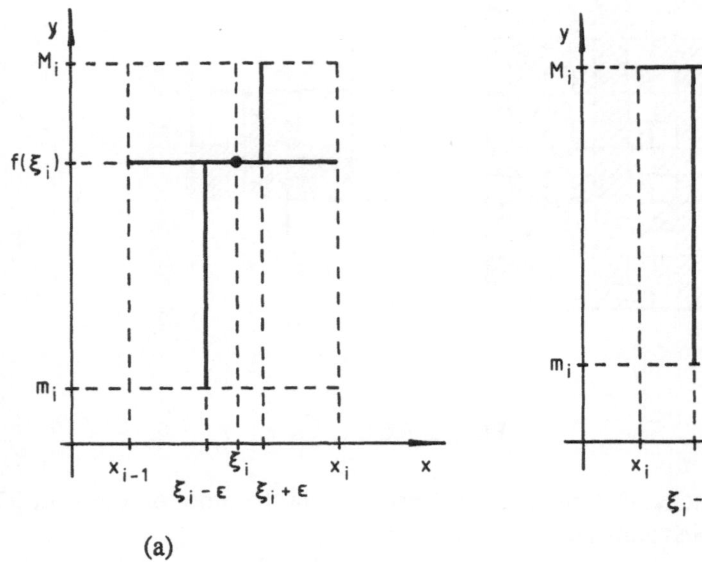

FIGURE 6.7

We obtain a different estimate if we choose knots $x_0, x_1, \ldots, x_{n+1}$ in Δ as follows: $x_i = a + 2i(b - a)/(2n + 1)$ for $i = 0, 1, 2, \ldots, n$ and $x_{n+1} = b$. Then $h = h_i = (b - a)/(2n + 1)$ for $i = 1, 2, \ldots, n$ and $h_{n+1} = h/2$. Hence, the set N_2 contains only one element, $N_2 = \{n + 1\}$, and $k = 1$. From Lemma 6.13, we now obtain

Corollary 6.8. For each positive integer n we have $d_{3n+1}(A_\Delta) \leq \dfrac{b-a}{(2n+1)\alpha}$.

Now we obtain estimates for the widths $d_n(A_\Delta)$ from below.

Lemma 6.12. Let L be a finite dimensional subspace of the space A_Δ with $\delta(A_\Delta; L) \le d$. Then for every $t \in \Delta$, there exist three functions φ_t, ψ_t and $\theta_t \in L$ with these properties:

(1) $\varphi_t(x) = 0$ for $|x - t| \ge 2\alpha d$, $\varphi_t(t) = 1$, and $0 \le \varphi_t(x) \le 1$ for $x \in \Delta$;

(2) $\psi_t(x) = 0$ for $|x - t| \ge \alpha d$, $0 \le \psi_t(x) \le 1$ for $x \in \Delta$, and there exists $\eta_t \in \Delta$ with $|\eta_t - t| \le \alpha d$ such that $\psi_t(\eta_t) = 1$;

(3) $\theta_t(x) = 0$ for $|x - t| \ge \alpha d$, $-1 \le \theta_t(x) \le 1$ for $x \in \Delta$, and there exist points ξ_t^- and $\xi_t^+ \in [t - \alpha d, t + \alpha d] \cap \Delta$ such that $\theta_t(\xi_t^-) = -1$ and $\theta_t(\xi_t^+) = 1$.

Proof. We consider for a prescribed $t \in \Delta$, $M > 0$, and a sufficiently small $\varepsilon > 0$ three functions $f_1(x)$, $f_2(x)$, and $f_3(x)$ defined by

$$f_1(x) = M\eta(\alpha d; x - t),$$

$$f_2(x) = M\delta(x - t),$$

$$f_3(x) = M(\delta(x - t - \varepsilon) - \delta(x - t + \varepsilon)),$$

where $\eta(h; x)$ and $\delta(x)$ are defined in (6.62). It follows from $\delta(A_\Delta; L) \le d$ that there exist functions $\varphi_t(M, \varepsilon; x)$, $\psi_t(M, \varepsilon; x)$, and $\theta_t(M, \varepsilon; x)$ in the subspace L for which

$$r(\alpha;\ \varphi_t(M, \varepsilon), f_1) \le d + \varepsilon,$$

$$r(\alpha;\ \psi_t(M, \varepsilon), f_2) \le d + \varepsilon,$$

$$r(\alpha;\ \theta_t(M, \varepsilon), f_3) \le d + \varepsilon.$$

The following properties follow from the definition of Hausdorff distance between functions and the above inequalities:

(1) for all points $(x,y(x))$ of the completed graph of the function $M^{-1}\varphi_t(M, \varepsilon)$,

$$|y(x)| \le (d + \varepsilon)M^{-1}, \quad |x - t| \ge 2\alpha(d + \varepsilon), \quad |y(t) - 1| \le dM^{-1}, \tag{6.70}$$

$$-dM^{-1} \le y(x) \le 1 + dM^{-1};$$

(2) for all points $(x,y(x))$ of the completed graph of the function $M^{-1}\psi_t(M, \varepsilon)$,

$$|y(x)| \le (d+\varepsilon)M^{-1}, \quad |x - t| \ge \alpha(d+\varepsilon), \quad -(d+\varepsilon)M^{-1} \le y(x) \le (d+\varepsilon)M^{-1}, \tag{6.71}$$

and there exists a point $\eta_t \in [t - \alpha(d+\varepsilon), t + \alpha(d+\varepsilon)] \cap \Delta$ such that $|y(\eta_t) - 1| \le (d+\varepsilon)M^{-1}$;

(3) for all points $(x,y(x))$ of the completed graph of the function $M^{-1}\theta_t(M, \varepsilon)$,

$$|y(x)| \le (d+\varepsilon)M^{-1}, \quad |x - t| \ge \alpha(d+\varepsilon), \quad |y(x) - 1| \le (d+\varepsilon)M^{-1}, \tag{6.72}$$

and there exist two points ξ_t^- and ξ_t^+ in $[t - \alpha(d+\varepsilon), t + \alpha(d+\varepsilon)] \cap \Delta$ for which $|y(\xi_t^-) + 1| \le (d+\varepsilon)M^{-1}$ and $|y(\xi_t^+) - 1| \le (d+\varepsilon)M^{-1}$.

Let us consider the segment functions

$$\bar{\varphi}_t(x) = \underset{\varepsilon\to 0,\, M\to\infty}{\text{Slim}} M^{-1}\varphi_t(M, \varepsilon; x),$$

$$\bar{\psi}_t(x) = \underset{\varepsilon\to 0,\, M\to\infty}{\text{Slim}} M^{-1}\psi_t(M, \varepsilon; x),$$

$$\bar{\theta}_t(x) = \underset{\varepsilon\to 0,\, M\to\infty}{\text{Slim}} M^{-1}\theta_t(M, \varepsilon; x).$$

Since L is a finite dimensional subspace of the space A_Δ, its unit sphere is compact with respect to pointwise convergence. Hence, there exist three functions φ_t, ψ_t and θ_t in L for which $\varphi_t(x) \in \bar{\varphi}(x)$, $\psi_t(x) \in \bar{\psi}(x)$, and $\theta_t(x) \in \bar{\theta}(x)$, respectively, for all $x \in \Delta$. Keeping in mind (6.70) - (6.71), it is straightforward to verify that φ_t, ψ_t and θ_t possess the asserted properties in the statement of the lemma.

Lemma 6.13. If L is a finite dimensional subspace of A_Δ and $\delta(A_\Delta; L) \le (b - a)/2\alpha n$, then the dimension of L is at least $3n + 2$.

Proof. Let $x_i = a + i(b - a)/n$ for $i = 0, 1, 2, \ldots, n$, and set $\xi_i = (x_{i-1} + x_i)/2$ for $i = 1, 2, 3, \ldots, n$. For $t = \xi_i$ the function φ_t of Lemma 6.12 will be denoted by φ_i, and for $t = x_i$, ψ_t and θ_t of Lemma 6.12 will be denoted by ψ_i and θ_i. We now consider the set of $3n + 2$ functions $\varphi_1, \varphi_2, \ldots, \varphi_n, \psi_0, \psi_1, \ldots, \psi_n, \theta_0, \theta_1, \ldots, \theta_n$. It suffices to show that these functions are linearly independent. Assume we have a linear combination of these functions summing to the zero function, i. e.,

$$\sum_{i=1}^{n} \lambda_i \varphi_i(x) + \sum_{i=0}^{n} (\mu_i \psi_i(x) + \nu_i \theta_i(x)) \equiv 0. \tag{6.73}$$

If we set $x = \xi_i$ in (6.73), then from conditions (1), (2), (3) of Lemma 6.12 and the compactness of the unit sphere of L, it follows that $\lambda_i = 0$, $i = 1, 2, 3, \ldots, n$. Now if in (6.73) we set consecutively $x = \eta_t$, $x = \xi_t^-$, and $x = \xi_t^+$ for $t = x_i$, we obtain analagously from properties (1), (2) and (3) of Lemma 6.12 the following system of equations:

$$\mu_i + \nu_i \theta_i(\eta_t) = 0, \qquad \mu_i \psi_i(\xi_t^-) + \nu_i = 0, \qquad \mu_i \psi_i(\xi_t^+) - \nu_i = 0,$$

where $t = x_i$. But since the function ψ_i is nonnegative, this system has a unique solution $\mu_i = \nu_i = 0$, $i = 0, 1, 2, \ldots, n$. Thus we have produced $3n + 2$ linearly independent functions in L.

Corollary 6.9. The inequality $d_{3n+1}(A_\Delta) \geq \dfrac{b - a}{2\alpha n}$ holds for every positive integer n.

Lemma 6.13. If L is a finite dimensional subspace of A_Δ and $\delta(A_\Delta; L) \leq (b - a)/\alpha(2n + 1)$, then the dimension of L is at least $3n + 3$.

Proof. Write $x_i = a + 2i(b - a)/(2n + 1)$ for $i = 0, 1, 2, \ldots, n$ and $x_{n+1} = b$, and set $\xi_i = a + (2i - 1)(b - a)/(2n + 1)$ for $i = 1, 2, \ldots, n + 1$. For this specification, $\xi_{n+1} = x_{n+1} = b$. As in the proof of Lemma 6.13, by φ_i we denote the function φ_t for $t = \xi_i$, $i = 1, 2, 3, \ldots, n + 1$, and for $t = x_i$, $i = 0, 1, 2, 3, \ldots, n - 1$, the functions ψ_t and θ_t will be denoted by ψ_i and θ_i. We add two more functions ψ_n and θ_n in L to these $3n + 1$ functions, satisfying the assumptions of Lemma 6.12 for $t = x_n$, and for which $\psi_n(b) = \theta_n(b) = 0$ as well. The existence of such functions in L follows from the compactness of the unit sphere in L with respect to pointwise convergence. For these $3n + 3$ functions in L, we can show as in the proof of Lemma 6.13 that they are linearly independent, completing the proof.

Corollary 6.10. The inequality $d_{3n+2}(A_\Delta) \geq \dfrac{b - a}{\alpha(2n + 1)}$ holds for every positive integer n.

Proof of Theorem 6.9 . The result follows directly from Corollaries 6.7 - 6.10. Indeed, keeping in mind the monotonicity of widths, we obtain from Corollary 6.7 and Corollary 6.9 that

$$\frac{b - a}{2\alpha n} \geq d_{3n}(A_\Delta) \geq d_{3n+1}(A_\Delta) \geq \frac{b - a}{2\alpha n},$$

i.e., $d_{3n}(A_\Delta) = d_{3n+1}(A_\Delta) = (b - a)/2\alpha n$. Similarly, from the other two corollaries, we have that $d_{3n+2}(A_\Delta) = (b - a)/\alpha(2n + 1)$. The theorem is proved.

T. Boyanov [1,3] studied in detail the widths of the set C_Ω of continuous functions defined on an arbitrary compact set Ω in m-dimensional Euclidean space with respect to Hausdorff distance. When this compact set is the m-dimensional unit cube D(m), the following result is obtained.

Theorem 6.10. (T. Boyanov and V. Popov [1]). If $C_{D(m)}$ is the set of continuous functions defined on the m-dimensional unit cube D(m), then for the widths $d_k(D(m))$ with respect to Hausdorff distance with parameter $\alpha = 1$, we have

$$d_k(C_{D(m)}) = \begin{cases} \dfrac{1}{2n + 1} & \text{if } 2(n + 1)^m + n(n + 1)^{m-1} \leq k < 3(n + 1)^m \\ \dfrac{1}{2n} & \text{if } 3n^m \leq k < 2(n + 1)^m + n(n + 1)^{m-1} \end{cases},$$

and hence $d_k(C_{D(m)}) \sim \frac{1}{2}\left(\frac{3}{n}\right)^{1/m}$ as $n \to \infty$.

The proof of Theorem 6.10 is similar to the proof of Theorem 6.9. T. Boyanov [1] also proved that the sequence $\langle d_n(C_\Omega)\rangle_1^\infty$ tends to zero if and only if Ω is locally connected.

Remarks . The ε-entropy of the set of continuous functions of several variables with respect to Hausdorff distance was considered by B. Penkov and Bl. Sendov [1].

Other questions connected with the computation of ε-entropy with respect to Hausdorff distance can be found in the papers of G. Clements [1], B. Boyanov [4], R. Dudley [1], and E. Bronstein [1].

Chapter 7

Approximation of Curves and Compact Sets in the Plane

In this chapter we shall consider the question of the approximation of compact sets in the plane various classes of point sets. The Hausdorff distance is quite natural in this situation. We note that interesting research on the approximation of point sets in the plane with respect to Hausdorff distance has been conducted by P. Davis, R. Vitale, and E. Ben-Sabar [1].

Maintaining the notation of Chapter 6, Z_D will denote the set of all compact connected subsets of the of the rectangle $D = \{(x,y) : a \le x \le b \text{ and } c \le y \le d\}$. The set Z_D becomes a compact metric space when equipped with Hausdorff distance:

$$(7.1) \qquad r(F, G) = \max\Big\{ \max_{(x,y)\in F} \inf_{(\xi,\eta)\in G} \max\{|x-\xi|, |y-\eta|\},\ \max_{(x,y)\in G} \inf_{(\xi,\eta)\in F} \max\{|x-\xi|, |y-\eta|\}\Big\}.$$

As means of approximation, we shall use the points sets γ in the plane which admit the following parametric representation :

$$(7.2) \qquad \gamma = (\varphi,\psi) = \{(x,y) : x = \varphi(t),\ \ y = \psi(t),\ \ t \in \Delta\},$$

where the functions φ and ψ belong to a given class. If $\varphi, \psi \in W$, then the curves in the plane with the representation of (7.2) will be called **W-curves** and will be denoted by W^2 in the sequel. We will be interested in these sets of W- curves :

C_Δ^2 = the set of continuous curves on $\Delta = [0, 1]$;

H_n^2 = the set of algebraic polynomial curves on $\Delta = [0, 1]$;

T_n^2 = the set of trigonometric polynomial curves on $\Delta = [0, 2\pi]$;

$S_{k,n}^2$ = the set of spline curves on $\Delta = [0, 1]$.

We will also consider two classes of piecewise monotone curves as well. By M_n^2, we denote those curves with a representation (7.2) where φ and ψ are piecewise monotone continuous functions of n-th order, i.e., $\varphi, \psi \in M_n$, for $\Delta = [0, 1]$. By $\mathbf{M}_n^2$, we denote the piecewise monotone point sets F having the representation

$$F = \{(x,y) : x \in \varphi(t), \quad y \in \psi(t), \quad t \in [0, 1]\},$$

where $\varphi, \psi \in \mathbf{M}_n$, i.e., φ and ψ are segment valued functions that are piecewise monotonic of nth-order (the reader may object to calling these curves).

It should be noted that the elements of the sets C_Δ^2, H_n^2, T_n^2, $S_{k,n}^2$ and M_n^2 are compact sets in the plane. The elements of $\mathbf{M}_n^2$ are also compact and may have thick sections, corresponding to those values of the parameter t for which the coordinate functions φ and ψ are multivalued simultaneously. The following inclusions are obvious: $C_\Delta^2 \supset H_n^2$, $C_\Delta^2 \supset T_n^2$, $C_\Delta^2 \supset S_{k,n}^2$ for $k = 1, 2, 3, \ldots$, and $C_\Delta^2 \supset M_n^2$. Also, if $F \in C_\Delta^2$, then $F \in Z_D$ for some rectangle D.

Let us also remind the reader that a continuous curve $\gamma \in C_\Delta^2$ may be represented parametrically by more than one pair (φ,ψ) of continuous functions on Δ.

Lemma 7.1. Suppose $\gamma_1 = (\varphi_1,\psi_1)$ and $\gamma_2 = (\varphi_2,\psi_2)$ are two continuous curves on $\Delta = [0, 1]$ with

$$\max \{|\varphi_1(t) - \varphi_2(t)| : t \in [0, 1]\} \le \varepsilon,$$

$$\max \{|\psi_1(t) - \psi_2(t)| : t \in [0, 1]\} \le \varepsilon.$$

Then $r(\gamma_1,\gamma_2) \le \varepsilon$.

Proof . It follows directly from the definition (7.1) for Hausdorff distance and the hypotheses of the lemma that

$$\max_{(x,y)\in\gamma_1} \min_{(\xi,\eta)\in\gamma_2} \max \{|x - \xi|, |y - \eta|\}$$

$$\le \max_{0\le t\le 1} \max \{|\varphi_1(t) - \varphi_2(t)|, |\psi_1(t) - \psi_2(t)|\} \le \varepsilon.$$

A similar inequality applies for the other component of Hausdorff distance, and the lemma is proved.

In view of Lemma 7.1, a number of questions on the approximation of point sets in the plane by curves of a given type could be reduced to the uniform approximation of the coordinate functions for certain parametric representations of these sets.

Let $F \in Z_D$ and let W^2 be a class of curves. The number

$$E(W^2; F) = \inf \{r(F,\gamma) : \gamma \in W^2\}$$

is naturally called the best approximation of F by curves in W^2. Our goal in the sequel is to find estimates for $E(W^2; F)$ subject to certain specifications for the point set F with respect to different means of approximation W^2, as in the case of best Hausdorff approximation of functions.

In order to find estimates for $E(W^2; F)$, we make use of some assertions connected with the concepts of labyrinths and corridors, introduced in Chapter 6. We divide the segment $[a, b]$ into p equal pieces, each of length $(b - a)/p$. This induces a partition of the square $D = [a, b] \times [a, b]$ into p^2 elementary squares by the straight lines $x = a + i(b - a)/p$, $i = 0, 1, 2, \ldots, p$ and $y = a + j(b - a)/p$. $j = 0, 1, 2, \ldots, p$. As before, each elementary square will be denoted by $\{x_i; y_j\}$, or more simply by $\{i; j\}$, where (x_i,y_j) is the left lower corner of the square. The elementary squares will be considered as closed point sets. The set of (p,p)-labyrinths (see Definition 6.3), consisting of elementary squares of a given partition of D into squares, will be denoted by $L^D_{p,p}$. The upper index D is required to distinguish the labyrinths of integer squares considered in the previous chapter from those introduced here, whose size depends on D.

We will also consider partitions of the rectangle $G = [0, 1] \times [a, b]$ divided into sp rectangles by the horizontal lines $x = \nu/s$, $\nu = 0, 1, 2, \ldots, s$ and the vertical lines $y = a + i(b - a)/p$, $i = 0, 1, 2, \ldots, p$. The set of all corridors (see Defintion 6.1) consisting of elementary rectangles of the given partition of G will be denoted by $K_{s,p}$.

We associate to each labyrinth in $L \in L^D_{p,p}$ two corridors in $K_{s,p}$.

Definition 7.1. We say that the labyrinth L in $L^D_{p,p}$ is decomposed into two corridors $K', K'' \in K_{s,p}$ if the following condition holds: whenever $F', F'' \in \mathbf{F}_\Delta$ and F' (resp. F'') belongs to the corridor K' (resp. K'') (see Definition 6.2), the set F given by

$$F = \{(x,y) : (t,x) \in F', \ (t,y) \in F'', \ t \in \Delta\}$$

belongs to the labyrinth L (see Defintion 6.6).

An elementary rectangle within a corridor in $K_{s,p}$ will be denoted as before by $\{\xi_i; \eta_j\}$, where (ξ_i,η_j) is the lower left corner of the rectangle. For simplicity, we write $\{i; j\}$ for $\{\xi_i; \eta_j\}$.

We associate to each corridor K in $K_{s,p}$ a 3 by s matrix $M(K)$ given by

$$M(K) = \begin{pmatrix} i_1 & i_2 & i_3 & \dots & i_s \\ j_1 & j_2 & j_3 & \dots & j_s \\ k_1 & k_2 & k_3 & \dots & k_s \end{pmatrix}$$

where $j_l \le i_l \le k_l$, $l = 1, 2, 3, \dots, s$, and the corridor K consists of just those squares $\{t_l; y_i\}$ for which $j_l \le i \le k_l$, $l = 1, 2, 3, \dots, s$, and $j_{l+1} \le i_l \le k_{l+1}$, $l = 1, 2, 3, \dots, s - 1$.

The corridor is called periodic if it is possible to choose i_s such that $j_1 \le i_s \le k_1$. The positive integer $\max\{k_l - j_l : 1 \le l \le s\}$ is called the thickness of the corridor K.

Lemma 7.2. If $L \in L^D_{p,p}$ is a labyrinth with closed passage of lenth s, then it is possible to decompose this labyrinth into two periodic corridors $K', K'' \in K_{s,p}$ such that each of them is of width at most 1.

Proof. Let the given closed passage be $\{\{x_i; y_j\} : (i,j) \in N\}$ where $N = \{(i_1,j_1), (i_2,j_2), \dots, (i_s,j_s)\}$. Denote $i'_l = \min\{i_l, i_{l+1}\}$ and $i''_l = \max\{i_l, i_{l+1}\}$, and $j'_l = \min\{j_l, j_{l+1}\}$, $j''_l = \max\{j_l, j_{l+1}\}$, where $i_{s+1} = i_1$ and $j_{s+1} = j_1$. We shall show that it is possible to decompose L into corridors K' and K'' with corresponding matrices

$$(7.3) \qquad M(K') = \begin{pmatrix} i_1 & i_2 & i_3 & \dots & i_s \\ i'_1 & i'_2 & i'_3 & \dots & i'_s \\ i''_1 & i''_2 & i''_3 & \dots & i''_s \end{pmatrix}, \qquad M(K'') = \begin{pmatrix} j_1 & j_2 & j_3 & \dots & j_s \\ j'_1 & j'_2 & j'_3 & \dots & j'_s \\ j''_1 & j''_2 & j''_3 & \dots & j''_s \end{pmatrix}.$$

It is obvious from the defintion of passage that the corridors K' and K'' so defined have width at most one and they are both periodic.

Let F' and F'' be two sets in F_Δ which belong to K' and K'', respectively. We shall show that the point set F given by

$$(7.4) \qquad F = \{(x,y) : (t,x) \in F', \ (t,y) \in F'', \ t \in \Delta\}$$

belongs to the labyrinth L. To this end, let us take an arbitrary point (x_0,y_0) of the set F. According to the representation (7.4), there exists $t_0 \in \Delta$ such that $(t_0,x_0) \in F'$, $(t_0,y_0) \in F''$. But since $F' \subset K'$ and $F'' \subset K''$, keeping in mind the matrix representation of a corridor in (7.3), we conclude that there is an index l which corresponds to the value t_0. Then the point (x_0,y_0) will belong to one of the following four rectangles: $\{i_l; j_l\}$, $\{i_l; j_{l+1}\}$, $\{i_{l+1}; j_l\}$, or $\{i_{l+1}; j_{l+1}\}$. But by the definition of a passage, we have $|i_l - i_{l+1}| + |j_l - j_{l+1}| \le 1$ so that either $i_l = i_{l+1}$ or $j_l = j_{l+1}$. Hence, either $i_l = i_{l+1}$ or $j_l = j_{l+1}$. This implies that the point (x_0,y_0) belongs to one of the two squares with indices

(i_l, j_l) and (i_{l+1}, j_{l+1}). But this means that (x_0, y_0) belongs to a square of the passage of L, i. e., it belongs to the labyrinth L.

It remains to show that each elementary square in L has a point in common with F. Let us take an arbitrary square $\{i_l; j_l\}$ of L, which participates in the passage under consideration. Consider the points $(t,x) \in F'$ and $(t,y) \in F''$ for $t \in [l/s, (l+1)/s]$. Since F' belongs to K' and F'' belongs to K'', then among these points there are those that belong to the elementary rectangles $\{l; i_l\}$, $\{l; i_{l+1}\} \subset F'$ and $\{l; j_l\}$, $\{l; j_{l+1}\} \subset F''$. Moreover, all the points under consideration belong to these four rectangles. By the definition of passage, either $i_l = i_{l+1}$ or $j_l = j_{l+1}$. We consider only the case $i_l = i_{l+1}$; the other case is similar. Since $t \in [l/s, (l+1)/s]$, all the points $(t,x) \in F'$ lie in the rectangle $\{l; i_l\}$ and there exists at least one point $(t_0,y) \in F''$ that belongs to the rectangle $\{l; j_l\}$ if $t_0 \in [l/s, (l+1)/s]$. Hence there is a point (x,y) belonging to the square $\{i_l; j_l\}$ for which $(t_0,x) \in F'$ and $(t_0,y) \in F''$, i.e., $(x,y) \in F$. The lemma is proved.

§ 7.1. Approximation by polynomial curves

Lemma 7.3. If K is an arbitrary corridor in $K_{s,p}$ with width at most 1 and $s \leq n/2\pi$, then there exists an algebraic polynomial of n-th degree whose graph belongs to the corridor K on the segment [0, 1].

Proof. Let the corridor K have the matrix representation

$$\begin{pmatrix} i_1 & i_2 & i_3 & \dots & i_s \\ j_1 & j_2 & j_3 & \dots & j_s \\ k_1 & k_2 & k_3 & \dots & k_s \end{pmatrix}.$$

On the segment [0, 1] we define a continuous function $\theta(t)$ which belongs to the corridor K that is linear on every subsegment $[(\nu - 1)/s, \nu/s]$, $\nu = 1, 2, 3, \dots, s$, and such that whenever $f \in C_{[0,1]}$ satisfies

$$\max\{|\theta(t) - f(t)| : t \in [0, 1]\} \leq \frac{b - a}{2p},$$

then f also belongs to the corridor K. It is easy to see that such a function θ can be constructed, and that it must be Lipschitz with constant $s(b - a)/p$, since the corridor K is of width at most 1.

By J. Favard's theorem [1,2], there exists an algebraic polynomial P of at most nth degree $(P \in H_n)$ such that

$$|\theta(t) - P(t)| \leq \frac{\pi s(b - a)}{pn}, \qquad t \in [0, 1].$$

Either of the following is a sufficient condition for the graph of P on [0, 1] to belong to the corridor K: $\pi s(b - a)/pn \leq (b - a)/2p$ or $s \leq n/2\pi$. The lemma is proved.

In view of Lemma 7.2 we obtain

Corollary 7.1. If L is a labyrinth in $L^D_{p,p}$ and L has a passage of length $s \leq n/2\pi$, then there exists a polynomial curve $\gamma \in H^2_n$ that belongs to the labyrinth L.

With the help of the above and Theorem 6.7, we can estimate the best approximation of arbitrary element in Z_D by polynomial curves of a given degree.

Theorem 7.1. If $F \in Z_D$, then for sufficiently large n we have

$$E(H^2_n ; F) \leq \frac{c}{n^{1/2}},$$

where $c = (b - a)(8\pi/3)^{1/2}$.

Proof. Consider a labyrinth $L \in L^D_{p,p}$ that contains the set F. By Theorem 6.7 the labyrinth L has a passage of length s, which is less than or equal to $4pp/3 + 2(p + p - 1)/3 \leq 4p(p + 1)/3$. Under the condition that $4p(p - 1)/3 \leq n/2\pi$, a polynomial curve $\gamma \in H^2_n$ belongs to L, and hence we may take $n = [8\pi(p - 1)^2/3]$. On the other hand, according to Lemma 6.8., we have $r(F,\gamma) \leq (b - a)/p$. Estimation of p from $n \leq 8\pi(p - 1)^2/3$ yields the assertion of the theorem.

Let c_1 be the following constant:

$$c_1 = \limsup_{n\to\infty} \frac{n^{1/2} E(H^2_n ; Z_D)}{b - a},$$

where $E(H^2_n ; Z_D) = \sup \{E(H^2_n ; F) : F \in Z_D\}$. We claim that

$$\frac{1}{4} \leq c_1 \leq \left(\frac{8\pi}{3}\right)^{1/2} = 2.894 \ldots . \tag{7.5}$$

The right-hand inequality of (7.5) follows from Theorem 7.1. To establish the left-hand inequality of (7.5), we construct a polygonal curve as follows. Take $p = [2\sqrt{n}]$ and connect successively by segments the centers of the following squares of the partition of D into p^2 pieces: {0; 0}, {1; 0}, {1; 1}, {0; 1}, {0; 2}, {1; 2}, . . ., {1; p - 1}, {2; p - 1}, {3; p - 1}, {3; p - 2}, {2; p - 2}, . . ., {p - 2; 0}, {p - 1; 0}. This snakelike path consisting of segments of length $(b - a)/p$ will be denoted by F^*. It is not hard to see that if an algebraic polynomial curve

$$\gamma = \{(x,y) : x = P(t),\ y = Q(t),\ t \in [0, 1]\},$$

where $P, Q \in H_n$ satisfies $r(F^*,\gamma) < (b - a)/2p$, then $P'(t)$ will have at least $(q - 1)[q/2]$ zeros. Consequently, $n > (q - 1)[q/2]$ holds, but for $n \geq 4$, we have

$$(q - 1)[q/2] \geq 2\sqrt{n}\,(\sqrt{n} - 1) = 2(n - \sqrt{n}) > n.$$

This leads to a contradiction, so that for each polynomial curve $\gamma \in H_n^2$, the inequality $r(F^*,\gamma) \geq (b - a)/4n^{1/2}$ must hold. This proves the left-hand inequality in (7.5).

The problem of finding the exact value of the constant c_1 remains unsolved. The following assertion for the trigonometric case is verified in a way similar to the proof of Theorem 7.1, this time using the corresponding theorems of N. Achieser and M. Krein [1].

Theorem 7.2. If $F \in Z_D$, then for sufficiently large n we have

$$E(T_n^2 ; F) \leq \frac{c}{n^{1/2}},$$

where $c = (b - a)(8/3)^{1/2}$.

With $E(T_n^2 ; Z_D)$ defined in the obvious way, the problem of finding the exact value of c_2 given by

$$c_2 = \limsup_{n\to\infty} \frac{n^{1/2} E(T_n^2 ; Z_D)}{b - a},$$

is also unsolved. It is only known that $1/4 < c_2 \leq 1.63\ldots$.

Now we look at the problem of the approximation of continuous curves of finite length by polynomial curves. For this purpose, we shall prove the next simple fact.

Lemma 7.4. If L is arbitrary labyrinth consisting of s elementary squares, then L has a closed passage whose length is at most $2(s - 1)$.

Proof. The assertion holds for a labyrinth consisting of two elementary squares. Let us assume that it holds true for labyrinths consisting of $s - 1$ squares. We take an arbitrary labyrinth consisting of s squares. Then we are always able to find an elementary square d such that $L' = L\backslash d$ is also a labyrinth, but of $s - 1$ squares. The passage of L may be obtained by adjoining two steps to the passage of L'. The theorem is proved.

Let Γ_l denote the set of all rectifiable continuous curves γ in the plane whose lengths are at most l.

Theorem 7.3. The following inequalities hold for sufficiently large n:

$$(7.6) \qquad \frac{1}{2n} \leq E(H_n^2; \Gamma_L) \leq \frac{8\pi l}{n},$$

and

$$(7.7) \qquad \frac{1}{2n} \leq E(T_n^2; \Gamma_L) \leq \frac{8l}{n}.$$

Proof . We just establish (7.6); the proof of (7.7) is similar and is left to the reader. Let $\gamma \in \Gamma_l$, i.e., γ is a rectifiable continuous curve of length at most l. It is possible to enclose the curve γ in a labyrinth $L \in L_{p,p}^D$ to which it belongs. If the sides of the elementary squares that form the labyrinth are of length d, then the labyrinth L that contains γ cannot have more than $2l/d - 4$ squares. Then by Corollary 7.1 and Lemma 7.4, we obtain the right-hand inequality in (7.6). The left-hand inequality in (7.6) is established in a manner similar to the left-hand inequality in (7.5).

If we designate

$$c_3 = \limsup_{n\to\infty} \frac{n\, E(H_n^2 \,;\, \Gamma_l)}{l}, \qquad c_4 = \limsup_{n\to\infty} \frac{n\, E(T_n^2 \,;\, \Gamma_l)}{l},$$

then by Theorem 7.3 we have $1/2 \leq c_3 \leq 8\pi$ and $1/2 \leq c_4 \leq 8$. The exact values of c_3 and c_4 are unknown.

Although the order of approximation of the entire class Γ_l is equal to $1/n$, each particular rectifiable curve is approximated faster. The following theorem holds.

Theorem 7.4. (Bl. Sendov and V. Popov [6]). If γ is a rectifiable curve, then $E(H_n^2; \gamma) = o(n^{-1})$ and $E(T_n^2; \gamma) = o(n^{-1})$.

First we prove some auxiliary statements, based on a result of N. Korneicuk and A. Polovina [1].

Lemma 7.5. If f is a Lipschitz function on the segment Δ, then there exist a sequence of piecewise linear functions $<\varphi_n>_1^\infty$, where φ_n has n components and interpolates f at the knots, i.e., φ_n is an interpolational spline of order (1,n), and a numerical sequence $<\lambda_n>_1^\infty$ with $\lim_{n \to \infty} \lambda_n = 0$ such that

$$|f(x) - \varphi_n(x)| \leq \frac{\lambda_n}{n}, \quad x \in \Delta, \quad n = 1, 2, 3, \ldots .$$

Using this result, we shall prove

Lemma 7.6. If $\gamma \in \Gamma_1$, then there exists a sequence $<\theta_n>_1^\infty$ of polygonal curves defined on [0, 1] inscribed into γ, where θ_n has n components, and a sequence of positive numbers $<\lambda_n>_1^\infty$ with $\lim_{n \to \infty} \lambda_n = 0$ such that $r(\gamma, \theta_n) \leq \lambda_n/n$.

Proof. Let $x = x(s)$, $y = y(s)$, $0 \leq s \leq 1$, be the parametrization of γ by arc length. The functions x(s) and y(s) satisfy a Lipschitz condition with constant 1 on [0, 1]. By Lemma 7.5, there exist a number sequence $<\lambda_n>_1^\infty$ with $\lim_{n \to \infty} \lambda_n = 0$ and sequences $<\varphi_n>_1^\infty$, $<\psi_n>_1^\infty$ of spline functions of order (1,n) such that

$$|x(s) - \varphi_n(s)| \leq \frac{\lambda_n}{n}, \quad |y(s) - \psi_n(s)| \leq \frac{\lambda_n}{n}, \tag{7.8}$$

for all $s \in [0, 1]$. In addition, we can assume that the functions φ_n and ψ_n have knots s_i, $i = 0, 1, 2, \ldots, n$, in common and that $x(s_i) = \varphi_n(s_i)$ and $y(s_i) = \psi_n(s_i)$ for $i = 0, 1, 2, \ldots, n$. For θ_n, we take the curve $x = \varphi_n(l_t)$ and $y = \psi_n(l_t)$, $0 \leq t \leq 1$. The conclusion of the lemma follows from (7.8) and Lemma 7.1.

The last result can be recast in the following form.

Corollary 7.2. If γ is a rectifiable curve in the plane, then $E(S_{1,n}^2; \gamma) = o(n^{-1})$.

Lemma 7.7. If θ_m is a polygonal curve in the plane with m knots and length l, i.e., $\theta_m \in S_{1,m}^2$, then

$$E(H_n^2\,;\,\theta_m) \le \frac{clm}{n^2},$$

where c is an absolute constant.

Proof. Let $x = x(s)$, $y = y(s)$, $0 \le s \le 1$, be the parametrization of θ_m by arc length. The functions $x(s)$, $y(s)$ are piecewise linear functions with m components, i.e., they belong to $S_{1,m}$. We shall present a different parametric representation for the curve θ_m. To this end, we define a function $\sigma(t)$ on $[0, 1]$. First, set $\alpha_i = \sqrt{s_i - s_{i-1}}$ for $i = 1, 2, \ldots, m$, $\beta_0 = 0$, and for $j = 1, 2, \ldots, m$, define β_j by

$$\beta_j = \frac{{}^1\sum_{i=1}^{j} \alpha_i}{\sum_{i=1}^{m} \alpha_i}.$$

The function σ is given by

$$\sigma(t) = A \int_0^t \int_0^{\xi} q(\tau)\, d\tau\, d\xi,$$

where the integrand q is given by

$$q(t) = \begin{cases} 1 & \text{if } \beta_j \le t < (\beta_j + \beta_{j+1})/2 \\ -1 & \text{if } (\beta_j + \beta_{j+1})/2 \le t < \beta_{j+1} \end{cases}, \tag{7.9}$$

and the constant A is determined form the condition $\sigma(l) = 1$.

Clearly $\sigma'(t)$ satisfies a Lipschitz condition with constant A on $[0, l]$. Let us estimate A. We have

$$\sigma(l) = \frac{Al^2}{4} \sum_{i=1}^{m} \alpha_i^2 \left(\sum_{i=1}^{m} \alpha_i\right)^{-2} = \frac{Al^3}{4} \left(\sum_{i=1}^{m} \alpha_i\right)^{-2} = 1. \tag{7.10}$$

Applying the Schwartz inequality under the condition $\sum_{i=1}^{m} \alpha_i^2 = l$, we see that $\sum_{i=1}^{m} \alpha_i$ is at most $(lm)^{1/2}$. Hence, (7.10) yields

(7.11) $$A \le \frac{4m}{l^2}.$$

Obviously, $\sigma(0) = 0$ and $\sigma(t)$ is a monotone continuous function on $[0, l]$. In addition, we have

(7.12) $$\sigma(\beta_j) = s_j, \quad \sigma'(\beta_j) = 0, \; j = 0, 1, 2, \ldots, m.$$

We now introduce our second parametric representation of the curve θ_m:

$$\theta_m = \{(x,y) : x = x(\sigma(t)), \; y = y(\sigma(t)), \; 0 \le t \le 1\}.$$

Since $x(s)$ and $y(s)$ belong to $S_{1,m}$ with knots s_i, $i = 0, 1, 2, \ldots, m$ and both satisfy a Lipschitz condition with constant 1 on $[0, l]$, it follows from (7.11) and (7.12) that $dx(\sigma(t)/dt$ and $dy(\sigma(t)/dt$ both satisfy a Lipschitz condition with constant $4ml^{-2}$ on $[0, 1]$.

From here, by Jackson's theorem on best uniform approximation of differentiable functions by algebraic polynomials, we see that there exist algebraic polynomials $P, Q \in H_n$ such that

(7.13) $$\max_{0 \le t \le 1} |x(\sigma(t) - P(t)| \le \frac{clm}{n^2}, \qquad \max_{0 \le t \le 1} |y(\sigma(t) - Q(t)| \le \frac{clm}{n^2},$$

hold, where c is an absolute constant.

Finally, consider the polynomial curve $\gamma_n \in H_n^2$ with parametric equations $x = P(t)$, $y = Q(t)$, $t \in [0, 1]$. From (7.13) and Lemma 7.1 we see that $r(\theta_m, \gamma_n) \le clmn^{-2}$, where c is an absolute constant. The lemma is proved.

Proof of Theorem 7.4. By virtue of Corollary 7.2 there exist a sequence $\langle\theta_n\rangle_1^\infty$ where $\theta_n \in S_{1,n}^2$ and θ_n is inscribed in γ and a sequence of positive numbers $\langle\lambda_n\rangle_1^\infty$ with $\lim_{n\to\infty} \lambda_n = 0$ such that

(7.14) $$r(\gamma, \theta_n) \le \frac{\lambda_n}{n}.$$

Let l be the length of a rectifiable curve γ, then the length of each θ_n is at most l. Then from Lemma 7.7, (7.14), and since $\theta_n \in \Gamma_l$, we obtain for fixed m and n

(7.15) $$E(H_n^2; \gamma) \le r(\gamma, \theta_m) + E(H_n^2; \theta_m) \le \frac{\lambda_m}{m} + \frac{clm}{n^2}.$$

Now if we choose λ_m carefully with respect to n, so that $m = n\lambda_m^{1/2}$ is an integer and $m \to \infty$ as $n \to \infty$, then we obtain from (7.15) that

$$E(H_n^2; \gamma) \le (1 - cl)\lambda_m^{1/2} n^{-1} = o(n^{-1}).$$

The proof for $E(T_n^2; \gamma)$ is analagous and is left to the reader.

§ 7.2. Characterization of best approximation in terms of metric dimension

The concept of metric dimension can be used to estimate the best approximation of point sets by polynomial curves (see A. Kolmogorov and V. Tihomirov [1, p. 21]).

Definition 7.2. Let A be a bounded complete metric space and let $H_\varepsilon(A)$ be the ε-entropy of the set A. The **upper and lower metric dimensions** $dm^+(A)$ and $dm^-(A)$ of the set A are given by

$$dm^+(A) = \limsup_{\varepsilon\to 0} \frac{H_\varepsilon(A)}{\log_2 \varepsilon^{-1}}, \qquad dm^-(A) = \liminf_{\varepsilon\to 0} \frac{H_\varepsilon(A)}{\log_2 \varepsilon^{-1}}.$$

If $dm^+(A) = dm^-(A)$, their common value is called the **metric dimension** of A, and will denoted by dm(A).

Theorem 7.5. If $F \in Z_D$, i.e. F is a compact connected set, and if $dm(F) = \sigma$, then for every positive δ, we have

$$E(H_n^2; F) = O(n^{-1/(\sigma+\delta)}) \quad \text{and} \quad E(T_n^2; F) = O(n^{-1/(\sigma+\delta)}).$$

Proof. We prove just the first of these equalities. We shall show that there exists a constant $c = c(\delta)$ such that for every positive integer p, a labyrinth $L \in L_{p,p}^D$ to which F belongs consists of at most $p^{\sigma+\delta}$ squares. From the definition of metric dimension, for sufficiently small $\varepsilon > 0$, we have

$$\frac{H_\varepsilon(F)}{\log_2 \varepsilon^{-1}} \le \sigma + \delta. \tag{7.16}$$

Let ν be an ε-covering of the set F which has a minimal number of elements. If $N_\varepsilon(F)$ is the number of elements in ν, then by definition $H_\varepsilon(F) = \log_2 N_\varepsilon(F)$, so that by (7.16) we obtain $N_\varepsilon(F) \le \varepsilon^{-\sigma-\delta}$. If we take $\varepsilon = 1/p$, then every element of ν may have points in common with at most four squares of the labyrinth L that contains F. As a result, the number of squares in L is at most $4p^{\sigma+\delta}$ for sufficiently large p. But then for each positive integer p, the number of squares in $L \in L^D_{p,p}$ to which F belongs is at most $c(\delta)p^{\sigma+\delta}$, where $c(\delta)$ is a positive constant, depending only on δ.

By Corollary 7.1 if

$$c(\delta)p^{\sigma+\delta} \le \frac{n}{2\pi}, \tag{7.17}$$

then

$$E(H^2_n ; F) \le \frac{b-a}{p_n}, \tag{7.18}$$

where p_n is the greatest positive integer p which satisfies the inequality (7.17), i.e.,

$$c(\delta)(p_n - 1)^{\sigma+\delta} > \frac{n}{2\pi}. \tag{7.19}$$

It follows from (7.19) that $1/p_n = O(n^{-1/(\sigma+\delta)})$, and the result follows from (7.18).

It should be noted that Theorem 7.5 is of interest for $1 \le \sigma \le 2$, since all subsets of D and in particular all elements of Z_D have metric dimension no higher than dm(D) = 2, and on the other hand, every connected set F for which dm(F) < 1 consists of a single point. It is natural to ask whether one can put $\delta = 0$ in Theorem 7.5. By Theorem 7.1., this is possible for $\sigma = 2$, but it turns out that this is impossible when $1 < \sigma < 2$.

Theorem 7.6. (T. Boyanov and Bl. Sendov [1]). For every $\sigma \in (1, 2)$ and for every $\delta > 0$ there exists a compact connected set $F_{\sigma,\delta}$ with metric dimension σ such that

$$\limsup_{n\to\infty} \frac{E(H^2_n; F_{\sigma,\delta})}{(\ln^{\delta} n/n)^{1/\sigma}} = \infty,$$

and

$$\limsup_{n\to\infty} \frac{E(T^2_n; F_{\sigma,\delta})}{(\ln^{\delta} n/n)^{1/\sigma}} = \infty.$$

Proof . Let $\sigma \in (1, 2)$ and $\delta > 0$ be fixed. We construct $F_{\sigma,\delta}$ as an intersection of a decreasing sequence of compact connected sets . We choose a positive integer k_0 such that $\sigma + (\log_2 k_0)/k_0 < 2$ and $\sigma + \log_2 (k_0 + 1) - \log_2 k_0 < 2$. Fix an integer $s > 0$ such that $s + 3 < \sigma s$ and $s(\sigma + \log_2 (k_0+ 1) - \log_2 k_0) \leq 2s - 1$. Such choices of k_0 and s are possible, since $1 < \sigma < 2$. Writing $\nu(k) = s(\sigma + \log_2 (k+ 1) - \log_2 k)$, we have $s + 3 \leq \nu(k) \leq 2s - 1$, provided $k \geq k_0$.

Now we shall consider a square with side $\varepsilon_k = 2^{-sk}$ which is divided into 2^{2s} squares each with side $\varepsilon_k 2^{-s}$. For $k \geq k_0$ we may extract $[2^{\nu(k)}]$ squares such that their union (a) forms a labyrinth; (b) is contained completely in the shaded set E shown in Figure 7.1 (because of $2^{\nu(k)} \leq 2^{2s}/2$); (c) contains the contour of the set E (because of $8 \cdot s^s \leq 2^{\nu(k)}$).

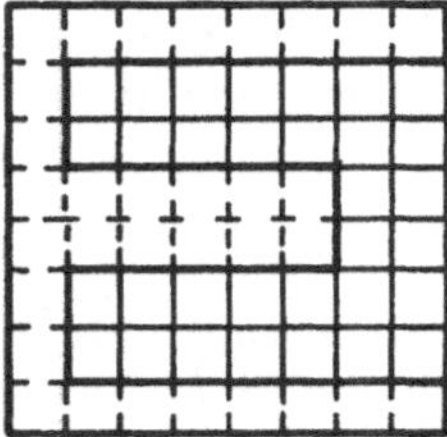

FIGURE 7.1

We now write $\lambda(k) = \frac{1}{k}\log_2 k$ and $N_k = 2^{sk(\sigma + \lambda(k))}$, and let F_{k_0} be a labyrinth of N_{k_0} squares, each one of them with side ε_{k_0}. If the set F_k of N_k squares with side ε_k is already defined, then F_{k+1} is obtained by choosing in the above manner in each square of F_k a totality of $[2^{\nu(k)}]$ squares with side $\varepsilon_k 2^{-s}$. As a result, F_{k+1} contains $N_{k+1} = 2^{\nu(k)} N_k$ squares with side ε_{k+1}.

We may now prescribe $F_{\sigma,\delta}$:

$$F_{\sigma,\delta} = \bigcap_{i=0}^{\infty} F_{k_i} .$$

Obviously, $F_{\sigma,\delta}$ is a connected compact set containing the corners of all squares that compose F_k for every $k > k_0$. Thus we obtain an ε_k-discrete subset of the set $F_{\sigma,\delta}$ consisting of N_k points.

We claim that $dm(F_{\sigma,\delta}) = \sigma$. For $\varepsilon > 0$ sufficiently small, there exists $k > k_0$ such that $\varepsilon_{k+1} < \varepsilon < \varepsilon_k$. Therefore

$$\frac{H_\varepsilon(F_{\sigma,\delta})}{\log_2 \varepsilon^{-1}} \leq \frac{\log_2 N_{k+1}}{\log_2 \varepsilon_k^{-1}} = \frac{s(k + 1)(\sigma + \lambda(k + 1))}{sk} ,$$

$$\frac{C_\varepsilon(F_{\sigma,\delta})}{\log_2 \varepsilon^{-1}} \geq \frac{\log_2 N_k}{\log_2 \varepsilon_{k+1}^{-1}} = \frac{sk(\sigma + \lambda(k))}{s(k-1)}.$$

Now as $\varepsilon \to 0$, we have $k \to \infty$ and $\lambda(k) \to 0$. We obtain from the definitions of dm^+ and dm^- that $dm^+(F_{\sigma,\delta}) \leq \sigma$ and $dm^-(F_{\sigma,\delta}) \geq \sigma$, so that $dm(F_{\sigma,\delta}) = \sigma$.

Let $\gamma_m \in H_m^2$ have the parametric representation

$$\gamma_m = \{(x,y) : x = P(t),\ y = Q(t),\ t \in [0, 1]\}.$$

Keeping in mind Figure 7.1, if the inequality $r(F_{\sigma,\delta}, \gamma_m) < \varepsilon_k/8$ holds, one can show that $P'(t)$ must have at least N_k zeros, one in each square of F_k. Therefore $m > N_k$.

Let $n = N_k$. It follows from the above analysis that $E(H_n^2; F_{\sigma,\delta}) \geq \varepsilon_k/8$, and then

$$\frac{E(H_n^2; F_{\sigma,\delta})}{(\ln^\delta n/n)^{1/\sigma}} \geq \frac{2^{-sk}}{8(|sk(\sigma + \lambda(k))|^\delta/2^{sk(\sigma + \delta(k))})^{1/\sigma}}$$

$$= \frac{1}{8}|s(\sigma + \lambda(k))|^{-\sigma/\delta}.k^{-\delta/\sigma}.\ 2^{sk\lambda(k)/\sigma},$$

which essentailly finishes the proof, since $2^{k\lambda(k)} = k$ and $k^s/k^\sigma \to \infty$. The second assertion for $E(T_n^2; F_{\sigma,\delta})$ is obtained in much the same way.

T. Boyanov and E. Karlov in unpublished work have obtained related results in the case $dm(F) = 1$, where F is a rectifiable curve.

§ 7.3. Approximation by piecewise monotone curves

A number of estimates on best approximation by polynomial curves have been obtained by counting zeros of the derivative of one of the polynomials in the parametric approximation. The perspective here is that $H_n^2 \subset M_n^2$ because $H_n \subset M_n$. Let us recall that $\gamma \in M_n^2$ provided

$$\gamma = \{(x,y) : x = \varphi(t),\ y = \psi(t),\ t \in [0, 1]\},$$

where $\varphi, \psi \in M_n$ and M_n is the set of continuous piecewise monotone functions with at most n sections of monotonicity.

The proof of the next result is similar to the proof of Theorem 7.1.

Theorem 7.7. The equality

$$\limsup_{n\to\infty} \frac{n^{1/2} E(M_n^2; Z_D)}{b - a} = c_5$$

holds, where c_5 is an absolute constant with $1/4 \le c_5 \le 0.816 \ldots$.

Proof. It is directly seen that the function f constructed in the proof of Lemma 7.3 is piecewise monotone and $f \in M_n$ where $n \le s/2$ when f belongs to the corridor $K \in K_{s,p}$. Let us recall that this corridor is of a special type. As a result, using Theorem 6.7, for each labyrinth $L \in L_{p,p}^D$ there exists $\gamma_n \in M_n$ such that γ_n belongs to the labyrinth L if $n \ge \frac{2}{3}p^2 - \frac{2}{3}p$. Hence, for every $F \in Z_D$ the inequality $E(M_n^2; F) \le (b - a)/p_n$ holds, where p_n is the largest positive integer which satisfies the inequality $2p^2 - 2p \le 3n$. Thus we obtain an upper bound for c_5: $c_5 \le \sqrt{2/3} = 0.816 \ldots$. The lower bound is obtained as in the case of polynomial curves. The theorem is proved.

The exact value of the constant c_5 has not been yet discovered. The determination of the exact value of

$$c_6 = \limsup_{n\to\infty} \frac{n\, E(M_n^2; \Gamma_l)}{l}$$

is also of some interest.

Since $H_n^2 \subset M_n^2$, from Theorem 7.4 one obtains

Corollary 7.3. If γ is a rectifiable curve, then $E(M_n^2; \gamma) = o(n^{-1})$.

The statements of Theorem 7.5 and Theorem 7.6 may be similarly modified.

§ 7.4. Other methods for the approximation of curves in the plane

We can obtain estimates for the best approximation of elements of Z_D (in particular, for rectifiable curves in the plane) using as approximation means the spline curves $S_{k,n}^2$. More precisely, using the methods of §7.2, we can estimate the constants c_k' and c_k'' given by

$$c_k' = \limsup_{n\to\infty} \frac{n^{1/2} E(S_{k,n}^2; Z_D)}{b - a} ,$$

$$c''_k = \limsup_{n\to\infty} \frac{nE(S^2_{k,n};\Gamma_1)}{l}.$$

The exact values of these constants remain unknown.

Let us note that $S^2_{1,n}$ consists of polygonal curves made up of at most $2n$ segments, if spline-functions with independent knots are used in the parametrization of the curve, or at most n segments, if spline functions with the same knots are considered.

The problem of the approximation of convex curves by polygons is of special interest. We denote the set of all convex n-gons in the plane by L_n. Clearly, $L_n \subset S^2_{1,n}$. We denote by Γ^0 the set of all convex closed curves in the plane. V. Popov [2,3,6] studied the problem of finding the best approximation of elements of Γ^0 by elements of L_n. We shall cite a few of these results without proof.

In consideration of such questions, we depart from a basic convention of this book, and now make use of Hausdorff distance as induced by the Euclidean distance in the plane rather than the box metric. In this sense, the Hausdorff distance $r_0(F, G)$ between F and G in Z_D is given by

$$r_0(F,G) = \max\Big\{ \max_{(x,y)\in F}\ \inf_{(\xi,\eta)\in G} \sqrt{(x-\xi)^2+(y-\eta)^2},\ \max_{(x,y)\in G}\ \inf_{(\xi,\eta)\in F} \sqrt{(x-\xi)^2+(y-\eta)^2}\Big\}.$$

The following result pertains to the best approximation $E_0(L_n;\gamma)$ of a convex curve $\gamma\in\Gamma^0$ by n-gons:

$$E_0(L_n;\gamma) = \inf\{r_0(P,\gamma) : P\in L_n\}.$$

Theorem 7.8. (V. Popov [3]). If γ is a closed convex curve of length 1, then

$$E_0(L_n;\gamma) \le \frac{1}{2n}\cdot\frac{\sin(\pi/n)}{1+\cos(\pi/n)}.$$

The following result shows that in a certain sense, this estimate cannot be improved; it is slightly better than a related result of H. Eggleston [1].

Theorem 7.9. (V. Popov [3]). For a regular n-gon P_n of length 1 we have

$$E_0(L_{n-1};P_n) \ge \frac{1}{2n}\cdot\frac{\sin(\pi/n)}{1+\cos(\pi/n)}.$$

An intriguing problem is determine that closed convex curve of length 1 that has the worst approximation by n-gons among all such curves. A reasonable conjecture is that this curve is the regular (n+1)-gon. R. Ivanov [1] has proved that the regualr (n+1)-gon has the worst approximation by n-gons among all the convex (n+1)-gons. This problem was considered by P. Kenderov [1], but with respect to another metric. Related questions are considered in P. Kenderov [2] and P. Gruber and P. Kenderov [1].

Using Theorem 7.8, we can address the approximation of closed convex curves by polynomial curves. The following assertion is true.

Theorem 7.10. (V. Popov [9]). If γ is a closed convex curve of length 1, then

$$E(H_n^2\,;\gamma) \geq c\ln^{-2}\ln^2 n, \tag{7.20}$$

where c is an absolute constant.

It should be noted that for each n, it is possible to find a closed convex curve γ_n of length 1 such that

$$E(H_n^2\,;\gamma_n) \geq c'\ln^{-2},$$

where c' is an absolute constant. In connection with this, there is no hope for the improvement of (7.20) by more than the factor $\ln^2 n$. Also, the exact asymptotic behavior of $E(H_n^2\,;\Gamma_1^0)$ has not been determined, where Γ_1^0 is the set of all closed convex curves of length 1.

Remarks . The approximation of point sets in the plane by polynomial curves has been studied by Bl. Sendov [10]. The improvability of Theorem 7.5 has been demonstrated by T. Boyanov and E. Karlov [1]. T. Boyanov [5] strengthened Theorem 7.6, replacing $\ln^\delta n$ by $\varphi(n)$ in the hypotheses of the theorem, where $\varphi(x) = o(x^\delta)$ as $x \to \infty$ for each $\delta > 0$. T. Boyanov [7] has considered an example that demonstrates the effect in approximation of different parametric representations of a plane curve.

Also related to the contents of this chapter are the papers of D. Rice [1], V. Martynjuk [4], N. A. Nasarenko [1], M. G. Nikolcheva [1], and E. Bronstein and L. Ivanov [1].

Chapter 8

Numerical Methods of best Hausdorff Approximation

Practical application of the polynomials of best Hausdorff approximation requires numerical methods for the approximate determination of these polynomials. As was already noted §4.1, the polynomial of best Hausdorff approximation need not be unique. Therefore it is possible to construct numerical methods for the determination of the polynomial of best Hausdorff approximation by analogy to the method of E. Remez [1] for the uniform metric only if the conditions for uniqueness considered in §4.1.1 are fulfilled: the function f has a unique polynomial of best Hausdorff approximation of degree n for sufficiently large n if f is λ-monotonic (see Definition 4.4).

The polynomial of best approximation is unique, if we use as distance only one of the componenets of Hausdorff distance, i. e., one-sided Hausdorff distance, from the approximating polynomial to the approximated function, provided the approximated function is continuous. The polynomial of best one-sided approximation coincides with the polynomial of best Hausdorff approximation when the uniqueness conditions of §4.4.1 are fulfilled. Thus, we consider numerical methods for the determination of the polynomials of best one-sided approximation.

§ 8.1. One-sided Hausdorff distance

One-sided Hausdorff distance was introduced for segment functions in §4.1.2.3 and was considered again in our study of widths in Chapter 6. We now consider such distances for segment functions with a parameter.

Definition 8.1. Let $f, g \in F_\Delta$ where $\Delta = [a, b]$. The **one-sided Hausdorff distance from f to g with parameter** $\alpha > 0$ is the number $h(\Delta, \alpha; f, g)$ given by

$$(8.1)\qquad h(\Delta, \alpha; f, g) = \max_{(x,y)\in f}\ \min_{(\xi,\eta)\in g}\ \max\{\alpha^{-1}|x-\xi|, |y-\eta|\}.$$

In the sequel, we will often just call this the one-sided distance from f to g. In view of (2.12), ordinary Hausdorff distance is expressed in the obvious way through both one-sided distances:

$$r(\Delta, \alpha;\ f, g) = \max\ \{h(\Delta, \alpha, f; g),\ h(\Delta, \alpha, g; f)\}.$$

Note that in the set $\mathbf{H}_\Delta$ and in particular in C_Δ the functional $h(\Delta, \alpha; f, g)$ satisfies all the properties of a metric except for symmetry. Such "distances" that do not possess symmetry are sometimes called **Δ-metrics**. Approximation with respect to Δ-metrics was considered by A. Andreev and V. Popov [1,2]. It is easy to see that for $f, g \in C_\Delta$ we have

$$\lim_{\alpha \to 0} h(\Delta, \alpha;\ f, g) = \|f - g\|_\Delta = \max_{x \in \Delta} |f(x) - g(x)|\ .$$

Definition 8.2. The best one-sided approximation $e(H_n, \Delta, \alpha; f)$ of f by algebraic polynomials of degree no higher than n with respect to one-sided Hausdorff distance with parameter α on the segment Δ is the number

$$e(H_n, \Delta, \alpha; f) = \inf\ \{h(\Delta, \alpha; P, f) : P \in H_n\}.$$

If for some $P_0 \in H_n$ we have $e(H_n, \Delta, \alpha; f) = h(\Delta, \alpha; P_0, f)$, then P_0 will be called a polynomial of best one-sided Hausdorff approximation of degree no higher than n for f.

We stress that best one-sided Hausdorff approximation is defined on the basis of one-sided distance from the approximated polynomial to the approximated function, and not vice versa.

8.1.1. EXISTENCE AND UNIQUENESS OF THE POLYNOMIAL OF BEST ONE SIDED APPROXIMATION

Theorem 8.1. For every function $f \in \mathbf{F}_\Delta^M$ and for every positive integer n, there exists a polynomial $P_0 \in H_n$ for which $e(H_n, \Delta, \alpha; f) = h(\Delta, \alpha; P_0, f)$.

Proof. From Definitions 8.1 and 8.2 it follows that for every positive integer m there exists $P_m \in H_n$ for which

$$h(\Delta, \alpha; P_m, f) = \max_{x \in \Delta}\ \min_{(\xi,\eta) \in f}\ \max\ \{\alpha^{-1}|x - \xi|, |P_m(x) - \eta|\} \tag{8.2}$$

$$\leq e(H_n, \Delta, \alpha; f) + m^{-1}.$$

But since $f \in \mathbf{F}_\Delta^M$, whence f is bounded, the sequence $\langle P_m \rangle_1^\infty$ is uniformly bounded by (8.2). Out of such a sequence it is possible to choose a uniformly convergent subsequence that will converge to a certain polynomial $P_0 \in H_n$. Then we obtain from (8.2) that

$$e(H_n, \Delta, \alpha; f) \geq h(\Delta, \alpha; P_0, f) \geq e(H_n, \Delta, \alpha; f),$$

or $h(\Delta, \alpha; P_0, f) = e(H_n, \Delta, \alpha; f)$. This completes the proof.

To see that the polynomial of best one-sided Hausdorff approximation need not be unique with respect to the class $\mathbf{F}_\Delta$, we present an example due to A. Andreev [1]. Let φ be this step function:

$$\varphi(x) = \begin{cases} -1/2 & \text{if } -1 \leq x < 0 \\ 2 & \text{if } 0 < x < 1 + \sqrt{3} \\ 8 + 4\sqrt{3} & \text{if } 1 + \sqrt{3} < x \leq 2 + \sqrt{3} \end{cases}$$

Clearly, $\varphi \in H_\Delta$, i.e., the completed graph of φ is in $\mathbf{H}_\Delta$, where $\Delta = [-1, 2 + \sqrt{3}]$. It can be shown that $e(H_2, [-1, 2 + \sqrt{3}], 1; \varphi) = 1$ and each polynomial of the form $P_\lambda(x) = x^2 + \lambda(x - \sqrt{2})(x - 1 - \sqrt{3})$ for λ positive and sufficiently small is a polynomial of best one-sided Hausdorff approximation of degree 2 for the function φ.

We note that polynomials of best uniform approximation also need not be unique in the case of discontinuous functions. It is a classical result of Chebyshev that uniqueness occurs with continuity. This fact remains valid for one-sided Hausdorff approximation, using the same tools as in the uniform case.

Theorem 8.2. Let $f \in C_\Delta$, $\Delta = [a, b]$. A necessary and sufficient condition for $P \in H_n$ to be a polynomial of best one-sided Hausdorff approximation for f is the existence of $n + 2$ points $\langle x_i \rangle_0^{n+1}$ where $a \leq x_0 < x_1 < \cdots < x_{n+1} \leq b$ such that

$$\min_{\xi \in \Delta} \max \{\alpha^{-1}|x_i - \xi|, |P(x_i) - f(\xi)|\} = h(\Delta, \alpha; P, f), \qquad i = 0, 1, \ldots, n+1$$

and $\operatorname{sgn}(f(x_i) - P(x_i)) = (-1)^i \varepsilon$ where $\varepsilon = \pm 1$, i.e., the existence of points of maximal deviation.

The proof of Theorem 8.2 is similar to the proof of the corresponding result for uniform distance (see I. Natanson [1, p. 1]). The uniqueness of the polynomial of best one-sided Hausdorff approximation follows directly from Theorem 8.2, provided the approximated function is continuous.

We now obtain an analogue of the Vallée-Poussin Theorem for best one-sided Hausdorff approximation.

Theorem 8.3. (B. Boyanov [1] and A. Andreev [1]). Let $f \in \mathbf{F}_{\Delta}^{M}$, where $\Delta = [a, b]$. Suppose for $P \in H_n$ there exist $n + 2$ points $\langle x_i \rangle_0^{n+1}$ such that $a \leq x_0 < x_1 < \cdots < x_{n+1} \leq b$ and $\operatorname{sgn}(f(x_i) - P(x_i)) = (-1)^i \varepsilon$, $\varepsilon = \pm 1$. If for each i we set

$$h_i = \min_{(\xi,\eta) \in f} \max \{\alpha^{-1}|x_i - \xi|, |P(x_i) - \eta|\},$$

then $e(H_n, \Delta, \alpha; f) \geq \min \{h_i : i = 0, 1, 2, \ldots, n + 1\}$.

Proof . First, an explanatory remark. In general $f(x_i) - P(x_i)$ is a segment, and by $\operatorname{sgn}(f(x_i) - P(x_i))$ we mean the sign of all y in the segment.

We first claim that if $x_0 \in \Delta$ and $y_0 > f(x_0)$, then for each positive λ, we have the inequality

$$\min_{(\xi,\eta) \in f} \max \{\alpha^{-1}|x_0 - \xi|, |y_0 - \eta|\} \leq \min_{(\xi,\eta) \in f} \max \{\alpha^{-1}|x_0 - \xi|, |y_0 - (\eta - \lambda)|\}. \tag{8.3}$$

Indeed, since $y_0 > f(x_0)$, then $y_0 > \eta$ for each $\eta \in f(x_0)$ and hence, subtracting the positive number λ from η can only increase the expression $\max \{\alpha^{-1}|x_0 - \xi|, |y_0 - \eta|\}$. Let us assume that the conclusion of the theorem is false, i.e.,

$$e(H_n, \Delta, \alpha; f) < \min \{h_i : i = 0, 1, 2, \ldots, n + 1\}. \tag{8.4}$$

This means that if P_0 is a polynomial of best one-sided Hausdorff approximation for f, then for $i = 0, 1, 2, \ldots, n + 1$,

$$\min_{(\xi,\eta) \in f} \max \{\alpha^{-1}|x_i - \xi|, |P_0(x_i) - \eta|\} \tag{8.5}$$

$$< \min_{(\xi,\eta)\in f} \max \{\alpha^{-1}|x_i - \xi|, |P(x_i) - \eta|\}.$$

If $P(x_i) > f(x_i)$, then $P_0(x_i) \le P(x_i)$, since otherwise (8.5) will fail by (8.3). Analogously, $P(x_i) < f(x_i)$ forces $P_0(x_i) \ge P(x_i)$. Consequently, the difference $P_0(x) - P(x)$ changes its sign at least $n + 1$ times, i.e., $P_0(x) \equiv P(x)$. This contradicts (8.4), and the theorem is proved.

Corollary 8.1. A sufficient condition for the $P \in H_n$ to be a polynomial of best one-sided Hausdorff approximation for $f \in \mathbf{F}_\Delta^M$, is the existence of $n + 2$ points $\langle x_i \rangle_0^{n+1}$ where $a \le x_0 < x_1 < \cdots < x_{n+1} \le b$ such that

$$\min_{(n,\xi)\in f} \max \{\alpha^{-1}|x - \xi|, |P(x_i) - \eta|\} = h(\Delta, \alpha; P, f)$$

and

$$\operatorname{sgn} (f(x_i) - P(x_i)) = (-1)^i \varepsilon, \text{ where } \varepsilon = \pm 1,\ i = 0, 1, 2, \ldots, n + 1.$$

Proof. With respect to the notation of Theorem 8.3 we have

$$e(H_n, \Delta, \alpha; f) \le h(\Delta, \alpha; P, f) = h_0 = h_1 = \cdots = h_{n+1} \le e(H_n, \Delta, \alpha; f),$$

that is, $e(H_n, \Delta, \alpha; f) = h(\Delta, \alpha; P, f)$.

§ 8.2. Coincidence of polynomials of best approximation with respect to one- and two-sided Hausdorff distance

In §4.4.1 we defined λ-monotonic functions within $\mathbf{H}_\Delta \subset \mathbf{F}_\Delta$ (see Definition 4.4). From Theorem 4.4 and Lemma 4.4 we obtain the following theorem.

Theorem 8.4. If $f \in \mathbf{F}_\Delta$ is λ-monotonic and $e(H_n, \Delta, \alpha; f) \le \lambda/\alpha$, then there exists a unique polynomial $P_0 \in H_n$ for which

$$r(\Delta, \alpha; f, P) = E(H_n, \Delta, \alpha; f) = e(H_n, \Delta, \alpha; f) = h(\Delta, \alpha; P, f),$$

i.e., the polynomial of best Hausdorff approximation (which is unique under these conditions) coincides with the polynomial of best one-sided Hausdorff approximation.

Proof . It suffices to note that the $n + 2$ points $\xi_0 < \xi_1 < \cdots < \xi_{n+1}$ in Lemma 4.4 satisfy exactly the conditions of Corollary 8.1.

By Theorem 8.4, finding the polynomial of best Hausdorff approximation for λ-monotone functions is reduced to the determination of the polynomial of best one-sided Hausdorff approximation .

§ 8.3. Numerical methods for calculating the polynomial of best one-sided approximation

Here we present a numerical algorithm of A. Andreev [1,2] for the approximate computation of the coefficients of the polynomial of best one-sided Hausdorff approximation . This algorithm is a modication of one of the methods of Remez [1].

The algorithm procedes as follows:

1. Construct a polynomial $P_{n,0} \in H_n$ such that there exist $n + 2$ points $\langle x_{k,0} \rangle_0^{n+1}$, for which $\operatorname{sgn} (P_{n,0}(x_{k,0}) - f(x_{k,0})) = (-1)^k \varepsilon$, where $\varepsilon = \pm 1$, $k = 0, 1, 2, \ldots, n + 1$. Set the index i equal to 0, and go to step 2.

2. Construct the function

$$\varphi_i(x) = \operatorname{sgn} (f(x) - P_{n,i}(x)) \left(\min_{(\xi,\eta)\in f} \max \{\alpha^{-1}|x - \xi|, |P_{n,i}(x) - \eta|\}\right).$$

3. Find $n + 2$ points $\langle z_{k,i} \rangle_0^{n+1}$ such that

$$\operatorname{sgn} \varphi_i(z_{k,i}) = (-1)^k \varepsilon, \text{ where } \varepsilon = \pm 1, \ k = 0, 1, 2, \ldots, n + 1.$$

4. If $\max \{ | |\varphi_i(z_{\nu,i})| - |\varphi_i(z_{\mu,i})| | : \nu, \mu = 0, 1, 2, \ldots, n + 1\} < \delta$, where $\delta > 0$ is a preassigned exactness, the polynomial $P_{n,i}$ is acceptable and the algorithm terminates. Otherwise go to step 5.

5. For the set of points $(z_{k,i}, \varphi_i(z_{k,i}))$ $k = 0, 1, 2, \ldots, n + 1$, construct the polynomial $Q_{n,i} \in H_n$ of best uniform approximation.

6. Form the polynomial $P_{n,i+1}(x) = P_{n,i}(x) + Q_{n,i}(x)$; then set i equal to $i + 1$, and return to step 2.

A Andreev [1] proved that the algogrithm described above converges for f Lipschitz. We present two examples of A. Andreev [2] of this algorithm in action.

Example 1. For the function

$$f_1(x) = \begin{cases} -1 & \text{if } -1 \leq x \leq -1/3 \\ 1 & \text{if } -1/3 < x \leq 1 \end{cases},$$

we have

$$0.2257 \leq e(H_5, [-1, 1], 1; f_1) = E(H_5, [-1, 1], 1; f_1) \leq 0.2266,$$

$$0.1413 \leq e(H_{10}, [-1, 1], 1; f_1) = E(H_{10}, [-1, 1], 1; f_1) \leq 0.1421,$$

and

$$P_5(f; x) = 0.22811x^5 + 3.18041x^4 - 1.07950x^3 - 3.96861x^2 + 1.85136x + 1.01437$$

$$P_{10}(f; x) = 1.17755x^{10} - 51.87041x^9 + 2.85984x^8 + 123.85465x^7 - 14.21173x^6$$
$$- 100.59484x^5 + 16.45178x^4 + 30.42548x^3 - 7.39353x^2 - 0.95663x$$
$$+ 1.11606,$$

where $P_5(f_1)$ and $P_{10}(f_1)$ are polynomials of best approximation of fifth and tenth degree respectively with respect to Hausdorff distance with parameter $\alpha = 1$. By Theorem 8.4, these polynomials coincide with the corresponding polynomials of best one-side approximation. The graphs of f_1 along with $P_5(f_1)$ and $P_{10}(f_1)$ appear in Figure 8.1.

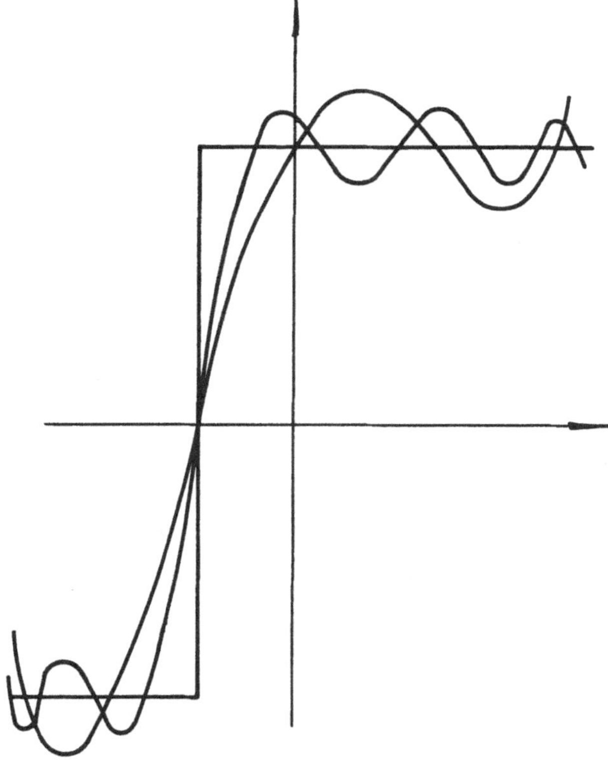

FIGURE 8.1

Example 2. For the function

$$f_2(x) = \begin{cases} -1 & \text{if } -1 \le x \le -1/3 \\ 0 & \text{if } -1/3 < x < 1/3 \\ 1 & \text{if } 1/3 \le x \le 1 \end{cases},$$

we have the follwing estimates for $\varepsilon_n = E(H_n, [-1, 1], 1; f_2) = e(H_n, [-1, 1], 1; f_2)$:

$$\varepsilon_1 = \varepsilon_2 = \frac{\sqrt{73} - 7}{6} = 0.25733\ldots, \qquad 0.1863 \le \varepsilon_5 = \varepsilon_6 \le 0.1871,$$

$$0.1481 \le \varepsilon_7 = \varepsilon_8 \le 0.1488, \qquad 0.0985 \le \varepsilon_{11} = \varepsilon_{12} \le 0.0995.$$

The graph of the function f_2 along with the graphs of its polynomials of best Hausdorff approximation of degree 1, 5, 7, and 11 are shown in Figure 8.2.

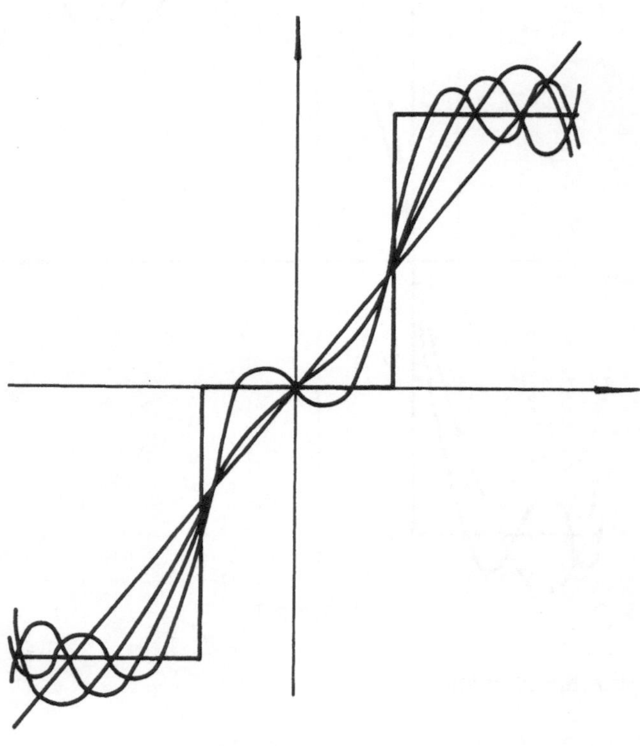

FIGURE 8.2

Remarks. A Andreev [3] considered the problem of stability of best approximations with respect to Hausdorff distance.

Numerical methods for the finding the polynomials of best uniform approximation of particular functions were considered by S. Markov and Bl. Sendov [1] and N. Kjurkciev and S. Markov [1]. P. G. Marinov and A. S. Andreev [1] considered one modification of the second algorithm of Remez producing rational functions of best Hausdorff approximation in some important cases.

The application of polynomials of best Hausdorff approximation to the design of antennas is studied in the following papers: H. Sinev, Sp. Tasev, N. Bankov, and N.

Kjurkciev [1,2], H. Sinev, N. Kjurkciev , M. Gacev, and S. Markov [1], and H. Sinev, N. Kjurkciev, and M. Gacev [1].

V. Veselinov [25] considered the problem of the asymptotic distribution of points of maximal deviation in the polynomials of best Hausdorff approximation.

The application of Hausdorff approximation to digital filters can be found in A. Andreev [4].

References

N. I. Achieser

1. Theory of approximation. Frederick Ungar Publ. Co., New York, 1956.

N. I. Achieser and M. G. Krein

1. On the best approximation of periodic differentiable functions by means of trigonometric sums. Doklady Akad. Nauk SSSR, 15 , 1937, 107-112 (in Russian).

A. S. Andreev

1. Hausdorff approximations and spline-interpolation. Thesis, Sofia, 1976 (in Bulgarian).
2. A numerical method for finding the polynomial of best Hausdorff approximation. C. r. Acad. Bulgare Sci., 29 , 1976, 163-166 (in Russian).
3. Stability of best approximations with respect to Hausdorff distance. In: Theory of Function Approximation. Moscow, 1977, 12-13 (in Russian).
4. Hausdorff distance and digital filters, D. Meyer-Ebrecht (hrsg.). ASST '87 Aachener Symposium für Signal-theorie, Proceedings, Springer-Verlag, 1987, 386-387.

A. S. Andreev and V. A. Popov

1. Approximation of functions with respect to a Δ-metric of Hausdorff type. God. Sofii. Univ. Mat. Fak., 64, 1971, 127-142 (in Bulgarian).
2. Approximation of continuous functions by rational function with respect to Δ-metric of Hausdorff type. God. Sofii. Univ. Mat. Fak., 65, 1973, 205-209 (in Bulgarian).

J.-P. Aubin and A. Cellina

1. Differential inclusions. Springer-Verlag, Berlin, 1984

E. A. Barbasin and Yu. I. Alimov

1. On the theory of relay differential equation. Izvestija vyss. ucebn. Zaved., Mat., 1962, No. 1, 3-13 (in Russian).

G. Beer

1. Upper semicontinuous functions and the Stone approximation theorem, J. Approximation Theory, 34, 1982, 1-11.
2. Cone lattices of upper semicontinuous functions, Proc. Amer. Math. Soc., 86, 1982 81-84.
3. Metric spaces on which continuous functions are uniformly continuous and Hausdorff distance, Proc. Amer. Math. Soc., 95, 1985, 653-658.
4. The approximation of real functions in the Hausdorff metric, Houston J. Math., 10, 1984, 325-338.

5. More on convergence of continuous functions and topological convergence of sets, Canad. Math. Bull., 28, 1985, 52-59.
6. Hausdorff distance and a compactness criterion for continuous functions, Canad. Math. Bull., 29, 1986, 463-468.
7. Complete subspaces of C(X,Y) with respect to Hausdorff distance, Math. Balkanica, 2, 1988, 78-84.

C. Berge
1. Topological spaces, Oliver and Boyd, Edinburgh, 1963.

S. N. Bernstein
1. Collected papers, vol. 1. Moscow, 1952 (in Russian).
2. Collected papers, vol. 2. Moscow, 1964 (in Russian).

V. I. Blagodatskih
1. On differentiability of solutions according to the initial conditions. Differencial'nye Uravnenija 9, 1973, 2136-2140 (in Russian).
2. Sufficient conditions for the optimality of differential inclusions. Izvestija Akad. Nauk. SSSR, Ser. mat., 38, 1974, 615-624 (in Russian).
3. Time optimal control problem for differential inclusions. Banach center publications, 1, 1976, 33-38.

L. N. Bondarenko
1. A note on a paper of Bl. Sendov. Ucenye Zapiski Pens. Politehn. Inst., 1970, No. 3, 24-28 (in Russian).

T. P. Boyadzieva and P. G. Boyadziev
1. On the polynomial approximation of sign (x - a), Constructive Function Theory '81, Sofia, 1983, 224-228.

B. D. Boyanov
1. On the polynomials of best approximation relative to Hausdorff distance. God. Sofii. Univ. Mat. Fak., 64, 1971, 161-170 (in Bulgarian).
2. On the connection between the variation and modulus of nonmonotonicity of a function. Izvestija Mat. Inst. BAN, 13, 1972, 99-103 (in Bulgarian).
3. On the approximation of functions by positive linear operators. Izvestija Mat. Inst. BAN, 14, 1973, 179-187 (in Bulgarian).
4. Estimates for the ε-entropy of the space of functions analytic in the unit circle. God. Sofii. Univ. Mat. Fak., 67, 1976, 191-194 (in Bulgarian).
5. Supplement to the paper of Lupas and Müller, Aequat. Math., 5, 1970, No. 1, 38-39.

B. D. Boyanov and V. M. Veselinov
1. A note on the approximation of functions in an infinite interval by linear positive operators. Bull. Math. Soc. Sci. Math. R. S. R., 14, 1970, No. 1, 9-13.
Bull. math. Soc. Sci. math. R. S. R., 14, 1970, No. 1, 9-13.

2. On the approximation of functions by Haar series. Mathematica (Cluj), 14 (37), 1972, No. 2, 189-192.
3. A general method generating estimates for the best Hausdorff approximation, C. r. Acad. Bulgare Sci., 29, 1976, 959-961.

T. P. Boyanov
1. On the widths of the space of continuous functions on a metric compactum. God. Sofii. Univ. Mat. Fak., 65, 1972, 25-32 (in Russian).
2. On the order of best approximation by algebraic polynomials with respect to a distance of Hausdorff type, C. r. Acad. Bulgare Sci. 23, 1970, 635-638 (in Russian).
3. Widths in Hausdorff metric. Thesis, Sofia, 1973 (in Bulgarian).
4. On the approximation of a function of class Lip α with respect to Hausdorff distance. C. r. Acad. Bulgare Sci., 27, 1974, 1629-1632 (in Russian).
5. Approximation of plane compacta by means of polynomial curves. Pliska 1, 1977, 134-136.
6. Exact asymptotics of the best Hausdorff approximation of classes of functions with given modulus of continuity, Serdica, 6, 1980, 84-97 (in Russian).
7. Approximation of the letter Γ by algebraic polynomial curves, Pliska, 5, 1983, 40-42 (in Russian).

T. P. Boyanov and L. Geshev
1. Hausdorff approximation in the class of functions with given modulus of continuity, Annuaire de l'Univ. de Sofia, Fac. de Mat., 72, 1978, 41-45 (in Russian).

T. P. Boyanov and E. Karlov
1. Polynomial approximations of nonrectifiable sets in the plane, Annuaire de l'Univ. de Sofia, Fac. de Mat., 72, 1978, 5-9.

T. P. Boyanov and V.A. Popov
1. On the widths of the space of continuous functions in the metric of Hausdorff. God. Sofii. Univ. Mat. Fak., 63, 1970, 167-185 (in Bulgarian).

T. P. Boyanov and Bl. Sendov
1. Metric dimensionality and approximation by polynomial curves in the plane. Serdica 2, 1976, 295-299 (in Russian).

E. M. Bronstein
1. ε-entropy of convex sets and functions. Sibir. Mat. Zurn., 17, 1976, 508-514 (in Russian).

E. M. Bronstein and L. D. Ivanov
1. On the approximation of convex sets by polytopes. Sibir. Mat. Zurn., 26, 1975, 1110-1112 (in Russian).

L. Cakalov
1. On a presentation of Newton's quotients in interpolation theory and its applications. God. Sofii. Univ. Fac. Mat. 34, 1938, 353-405 (in Bulgarian).

Z. A. Canturia

1. Modulus of variation of functions and its application in Fourier series. Doklady Akad. Nauk SSSR, 214, 1974, 63-66 (in Russian).

J. W. S. Cassels

1. Introduction to diophantine approximation. Cambridge Tracts in Mathematics and Mathematical Physics, No. 45, Cambridge, 1957.

C. Castaing and M. Valadier

1. Convex analysis and measurable multifunctions, Lecture Notes in Mathematics, v. 580, Springer-Verlag, Berlin, 1975.

A. Cellina

1. Approximation of set valued functions and fixed point theorems, Ann. Mat. Pura Appl., 82, 1969, 17-24
2. A further result on the approximation of set-valued mappings, Rend. Acc. Naz. Lincei, 48, 1970, 412-416.

G. P. Clements

1. Entropies of sets of functions of bounded variation. Can. J. Math., 15, 1963, 422-432.

P. J. Davis, R. A. Vitale, and E. Ben-Sabar

1. On the deterministic and stochastic approximation of regions. J. Approximation Theory, 21, 1977, 60-88.

F. De Blasi

1. Characterizations of certain classes of semicontinuous multifunctions by continuous approximations, J. Math. Anal. Appl. 106, 1985, 1-18.

F. De Blasi and J. Myjak

1. On continuous approximations for multifunctions, Pacific J. Math., 123, 1986, 9-31.

F. Deutsch and P. Kenderov

1. Continuous selections and approximate selections for set-valued mappings and applications to metric projections, SIAM J. Math. Anal., 14, 1983, 185-194.

D. G. Dimitrov and V. A. Popov

1. A generalization of Timan's theorem for approximation of functions by algebraic polynomials. Serdica, 6, 1980, 9-15.

J. Dieudonné

1. Foundations of modern analysis. Academic Press, New York, 1960.

H. N. Djidjev

1. On the Hausdorff convergence of summative formulae, Constructive Theory of Functions '84, Sofia, 1984, 291-297.

A. L. Donchev

1. Perturbations, approximations, and sensitivity analysis of optimal control systems, Lecture Notes in Control and Inf. Sciences, v. 52, Springer-Verlag, Berlin, 1983.

A. L. Donchev and E. M. Farkhi

1. An average modulus of continuity for multivalued maps and its applications to differential inclusions, Proc. International Conf. on Constructive Theory of Functions, Varna, 1987, Publ. House of BAS, Sofia, 1988, 127-131.

A. L. Donchev and V. M. Veliov

1. Singular perturbations in Meyer's problem for linear systems, SIAM J. Control and Optimization, 21, 1983, 566-581.
2. Singular perturbations in linear control systems with weakly coupled stable and unstable fast subsystems, J. Math. Anal. Appl., 110, 1985, 1-30.

A. S. Dzafarov and G. M. Gasanov

1. Almost nonperiodic functions with respect to Hausdorff metric and their properties. Izvestija Akad. Nauk. Azerbaidz. SSR, 1977, No. 1, 57-62 (in Russian).

E. P. Dolzenko

1. Some new converse theorems of approximation theory. In: Constructive Function Theory. Sofia, 1980 (in Russian).
2. Some metric properties of algebraic hypersurfaces. Izvestija Akad. Nauk. SSSR, ser. mat., 27, 1963, 241-252 (in Russian).

E. P. Dolzenko and E. A. Sevastianov

1. On the approximation of functions in Hausdorff metric. Doklady Akad. Nauk SSSR, 226, 1976, 768-770 (in Russian).
2. On the approximation of functions in Hausdorff metric by means of piecewise (in particular rational) functions. Mat. Sbornik, 101, 1976, 508-541 (in Russian).
3. Approximations in Hausdorff metric. In: Theory of Function Approximation. Moscow, 1977, 175-182 (in Russian).
4. On the dependance of the properties of functions on the rate of their approximation by polynomials. Izvestija Akad. Nauk. SSSR, ser. mat., 42, 1978, 270-304 (in Russian).

E. P. Dolzenko and P. L. Ul'janov

1. On certain questions of functions theory. Vestnik Moskov. Univ. Ser. I, 1980, No. 1, 3-13.

R. Dudley

1. Metric entropy of some classes of sets with differentiable boundaries. J. Approximation Theory 10, 1974, 227-236.

H. G. Eggleston

1. Problems in Euclidean space. London, 1957.

A. I. Ermakov

1. Best Hausdorff approximation by algebraic polynomials and continuity of functions, Mat. Zametki, 28, 1980, 843-858 (in Russian).

J. Favard

1. Sur les meilleurs procédés d'approximation de certain classes des fonctions par polynomes trigonométriques. Bull. Sci. Math. 61, 1937, 209-224.
2. Sur l'approximation des fonctions, Bull. Sci. Math. 63, 1938, 338-352.

C. H. Fitzgerald and L. L. Schumaker

1. A differential equation approach to interpolation at extremal points. J. Analyse Math., 22, 1969, 117-134.

G. Freud and Bl. Sendov

1. On a method of approximation of periodic functions by trigonometric polynomials. Magyar Tud. Akad. Mat. kütato inter. köslemenyei, Ser. A, 9, 1964, 491-494 (in Russian).

G. M. Gasanov

1. On the order of convergence of Hermite-Fejer interpolational processes in Hausdorff metric. Izvestija Akad. Nauk. Azerbaidz. SSR, 1970, No. 3, 3-7 (in Russian).
2. On quadrature processes in the class of bounded and integrable functions. Izvestija vyss. ucebn. Zaved. Mat., 1974, No. 8, 22-25 (in Russian).

G. M. Gasanov and V. A. Popov

1. Approximation of locally monotone functions by positive linear operators in L and its application for the estimate of error in the collocation method. Serdica, 2, 1976, 75-81 (in Russian).

G. M. Gasanov and S. P. Suleimanov

1. On the order of convergence of quadrature processes in the class of bounded functions. Izvestija Akad. Nauk. Azerbaidz. SSR, 1973, No. 2, 61-66 (in Russian).

T. R. Gicev

1. On the realization of the time optimal linear control. Izvestija Mat. Inst. BAN, 13, 1972, 227-246 (in Bulgarian).
2. β-continuity of the optimal control as a function of the initial stage. Izvestija Mat. Inst. BAN, 14, 1973, 35-49 (in Bulgarian).
3. Certain questions of correctness of the linear problem of optimal control with minimal impulse, II. Differencial'nye Uravnenija, 9, 1973, 1561-1751 (in Russian).
4. Certain questions of correctness of the linear problem of optimal control with minimal impulse, I. Differencial'nye Uravnenija, 9, 1973, 1383-1392 (in Russian).
5. Correctness of linear problem of optimal control. Thesis, Moscow, 1973. (in Russian).
6. A test for convergence of Stieltjes' integral. Vestnik Moskov. Univ., Ser. I, 30, 1975, 18-22 (in Russian).
7. Correctness of continuous games in a square. Serdica, 2, 1976, 219-34 (in Russian).

T. R. Gicev and N. H. Rosov
1. Nonlinear controlled impulse systems, I. The influence of certain perturbations. Differencial'nye Uravnenija, 15, 1979, 1933-1939 (in Russian).
A. A. Goncar
1. Estimates of the growth of rational functions with certain applications. Mat. Sbornik, 72, 1967, 489-503 (in Russian).
2. On the rate of rational approximation of continuous functions with characteristic singularities. Mat. Sbornik, 73, 1967, 630-638 (in Russian).
P. Gruber and P. Kenderov
1. Approximation of convex bodies by polytopes, Rend. Ciricolo Mat. Palermo, Ser. 2, 31, 1982, 195-225.
A. I. Guseinov and G. M. Gasanov
1. On an estimate of error of the approximate solutions of an integral equation. Doklady Akad. Nauk SSSR, 211, 1973, 1270-1272 (in Russian).
F. Hausdorff
1. Mengenlehre. W. Gruyter & Co., Berlin, 1927.
H. Hermes
1. The generalized differential equation, Adv. Math., 4, 1970, 149-169.
H. Hitov
1. The number of corridors in a rectangle. Izvestija Mat. Inst. BAN, 13, 1972, 265-276 (in Bulgarian).
H. Hörmander
1. Sur la fonction d'appui des ensembles convexes dans une espace localement convexe. Arkiv. Mat., 3, 1954, 181-186.
V. H. Hristov
1. Criteria of the Dini-Lipschitz tyupe for the uniform convergence of interpolational polynomials. Pliska, 1, 1977, 128-133 (in Russian).
2. On convergence criteria for Fourier series. Thesis, Sofia, 1977 (in Bulgarian).
V. H. Hristov and P. P. Petrusev
1. An improvement of the Dini-Lipschitz criterion on uniform convergence of Fourier series. C. r. Acad. Bulgare Sci., 29, 1976, 1579-1582 (in Russian).
2. Sufficient conditions for the convergence of Fourier series. In: Theory of the Function Approximation. Moscow, 1977, 392-396 (in Russian).
M. Hukuhara, Sur l'application semi-continue dont la valeur est un compact convexe, Funckcial. Ekvac., 10, 1967, 43-66.
R. Ivanov
1. Approximation of convex n-gons by inscribed (n - 1)-gons. In: Mathematics and Education in Mathematics, 1973. Sofia, 1974, 115-122 (in Bulgarian).

G. L. Iliev

1. Approximation of functions with respect to a new metric. C. r. Acad. Bulgare Sci., 28, 1975, 299-301 (in Russian).
2. Approximation of functions with respect to certain A-distances. In: Theory of Function Approximation. Moscow, 1977, 204-205 (in Russian).
3. Estimation of Hausdorff approximation by Müntz polynomials. C. r. Acad. Bulgare Sci., 29, 1976, 1253-1256.
4. Exact estimation under the partially monotone approximation and interpolation. C. r. Acad. Bulgare Sci., 30, 1977, 491-494.

G. L. Iliev and A. S. Andreev

1. Numerical methods for finding the polynomial of best approximation with respect to a new metric in the space of continuous functions. Serdica, 2, 1976, 360-364 (in Russian).

G. L. Iliev and V. A. Popov

1. An analogue of Jackson's theorem for G-distance in the space of continuous functions. C. r. Acad. Bulgare Sci., 28, 1975, 1159-1162.

M. Jacobs

1. Attainable sets in systems with unbounded controls. J. Diff. Equations, 4, 1968, 408-423.

E. Kamke

1. Differentialgleichungen. 1. Gewöhnliche differentialgleichungen. Leipzig, 1959.

L. V. Kantorovic and G. P. Akilov

1. Functional analysis in normed spaces. Pergamon Press, Oxford, 1964.

M. S. Kasciev

1. On the approximation of a class of complex functions with respect to a metric of Hausdorff type. Izvestija Mat. Inst. BAN, 14. 1973, 105-113 (in Bulgarian).

B. Kasimov

1. On the polynomial approximation of the Zygmund class in Hausdorff metric, Mat. Zamettki, 28, 1980, 335-342 (in Russian).

P. S. Kenderov

1. On an optimal property of the square. C. r. Acad. Bulgare Sci., 26, 1973, 1143-1146 (in Russian).
2. Approximation of plane convex compacta by polygons, C. r. Acad. Bulgare Sci., 33, 1980, 889-891.
3. Approximation of plane convex compacta by polygons, Constructive Function Theory '81, Sofia, 1983, 376-381.

N. V. Kjurkciev and S. M. Markov

1. On the numerical approximation of the "cross" set, God. Sofii. Univ. Mat. Fak., 66, 1974, 19-25 (in Bulgarian).

N. V. Kjurkciev and Bl. Sendov
1. Approximation of a class of functions by algebraic polynomials with respect to Hausdorff distance. God. Sofii. Univ. Mat. Fak., 67, 1976, 543-579 (in Bulgarian).
E. Klein and A. Thompson
1. Theory of Correspondences, Wiley-Interscience, Montreal, 1984.
H.-B. Knoop and P. Pottinger
1. Ein satz vom Korovkin-typ für C^k-raum. Math. Z., 148, 1976, 23-32.
A. N. Kolmogorov
1. Asymptotic characteristics of certain totally bounded metric spaces. Doklady Akad. Nauk SSSR, 108, 1956, 385-389 (in Russian).
2. On the convergence of A. V. Skorohod. Teor. Verojatn. Primen., 1, 1956, 239-247 (in Russian).
3. Über die beste annäherung von functionen einer funktionsklasse. Math. Ann., 37, 1936, 107-111.
A. N. Kolmogorov and V. M. Tihomirov
1. ε-entropy and ε-capacity of sets in function spaces. Uspehi Mat. Nauk, 14, 1959, No. 2, 3-86.
N. P. Korneicuk
1. Exact constants in approximation theory, Nauka, Moscow, 1987 (in Russian).
N. P. Korneicuk and A. I. Polovina
1. On the approximation of continuous functions by algebraic polynomials on a closed interval. Doklady Akad. Nauk SSSR, 166, 1966, 281-283 (in Russian).
P. P. Korovkin
1. Linear operators and approximation theory. Hindustan, India, 1960.
2. An axiomatic approach to certain questions in the constructive approximation theory of functions of one variable. Ucenye Zapiski Kaliningr. gosudarst. ped. Inst., 69, 1969, 91-109 (in Russian).
3. An axiomatic approach to certain questions of constructive approximation. In: Constructive Function Theory. Sofia 1972, 55-63 (in Russian).
K. Kuratowski
1. Topology, v. 1, Academic Press, New York, 1966.
S. E. Laveuve
1. Definition einer Kahan-arithmetik und ihre implementierung. In: Interval Mathematics., Lecture Notes in Computer Sciences vol.29, Springer-Verlag, Berlin, 1975, 236-245.
P. Lévy
1. Théorie de l'addition des variables aléatoires. G. Villard, Paris, 1937.
G. G. Lorentz
1. Zur theorie der polynome von S. Bernstein. Mat. Sbornik 2, 1937, 543-556.

2. Berstein polynomials. Univ. Toronto Press, Toronto, 1953.
3. Approximation of functions. Holt, Rinehart, and Winston, New York, 1966.

G. G. Lorentz and K. Zeller
1. Degree of approximation by monotone polynomials, I. J. Approximation Theory, 1, 1968, 501-504.

K. N. Lungu
1. On the best approximations by rational functions with respect to Hausdorff distance. Mat. Zametki, 11, 1972, 491-498 (in Russian).

A. Lupas
1. Die folge der betaoperatoren. Univ. Stuttgart, 1972.

A. Lupas and M. W. Müller,
1. Approximation properties of the M_n-operators. Aequat. Math., 5, 1970, 19-37.

V. L. Makarov and A. M. Rubinov
1. The mathematical theory of economic dynamics and equilibrium. Moscow, 1973 (in Russian).

P. G. Marinov and A. S. Andreev
1. A modified Remez algorithm for approximate determination of the rational function of best approximation in Hausdorff metric, C. r. Acad. Bulgare Sci., 40, 1987, 13-16.

S. M. Markov
1. Extended interval arithmetic. C. r. Acad. Bulgare Sci., 30, 1977, 1239-1242.
2. A differential calculus for interval-valued functions based on extended interval arithmetic. C. r. Acad. Bulgare Sci., 30, 1977, 1377-1380.
3. Relations between the integral and Hausdorff distance with applications to differential equations. Pliska 1, 1977, 112-121.
4. One-sided approximation of functions with respect to Hausdorff distance. In: Constructive Function Theory. Sofia, 1972, 69-75 (in Russian).

S. M. Markov and Bl. Sendov
1. On the numerical evaluation of a class of polynomials of best approximation. God. Sofii. Univ. Mat. Fak., 61, 1968, 17-27 (in Bulgarian).

V. T. Martynjuk
1. On linear methods of approximation of bounded functions of two variables with respect to a metric of Hausdorff type. Izvestija vyss. ucebn. Zaved. Mat., 1968, No. 7, 42-50 (in Russian).
2. Linear methods of approximation of bounded functions of two variables with respect to a metric of Hausdorff type. Thesis, Dnepropetrovsk, 1968 (in Russian).
3. On the approximation of V. A. Steklov's function in the Hausdorff metric. Mat. Zametki, 5, 1969, 21-30 (in Russian).
4. On the approximation of curves defined by parametric equations by polygonal lines in Hausdorff metric. Ukrain. Mat. Zurn., 28, 1976, 87-91 (in Russian).

V. T. Martynjuk and V. F. Storcai
1. Approximation by polyhedral functions in Hausdorff metric. Ukrain. Mat. Zurn., 25, 1973, 115-120 (in Russian).
E. Michael
1. Topologies on spaces of subsets. Trans. Amer. Math. Soc., 71, 1951, 152-182.
R. M. Minkova
1. On the convergence of derivatives of linear operators, C. r. Acad. Bulgare Sci., 23, 1970, 627-629 (in Bulgarian).
2. Convergence of derivatives of linear operators. Izvestija vyss. ucebn. Zaved. Mat., 1976, No. 8, 52-59 (in Russian).
R. E. Moore
1. Interval analysis. Prentice-Hall, Englewood Cliffs, N. J., 1966.
R. E. Moore and C. T. Yang
1. Interval analysis. LMSD-285875, Lockheed Missiles and Space Co., Palo Alto, California, 1959.
E. S. Moskona
1. Rational approximation of functions with bounded variation in Hausdorff metric, C. r. Acad. Bulgare Sci., 40, 1987, 33-36.
M. W. Müller
1. Approximation unbeschränkter functionen bezüglich einer Korovkin-metric. In: Theory of Function Approximation. Moscow, 1977, 269-272.
M. W. Müller and H. Walk
1. Konvergenz and güteaussagen für die approximation durch folgen linearer positiver operatoren. In: Constructive Function Theory. Sofia, 1972, 221-234.
J. Mycielski
1. Problem 6096 (1976, 489). Amer. Math. Monthly, 85, 1978, 124.
S. Nadler
1. Hyperspaces of sets, Dekker, New York, 1978.
S. Naimpally
1. Graph topology for functions spaces, Trans. Amer. Math. Soc. 268 (1966), 127-142.
N. A. Nasarenko
1. Approximation of curves by means of parametrized splines, Constructive Function Theory '81, Sofia, 1983, 111-114 (in Russian).
I. P. Natanson
1. Konstructive functionen theorie. Akademie-Verlag, Berlin, 1955.
2. Constructive theory of functions. AEC-tr-4503 (i,ii), United States Atomic Energy Commission, 1961.
3. Constructive function theory, vol. I, II, III. Frederick Ungar Publ. Co., New York, 1964-1965.

D. J. Newman

1. Rational approximation to |x|. Mich. Math. J., 11, 1964, 11- 14.

D. Newman, E. Passov, and L. Raymon

1. Piecewise monontone polynomial approximation. Trans. Amer. Math. Soc., 172, 1972, 465-472.

M. G. Nikolcheva

1. Approximation of curves in the plane, Constructive Function Theory '81, Sofia, 1983, 115-117 (in Russian).

S. M. Nikol'skii

1. On the best approximation of Lipschitz functions by polynomials. Izvestija Akad. Nauk SSSR, Ser. mat., 10, 1946, 295-318 (in Russian).

C. Olech

1. Extremal solutions of a control system, J. Differential Equat., 2, 1966, 74-101.

H. J. Ortolf

1. Eine verallgemeinerung der intervallaithmetik. Ber. Ges. Math. u. Datenverarbeitung (Bonn), 1969, No. 11.

A. A. Panov

1. On the number of (p,q)-corridors and (p,q)-labyrinths. Uspehi Mat. Nauk, 30, 1975, 221-222 (in Russian).
2. Evaluation of ε-entropy and ε-capacity of continuous function spaces and continuous curve spaces in Hausdorff metric. Thesis, Moscow, 1975 (in Russian).
3. Evaluation of number of (p,q)-labyrinths. Serdica, 1, 1975, 399-408 (in Russian).
4. Evaluation of ε-entropy of continuous function spaces with Hausdorff metric. Mat. Zametkik, 21, 1977, 39-50 (in Russian).

B. Penkov and Bl. Sendov

1. Entropy of continuous functions set of many variables, C. r. Acad. Bulgare Sci., 17, 1964, 335-337.
2. Hausdorffsche metrik und approximationen, Numer. Math., 9, 1966, 214-226.
3. Hausdorff metric and its application. In: Numerische Methoden der Approximationstheorie, Band 1. Basel, 1972, 127-146.

P. P. Petrusev

1. On the rational approximation of functions with convex derivative. C. r. Acad. Bulgare Sci., 29, 1976, 1249-1252 (in Russian).
2. On the rational approximation of functions. C. r. Acad. Bulgare Sci., 29, 1976, 1405-1408.
3. Approximation of certain clasees of real functions in uniform and Hausdorff distances. Thesis, Sofia, 1977 (in Bulgarian).
4. On rational approximation of functions with unbounded variation. Serdica, 2, 1976, 149-153.

5. Best rational approximation in Hausdorff metric, Serdica, 6, 1980, 29-41 (in Russian).
6. Some new characterizations in the theory of rational approximations, Constructive Function Theory '81, Sofia, 1983, 121-124 (in Russian).

P. P. Petrusev and V. H. Hristov

1. Convergence of Fourier series in Hausdorff metric. Pliska 1, 1977, 21-36 (in Russian).
2. On approximation by Müntz polynomials in Hausdorff metric. C. r. Acad. Bulgare Sci., 29, 1976, 955-958 (in Russian).

P. P. Petrusev and V. A. Popov

1. Rational approximation of real functions, Encyclopedia of Math. and its Appl., Cambridge Univ. Press, Cambridge, 1988.

P. P. Petrusev and S. Tasev

1. Some converse theorems in Hausdorff metric. C. r. Acad. Bulgare Sci., 29, 1976, 1721-1724 (in Russian).

A. P. Petukhov

1. On the approximation of discontinuous functions in Hausdorff metric, Mat. Zametki, 37, 1985, 25-40 (in Russian).
2. Approximation of functions by singular integrals in the Hausdorff metric, Mat. Sbornik (new series), 135 (177), 1988, 235-252.

R. R. Phelps

1. Convex functions, monotone operators, and differentiability, Lecture Notes in Mathematics v. 1364, Springer-Verlag, Berlin, 1989.

V. A. Popov

1. Approximation of functions of several variables by algebraic polynomials in Hausdorff metric. Thesis, Sofia Univ., 1965 (in Bulgarian).
2. Approximation of convex figures. C. r. Acad. Bulgare Sci., 21, 1968, 993-995 (in Russian).
3. Approximation of convex sets. Izvestija Mat. Inst. BAN, 11, 1969, 67-80 (in Bulgarian).
4. Approximation of convex functions by polygons. C. r. Acad. Bulgare Sci., 23, 1970, 643-645 (in Russian).
5. Approximation of convex functions by polygons. Izvestija Mat. Inst. BAN, 11, 1970, 117-126.
6. Convex approximations. Thesis, Sofia, 1970.
7. Some problems connected with convergence in Hausdorff metric. Izvestija Mat. Inst. BAN, 12, 1970, 87-96 (in Bulgarian).
8. On a converse problem of approximation theory in Hausdorff metric. God. Sofii. Univ. Mat. Fak., 65, 1972, 201-204 (in Russian).

9. Parametric approximation of convex curves by polynomial curves. God. Sofii. Univ. Mat. Fak., 67, 1976, 333-341 (in Russian).
10. On approximation of functions of bounded variation by rational functions. Serdica 1, 1975, 96-103 (in Russian).
11. Local approximation of a function. Mat. Zametki, 17, 1975, 369-382 (in Russian).
12. Local approximation of functions by linear operators. In: Mathematics and Mathematical Education, 1973. Sofia, 1974, 183-191 (in Bulgarian).
13. Direct and converse theorems in approximation theory. Thesis, Sofia, 1976 (in Bulgarian).
14. Some characteristics of functions and their application to spline approximation and rational approximation. In: Theory of Function Approximation. Moscow, 1974, 286-293 (in Russian).
15. Approximation de fonctions d'un grand nombre de vairaibles indépendantes au moyen de polynòmes dans la métrique de Hausdorff. C. r. Acad. Bulgare Sci., 19, 1966, 561-564.
16. On the connection between rational and spline approximation. C. r. Acad. Bulgare Sci., 27, 1974, 623-626.
17. Approximation of convex functions by algebraic polynomials in Hausdorff's metric. Serdica, 1, 1975, 386-398.
18. Rational uniform approximation of the class V_r and its applications. C. r. Acad. Bulgare Sci., 29, 1976, 791-794.
19. Uniform rational approximation of the class V_r and its applications. Acta Math. Acad. Sci. Hung., 29, 1977, 119-129.

V. A. Popov and A. S. Andreev

1. On the convergence of linear operators with respect to Hausdorff distance. God. Sofii. Univ. Mat. Fak., 62, 1969, 215-223 (in Bulgarian).

V. A. Popov and V. M. Veselinov

1. On the best approximation by algebraic and trigonometric polynomials with respect to Hausdorff metric. Izvestija Mat. Inst. BAN, 10, 1969, 213-221 (in Bulgarian).
2. Some remarks on the derivatives of linear positive operators. God. Sofii. Univ. Mat. Fak., 64, 1971, 143-152 (in Russian).
3. One generalization of Popoviciu's theorem for Bernstein polynomials. Mathematica (Cluj), 16, 1974, 159-172.

V. A. Popov and Bl. Sendov

1. On approximation of functions by spline functions and rational functions. In: Constructive Function Theory. Sofia, 1972, 89-94 (in Russian).
2. Approximation of monotone functions by monotone polynomials in Hausdorff's metric. Rev. Anal. Numér. Théorie Approx., 3, 1974, 79-88.

V. A. Popov and S. L. Troyanski
1. On approximation of abstract functions in Hausdorff metric. Math. Balkanica, 1, 1971, 190-194 (in Russian).
T. Popoviciu
1. Sur l'approximation des fonctions convexes d'ordre supérieur. Mathematica (Cluj), 10, 1934, 49-54.
H. Poppe
1. Über graphentopologien für abbildungsraüme I, Bull. Acad. Pol. Sci., Ser. Sci. Math. Astron. Phys., 15, 1967, 71-80.
P. Pottinger
1. Zur linearen approximation im raum $C^k(I)$. Habilitationsschrift, Duisburg, 1976.
I. I. Privalov
1. Boundary properties of analytic functions. Moscow, 1950 (in Russian).
Yu. V. Prohorov
1. Convergence of random processes and limit theorems in probability theory. Teor. Verojatn. Primen. 1, 1956, 177-238 (in Russian).
H. Ratschek
1. Die subdistributivität der intervallarithmetik. Z. Angew. Math. u. Mech., 51, 1971, 189-192.
H. Ratschek and G. Schröder
1. Über die ableitung von intervallwertigen functionen. Computing, 7, 1971, 172-187.
E. Ya. Remez
1. General computing methods of Chebyshev approximation. Moscow, 1957 (in Russian).
J. R. Rice
1. General purpose curve fitting, Approximation Theory Proc. Symp. Lancaster, July 1969. London, 1970, 191-204.
R. T. Rockafellar
1. Convex analysis, Princeton University Press, Princeton, 1970.
J. Roulier
1. Polynomials of best approximation, which are monotone. J. Approximation Theory, 9, 1973, 212-217.
O. S. Sabazov
1. On an estimate of approximation of functions by rational operators of Bernstein polynomial type. Doklady Akad. Nauk Tadzik. SSR, 13, 1970, 3-6 (in Russian).
G. Schmid
1. Approximation unbeschränkter funktionen. Univ. Stuttgart. Stuttgart, 1972.
G. Schröder
1. Charakterisierung des quasilinearen räumes I(R) und klassifizierung der quasilinearen räume der dimension 1 und 2. Computing, 10, 1972, 111-120.

I. J. Schoenberg

1. On spline functions. Inequalitites. New York, 1957, 255-292.

Bl. Sendov

1. Approximation of functions by algebraic polynomials with repect to a metric of Hausdorff type. God. Sofii. Univ. Mat. Fak., 55, 1962, 1-39 (in Bulgarian).
2. On the best approximation by algebraic polynomials with respect to Hausdorff distance. God. Sofii. Univ. Mat. Fak., 56, 1963, 195-207 (in Bulgarian).
3. On an estimate of approximation of functions by Bernstein polynomials. Mathematica (Cluj), 5, 1963, 145-154 (in Russian).
4. Linear methods of approximation of periodic functions with respect to a metric of Hausdorff type. Doklady Akad. Nauk SSSR, 160, 1965, 1023-1025 (in Russian).
5. On a certain linear method of approximation of periodic functions with respect to Hausdorff distance. God. Sofii. Univ. Mat. Fak., 58, 1965, 107-140 (in Bulgarian).
6. On the interpolation process of Fejer. Izvestija Mat. Inst. BAN, 9, 1966, 133-145 (in Bulgarian).
7. On the best approximation with respect to Hausdorff distance. God. Sofii. Univ. Mat. Fak., 59, 1966, 85-103 (in Bulgarian).
8. Approximation of stepwise functions with respect to Hausdorff distance. Mat. Zametki, 2, 1967, No. 1, 61-70 (in Russian).
9. On convergence of sequences of linear positive operators. God. Sofii. Univ. Mat. Fak., 60, 1967, 279-296 (in Bulgarian).
10. Approximation of point sets with polynomial curves in the plane. God. Sofii. Univ. Mat. Fak., 60, 1967, 211-222 (in Bulgarian).
11. On P. Korovkin's theorems for convergence of sequences of linear positive operators, Doklady Akad. Nauk SSSR, 177, 1967, 518-520 (in Russian).
12. Approximation with respect to Hausdorff distance. Thesis, Moscow, 1967 (in Russian).
13. Approximation with respect to Hausdorff distance. Mat. Zametki, 3, 1968, 481-494 (in Russian).
14. Certain questions in the theory of approximation of functions and sets in the Hausdorff metric. Uspehi Mat. Nauk 24, 1969, No. 5, 141-178 (Russian Math. Surveys, 24, 1969, No. 5, 143-183).
15. Order of best Hausdorff approximation for a class of analytic functions. C. r. Acad. Bulgare Sci., 27, 1974, 1621-1623 (in Russian).
16. Approximation of a semicircle by algebraic polynomials. Trudy Mat. Inst. Steklov, 134, 1975, 278-282 (Proc. Steklov Inst. Math., 134, 1975, 315-320).
17. On a modification of Goncar's theorem on the rate of rational approximation. Mat. Zametki, 17, 1975, 383-390 (in Russian).

18. Certain problems of Hausdorff approximation. In: Theory of function approximation. Moscow, 1977, 322-329 (in Russian).
19. Best Hausdorff approximations by spline functions with equidistant knots. Pliska, 1, 1977, 79-92.
20. Approximation bezüglich einer Hausdrff'chen metrik mittels algebraisher polynome. Internationationales kolloquium über aktualle probleme der rechentechnik, II. Dresden, 1962. Vorträge, Dresden, 1963, 121-122.
21. Approximation relative to Hausdorff distance. Approximation theory, London, 1970, 101-108.
22. Approximation of analytic functions in Hausdorff metric. In: Functional analysis and its applications, Lecture notes in mathematics, vol. 399, Springer-Verlag, Berlin, 1974, 490-500.
23. Convergence of Vallée-Poussin sums in Hausdorff distance. C. r. Acad. Bulgare Sci., 26, 1973, 1431-1432.
24. Hausdorff approximation of functions and point sets. Topics in numerical analysis II. New York, 1975, 175-184.
25. Segment arithmetic and segment limit. C. r. Acad. Bulgare Sci., 30, 1977, 955-958.
26. Segment derivatives and Taylor's formula. C. r. Acad. Bulgare Sci., 30, 1977, 1093-1096.
27. Order of best Hausdorff polynomial approximation of certain functions. Serdica, 1, 1975, 77-87.
28. Best Hausdorff approximation with equidistant-knot spline functions. C. r. Acad. Bulgare Sci., 29, 1976, 1717-1719.
29. Exact estimation for the best Hausdorff spline approximation. C. r. Acad. Bulgare Sci., 30, 1977, 187-190.
30. Convergence of sequences of monotonic operators in A-distance. C. r. Acad. Bulgare Sci., 30, 1977, 657-660.

Bl. Sendov and B. D. Boyanov

1. On a property of a class of linear positive operators. God. Sofii. Univ. Mat. Fak., 64, 1971, 115-117.

Bl. Sendov and T. P. Boyanov

1. A problem of labyrinths. C. r. Acad. Bulgare Sci., 25, 1972, 583-585.

Bl. Sendov, S. Dimiev and B. Penkov

1. ε-entropy and ε-capacity of the set of continuous curves. Vestnik Moskov. Univ. Ser. I, 3, 1962, 20-23 (in Russian).

Bl. Sendov and B. Penkov

1. ε-entropy and ε-capacity of the set of continuous functions. Vestnik Moskov. Univ., Ser. I, 3, 1962, 15-19 (in Russian).
2. ε-entropy and ε-capacity of the space of continuous functions. Izvestija Mat. Inst. BAN, 6, 1962, 27-50 (in Bulgarian).

3. On the widths of the set of continuous functions. Izvestija Mat. Inst. BAN, 10, 1969, 5-15 (in Bulgarian).
4. On widths of the space of continuous functions. C. r. Acad. Bulgare Sci., 17, 1964, 689-691.

Bl. Sendov, B. Penkov, V. A. Popov and S. M. Markov

1. Hausdorff derivatives in F_Δ. Serdica, 2, 1976, 131-137.

Bl. Sendov and V. A. Popov

1. On certain properties of Hausdorff metric. Mathematica (Cluj), 8, 1966, 163-172 (in Russian).
2. On convergence of the derivatives of linear positive operators. C. r. Acad. Bulgare Sci., 22, 1969, 507-509 (in Russian).
3. On the approximation by spline functions. C. r. Acad. Bulgare Sci., 23, 1970, 755-758 (in Russian).
4. Convergence of the derivatives of linear positive operators. Izvestija Mat. Inst. BAN, 11, 1969, 107-115 (in Russian).
5. Approximation of functions of several variables by algebraic polynomials in a Hausdorff type metric. God. Sofii. Univ. Mat. Fak., 63, 1970, 61-76.
6. Approximation of curves in the plane by polynomial curves. C. r. Acad. Bulgare Sci., 23, 1970, 639-642 (in Russian).
7. An analogue of Nikol'skii's theorem for the approximation of functions by algebraic polynomials in the Hausdorff metric. In: Constructive Function Theory. Sofia, 1972, 95-103.
8. General estimates for the approximation of functions by linear positive operators, God. Sofii. Univ. Mat. Fak., 65, 1972, 191-200 (in Russian).
9. The exact asymptotic behavior of the best approximation by algebraic and trigonometric polynomials in the Hausdorff metric. Mat. Sbornik, 89, 1972, 138-147 (Mat. USSR Sbornik, 18, 1972, 139-149).
10. On a generalization of Jackson's theorem for best approximation. J. Approximation Theory. 9, 1973, 102-111.
11. Averaged moduli of smoothness, Bulg. Acad. Sci. Sofia, 1983.

Bl. Sendov, S. P. Tashev, and P. Petrushev

1. Characterization of the S-derivatives of Lipschitz functions, Serdica, 4, 1978, 260-266 (in Russian).

H. Sinev, N. Kjurkciev and M. Gacev

1. Design of a linear antennal lattice with differential diagram of Hausdorff type. Izvestija VMEI Sofia, 35, 1976, 19-24 (in Bulgarian).

H. Sinev, N. Kjurkciev, M. Gacev and S. Markov

1. Application of a certain class of polynomials of best approximation to the design of linear antennal lattices. Izvestija VMEI Sofia, Radioelektronika, 34, 1975, No. 1, 1-6 (in Bulgarian).

H. Sinev, S. Tasev, N. Bankov and N. Kjurkciev
1. Design of linear antennal lattices with optimal diagrams of directedness. Izvestija VMEI Sofia, Radioelektronika i Elektroakustika, 4, 1978, 1-19 (in Bulgarian).
O. Shisha
1. Monotone approximation. Pacific J. Math., 15, 1965, 667-671.
R. Smithson
1. Multifunctions, Nieuw Archief voor Wiskunde, 20, 1972, 31-53.
V. S. Spiridonov
1. On the convergence of an interpolation process with regard to Hausdorff metric. Izvestija Mat. Inst. BAN, 9, 1966, 219-227 (in Bulgarian).
S. Stoinski
1. H-almost periodic functions. Functiones et Approximatio. Comentarii Mathematici (Poznan), 1, 1974, 114-122.
2. Periodic functions by means of certain linear operators with regard to Hausdorff metric. Functiones et Approximatio. Comentarii Mathematici (Poznan), 1, 1974, 123-131.
S. P. Suleimanov
1. On the order of approximation of functions by integral operators in an infinite region with respect to Hausdorff distance. Doklady Akad. Nauk Azerbaidz. SSR, 28, 1972, No. 2, 14-18 (in Russian).
S. P. Suleimanov and G. M. Gasanov
1. On kthe order of approximation of summation formulas on the entire axis in Hausdorff metric. Ucenye Zapiski Azerbaidz. gos. Ped. Inst., 11, 1971, No. 2, 108-112 (in Russian).
T. Sunaga
1. Theory of an interval algebra and its application to numerical analysis. RAAG Memoirs II, Tokyo: Gekujutsu Bunken Fukuykai, 1958, 547-564.
S. Tasev
1. Approximation of bounded sets in the plane in Hausdorff metric, C. r. Acad. Bulgare Sci., 29, 465-468.
2. On the existence of the solution of differential equations with multivalued right hand side, Serdica, 6, 1980, 98-104.
3. On the distribution of the points of maximal devaition for the polynomials of best Chebyshev and Hausdorff approximations, PWN and North-Holland Publishers, 1981, 791-799.
4. New estimates for the Hausdorff and local approximation of functions, Constructive Function Theory '81, Sofia, 1983, 551-557.
V. M. Tihomirov
1. On n-dim widths of certain function classes. C. r. Acad. Bulgare Sci., 130, 1960, 734-738 (in Russian).

2. Widths of sets in function spaces and the theory of best approximation. Uspehi Mat. Nauk, 15, 1960, 81-120 (in Russian).

V. M. Veliov

1. Estimations of the influence of some regular perturbations and inertness in time optimal control problem, Univ. Annual Appl. Math., Sofia, 15, 1979, 83-98 (in Russian).
2. An estimate for the influence of inertness of optimal control in the linear time optimal control problem, Proc. 19th Conf. on Math. and Math. Educ., Golden Sands, 1980, Bul. Acad. Sci 1980, Sofia, 121-127 (in Russian).

V. M. Veliov and A. L. Donchev

1. Continuity of the family of trajectories of linear control systems with singular perturbations, Doklady Acad. Nauk SSSR, 293, 1987, 274-278 (in Russian).

V. Veselinov

1. Approximation by trigonometric polynomials with respect to Hausdorff distance. Thesis, Sofia Univ., Sofia, 1965 (in Bulgarian).
2. On the best approximation of periodic totalities by trigonometric polynomials in Hausdorff metric. God. viss. ucebn. Zaved., prilozna Mat., 4, 1967, No. 1, 69-86 (in Bulgarian).
3. Approximation of functions by trigonometric polynomials in a Hausdorff type metric. Mathematica (Cluj), 9, 1967, No. 1, 185-199 (in Russian).
4. On certain estimates of best approximation by means of algebraic polynomials in Hausdorff metric. C. r. Acad. Bulgare Sci., 21, 1968, 5-8 (in Russian).
5. Approximation of unbounded functions by positive linear operators in Hausdorff metric. God. viss. ucebn. Zaved., prilozna Mat., 5, 1969, No. 2, 37-45 (in Russian).
6. Approximation of unbounded functions by positive linear operators in Hausdorff metric. C. r. Acad. Bulgare Sci., 22, 1969, 499-502 (in Russian).
7. On certain estimates and asymptotic formulas for operators on an infinite interval. God. Sofii. Univ. Mat. Fak., 64, 1971, 153-159 (in Bulgarian).
8. On the convergence of sequences of linear operators. Mathematica (Cluj), 12, 1970, No. 2, 377-382 (in Russian).
9. On the order of approximation of semicontinuous functions by nonlinear positive operators. C. r. Acad. Bulgare Sci., 24, 1971, 705-708 (in Russian).
10. On the extension of the modulus of nonmonotonicity on an infinite interval and the approximation of unbounded functions in Hausdorff metric. C. r. Acad. Bulgare Sci., 24, 1971, 1153-1154 (in Russian).
11. Approximation of functions by linear operators with respect to Hausdorff and uniform metrics. Thesis, Sofia, 1972 (in Bulgarian).
12. On the exact order of approximation of functions by Bernstein polynomials in Hausdorff metric. Mat. Zametki, 12, 1972, 501-510 (in Russian).

13. On approximation of functions by positive linear operators on an infinite interval. In: Constructive Function Theory, Sofia, 1972, 31-34 (in Russian).
14. On the weighted approximation of functions on the entire axis by linear operators. C. r. Acad. Bulgare Sci., 26, 1973, 1151-1154 (in Russian).
15. Some new estimates for Bernstein polynomials and Vallée-Poussin integrals. C. r. Acad. Bulgare Sci., 27, 1974, 747-750.
16. On a theorem of Baire. God. Sofii. Univ. Mat. Fak., 65, 1972, 211-218 (in Russian).
17. On some estimates on the approximation of functions by de la Vallée-Poussin and Landau operators. God. Sofii. Univ. Mat. Fak., 66, 1974, 153-158 (in Russian).
18. On weighted approximations by entire functions of exponential type. Mat. Zametki, 16, 1974, 185-192 (in Russian).
19. The exact constants in the theory of approximation by Bernstein polynomials in the Hausdorff metric. C. r. Acad. Bulgare Sci., 27, 1974, 1183-1186.
20. Best weighted approximation by polynomials in Hausdorff metric. C. r. Acad. Bulgare Sci., 28, 1975, 1019-1021 (in Russian).
21. Converse theorems for Hausdorff approximation of functions by linear operators. C. r. Acad. Bulgare Sci., 29, 1976, 159-162 (in Russian).
22. On the convergence of a sequence of nonlinear operators to semicontinuous functions. C. r. Acad. Bulgare Sci., 30, 1977, 1537-1540 (in Russian).
23. Weighted approximations in Hausdorff metric. In : Theory of Function Approximation. Moscow, 1977, 80-82 (in Russian).
24. Best approximation of functions by polynomials with positive coefficients in the Hausdorff metric. C. r. Acad. Bulgare Sci., 27, 1974, 599-602.
25. Asymptotic distribution of points of maximal deviation in the polynomials of best Hausdorff approximation. Proc. of the Colloquium on Fourier Analysis and Approximation Theory. August 16-21, 1976, Budapest.

V. M. Veselinov and F. V. Buong

1. The best constants for the approximation of functions by certain singular integrals in the Hausdorff metric. God. Sofii. Univ. Mat. Fak., 68, 1977, 227-234 (in Russian).

V. M. Veselinov and I. Kirkorov

1. On an estimate of function approximation by generalized polynomials of Landau. C. r. Acad. Bulgare Sci., 23, 1970, 647-650 (in Russian).

N. V. Vladov

1. Exact asymptotics for the rational approximation of sign (x) in Hausdorff distance, C. r. Acad. Bulgare Sci., 39, 1986, 37-40.

H. Walk

1. Approximation unbeschränkter funktionen durch lineare positive operatoren. Habilitationsschrift. Stuttgart, 1970.

2. Approximation von ableitungen unbeschränkter funktionen durch lineare operatoren. L'analyse Numer. Théorie Approx., 6, 1977, 99-105.

S. M. Yakov

1. Some questions in the theory of approximation by algebraic polynomials. Sofia Univ. Thesis. Sofia, 1963 (in Bulgarian).

V. M. Zolotarev and V. V. Senatov

1. Two-sided estimates of the Levi metric. Teor. Verojatn. Primen., 20, 1975, 239-250.

Author Index

Notation Index

Subject Index